Technologies in Plant Biotechnology and Breeding of Field Crops

Kamaluddin • Usha Kiran • M. Z. Abdin
Editors

Technologies in Plant Biotechnology and Breeding of Field Crops

Springer

Editors
Kamaluddin
Department of Genetics and Plant Breeding
Banda University of Agriculture and Technology
Banda, India

Usha Kiran
Vanercia Institute of Technical Education
Gautam Buddh Nagar, Uttar Pradesh, India

M. Z. Abdin
Department of Biotechnology
Jamia Hamdard University
New Delhi, India

ISBN 978-981-16-5769-6 ISBN 978-981-16-5767-2 (eBook)
https://doi.org/10.1007/978-981-16-5767-2

This Springer imprint is published by the registered company Springer Nature Singapore Pte Ltd.
The registered company address is: 152 Beach Road, #21-01/04 Gateway East, Singapore 189721, Singapore

Preface

Agriculture productivity is adversely affected due to uneven weather conditions, increased temperature, and less availability of irrigational water. Global climate change together with the rapidly increasing population is escalating considerable pressure on the agriculture sector to produce more food from less land. Food production may decline, in both tropical and temperate regions, due to anticipated increases in temperature and erratic climatic conditions.

In the current century, plant breeders and scientists have a difficult task to develop crop varieties which could produce more food on less acreage of arable land and using less amount of water. They are under extreme pressure to boost the genetic production potential of crops with the aim to feed the growing human population and provide a livelihood to a large section of society. They should develop tools to save scarce land resources and preserve natural habitats.

Researchers and students involved in crop improvement and plant breeding are fascinated to know the details of what and how the achievements in crop improvement were made. The conventional and molecular approaches, together, are being used to improve varieties and hybrids of major crops in light of their origin, evolution, taxonomy, production, and productivity and need by human civilization.

The book consists of 12 chapters, and each chapter is articulated by more than one eminent scientists, across the globe, to ensure distinctiveness and applicability.

Chapter 1, **Conventional Plant Breeding to Modern Biotechnological Approaches in Crop Improvement**, describes the history and the current status of conventional breeding methods along with the use of modern biotechnological tools for crop improvement. Several conventional and advanced biotechnological approaches including plant tissue culture, genetic transformation, hybrids generation, tilling, RNAi, genome editing, and nano-biotechnology have been discussed in this chapter, which helps to transfer knowledge from scientist to farmer field.

Chapter 2, **Marker-Assisted Selection for Value Addition in Crop Plants**, describes the importance of marker-assisted selection, as an important modern molecular tool for selecting plants having desirable genotype, efficiently at an early stage, reducing the time of development of improved cultivars. The role of marker-assisted selection in the identification of cultivars, assessment of genetic diversity and purity, selection of parents, the study of heterosis, etc. has been review in the chapter. Different categories of MAS are listed and described in the chapter.

Chapter 3, **Application of Gene Mining and Editing Technologies for Agricultural Research and Breeding**, highlights the dependence of agriculture and plant breeding on genetic engineering technologies that have become an attractive tool for rapidly improving plant tolerance to abiotic and biotic stresses. This chapter also describes gene mining tools that led to the selection of genes for engineering plant tolerance to abiotic and biotic stresses in the four major food crop plant families.

Chapter 4, **Germplasm Conservation for Biotechnology and Plant Breeding**, describes the use of a wider range of plant genetic diversity throughout the world for plant genetic resources (PGR) conservation programs for food and agriculture. The first part of the chapter deals with conservation methods and approaches for characterizing germplasm, and the second part of the chapter focuses on the exploitation of genetic resources in biotechnology and the selection of genotypes with good agronomic performance, under environmental constraints.

Chapter 5, **Recent Advances in Rice Breeding Using Biotechnological and Genomics Tools**, highlights the development of technologies in the area of rice breeding, biotechnology, and sequencing, which are the major driving force for boosting rice production.

Chapter 6, **Application of Genomics and Breeding Technologies to Increase Yield and Nutritional Qualities of Rapeseed-Mustard and Sunflower**, describes the breeding objectives of the oilseed crop. The breeding objectives are to increase the yield of seeds and oil, improve the quality of the oil and meal according to its use, and develop stable biotic and abiotic resistant/tolerant varieties. The chapter aims to provide an overview of the various advanced approaches for genomics-assisted breeding to enhance genetic gain in yield and nutritional quality.

Chapter 7, **Accelerated In Vitro Propagation of Sweet Potato Clones (*Ipomoea batatas* L.)**, is a protocol-styled chapter, which describes a low-input suspension technique that combines the use of the liquid and solid medium and permits the successful propagation of genetically diverse sweet potato genotypes [*Ipomoea batatas* (L.) Lam.] with a high multiplication rate. Sweet potato is an important staple crop in low-income/technology countries of Africa, Asia, and South America; the described method may find valuable application for the breeding programs in these regions.

Chapter 8, **Emerging Technologies in Plant Breeding for Fiber Crops, Cotton and Sunn Hemp**, describes the major breeding approaches in fiber crops to breed the cultivars with increased yield and enhance the quantity of fiber and better tolerance to various abiotic and biotic.

Chapter 9, **Recent Advances in Omics Approaches for Mungbean Improvement**, highlights the recent advancements and significant achievements made in new mungbean variety development, post mungbean genome sequence decoding.

Chapter 10, **Genomics-Assisted Breeding Approaches in Lentil (*Lens culinaris* Medik)**, describes the functional genomics approaches with an aim of tagging the genes controlling the traits of interest. The chapter also describes the comparative genomics approaches to improve the cross-species transferability of genetic information among legume species. MAS, DNA-based selection of desirable plant, a remarkable way to introduce QTLs/genes associated with yield enhancement, and biotic or abiotic traits, have also been discussed.

Chapter 11, **Transgenic and Molecular Approaches for Pigeon Pea and Chickpea Improvement**, describes the importance of genomic resources, molecular markers, high-density genetic maps, and transcriptomic resources for the development of transgenic plants for pigeon pea and chickpea.

Chapter 12, **Intellectual Property Rights in Plant Biotechnology and Breeding**, attempts to explain the forms and scope of intellectual property rights relevant to plant biotechnology, plant variety, biosecurity, and biopiracy.

We extend our heartfelt gratitude to the contributing authors for their efforts to review the published research and present the subject topic in an unpretentious manner. Their patience and cooperation in meeting deadlines are also appreciated. Compiling and editing this book, with full-time research jobs, has been very demanding. We are thankful to our family, friends, and students for their patience in putting up with these demands.

I hope the content of the book would be a valuable resource for graduate and postgraduate students, teachers, plant breeders and researchers, plant scientists, seed producers, and policymakers interested in agriculture, particularly in modern breeding technologies.

Banda, India Kamaluddin
Gautam Buddh Nagar, India Usha Kiran
New Delhi, India M. Z. Abdin

Contents

Editors and Contributors

About the Editors

Kamaluddin is an associate professor in the Department of Genetics and Plant Breeding, Banda University of Agriculture and Technology, Banda, India. He obtained his Ph.D. degree from BHU, Varanasi. He has expertise in plant breeding and molecular breeding. Dr. Kamaluddin has published original research articles, review articles, book chapters, and abstracts in different journals of national and international repute. Presently, he is involved in the development of heat- and drought stress-tolerant pulse cultivars, using marker-assisted selection. He has published *Plant Biotechnology: Principles and Application* (Springer) and others in the capacity of coeditor.

Usha Kiran is currently working as director, Vanercia Institute of Technical Education, Gautam Buddh Nagar, UP, India. She obtained her Ph.D. in biotechnology from Jamia Hamdard, New Delhi, India. Her research is focused on signaling modulators used by plant cell to evoke protective cellular response, especially proline biosynthesis and degradation during stress conditions. Her work involves extensive use of bioinformatic tools for analyses. Her research articles, review articles, book chapters, and abstracts have been published in journals of national and international repute. In her pursuit to bring technologies and their application to fellow scientists and research scholars, she has published books on applications of biotechnology in plant science such as *Plant Biotechnology: Principles and Application*, Springer.

M. Z. Abdin is currently working as head of the Department of Biotechnology, School of Chemical and Life Sciences, Jamia Hamdard, New Delhi, India. He has published four books in the capacity of editor/coauthor. Dr. Abdin has been working in the field of plant biotechnology with a focus on metabolic engineering of medicinal plants and crops for better yield, quality, and tolerance to abiotic/biotic stresses. He is also working on the modulation of secondary plant metabolites synthesis employing RNAi technology. Further, he is also working for the development and use of DNA fingerprints for authentication of herbal/Unani drugs. Dr. Abdin has published extensively in journals of national and international repute.

Contributors

M. Z. Abdin Department of Biotechnology, School of Chemical and Life Sciences, Jamia Hamdard, India

Javed Akhatar Punjab Agricultural University, Ludhiana, Punjab, India

N. L. Anglin International Potato Center (CIP), Lima, Peru
United States Department of Agriculture (USDA), Aberdeen, ID, USA

Anu CCS Haryana Agricultural University, Hisar, Haryana, India

L. Arroyo International Potato Center (CIP), Lima, Peru

Madhuri Arya Department of Plant Breeding and Genetics, Tirhut College of Agriculture, Dholi, Bihar, India

Mounawer Badri Laboratory of Extremophile Plants, Centre of Biotechnology of Borj Cedria, Hammam-Lif, Tunisia

Olalekan Bakare Environmental Biotechnology Laboratory, Department of Biotechnology, University of the Western Cape, Bellville, South Africa

Pratik Istari Bhoyar Indian Agricultural Research Institute, Regional Station, Indore, Madhya Pradesh, India

Mogamat Fahiem Carelse Environmental Biotechnology Laboratory, Department of Biotechnology, University of the Western Cape, Bellville, South Africa

A. K. Choubey Department of Soil Science and Agriculture Chemistry, College of Agriculture, Banda University of Agriculture and Technology, Banda, Uttar Pradesh, India

Indrapreet Dhaliwal Department of Plant Breeding and Genetics, Punjab Agricultural University, Ludhiana, Punjab, India

D. Ellis International Potato Center (CIP), Lima, Peru

J. Espirilla International Potato Center (CIP), Lima, Peru

G. Flores International Potato Center (CIP), Lima, Peru

Arun Gokul Department of Plant Sciences, University of Free State, Phuthaditjhaba, South Africa

D. K. Janghel Department of Genetics and Plant Breeding, CCS Haryana Agricultural University, Hisar, Haryana, India

Kamaluddin Department of Genetics and Plant Breeding, College of Agriculture, Banda University of Agriculture and Technology, Banda, Uttar Pradesh, India

Harjeevan Kaur Punjab Agricultural University, Ludhiana, Punjab, India

Marshall Keyster Environmental Biotechnology Laboratory, Department of Biotechnology, University of the Western Cape, Bellville, South Africa
DST-NRF Centre of Excellence in Food Security, University of the Western Cape, Bellville, South Africa

Usha Kiran Vanercia Institute of Technical Education, Gautam Buddh Nagar, Uttar Pradesh, India

Ashwil Klein Plant Omics Laboratory, Department of Biotechnology, University of the Western Cape, Bellville, South Africa

J. Kreuze International Potato Center (CIP), Lima, Peru

Hitesh Kumar Department of Genetics and Plant Breeding, College of Agriculture, Banda University of Agriculture and Technology, Banda, Uttar Pradesh, India

Sarita Kumari Dr. Rajendra Prasad Central Agricultural University, Pusa Samastipur, Bihar, India

Mukul Kumar Department of Genetics and Plant Breeding, College of Agriculture, Banda University of Agriculture and Technology, Banda, Uttar Pradesh, India

Rahul Kumar Division of Genetics, Indian Agricultural Research Institute, New Delhi, India

Rajesh Kumar Dr. Rajendra Prasad Central Agricultural University, Pusa Samastipur, Bihar, India

Ndiko Ludidi Plant Biotechnology Research Group, Department of Biotechnology, University of the Western Cape, Bellville, South Africa
DSI-NRF Centre of Excellence in Food Security, University of the Western Cape, Bellville, South Africa

S. B. Mishra Department of Plant Breeding and Genetics, Tirhut College of Agriculture, Dholi, Bihar, India

H. S. Negi Department of Plant Pathology, College of Agriculture, Banda University of Agriculture and Technology, Banda, Uttar Pradesh, India

Lee-Ann Niekerk Environmental Biotechnology Laboratory, Department of Biotechnology, University of the Western Cape, Bellville, South Africa

Nalini Kant Pandey Mitakshara IP Services, Ghaziabad, Uttar Pradesh, India

Durga Prasad Department of Plant Pathology, Banda University of Agriculture and Technology, Banda, Uttar Pradesh, India

Aditya Pratap Division of Crop Improvement, ICAR-Indian Institute of Pulses Research, Kanpur, Uttar Pradesh, India

Rizwana Rehsawla CCS Haryana Agricultural University, Hisar, Haryana, India

A. Rojas International Potato Center (CIP), Lima, Peru

J. C. Sánchez International Potato Center (CIP), Lima, Peru

Vijay Sharma Banda University of Agriculture and Technology, Banda, Uttar Pradesh, India

Bhupendra Kumar Singh Department of Entomology, Banda University of Agriculture and Technology, Banda, Uttar Pradesh, India

Chandra Mohan Singh Dr. Rajendra Prasad Central Agricultural University, Pusa Samastipur, Bihar, India
Department of Genetics and Plant Breeding, Banda University of Agriculture and Technology, Banda, Uttar Pradesh, India

Smita Singh Department of Genetics and Plant Breeding, Banda University of Agriculture and Technology, Banda, Uttar Pradesh, India

Sumeet Kumar Singh Dr. Rajendra Prasad Central Agricultural University, Pusa Samastipur, Bihar, India

Vivek K. Singh CCS Haryana Agricultural University, Hisar, Haryana, India

Preeti Sonkar Department of Genetics and Plant Breeding, College of Agriculture, Banda University of Agriculture and Technology, Banda, Uttar Pradesh, India

Sandeep Kumar Suman Dr. Rajendra Prasad Central Agricultural University, Pusa Samastipur, Bihar, India

R. Vollmer International Potato Center (CIP), Lima, Peru

Suresh Yadav Division of Genetics, Indian Agricultural Research Institute, New Delhi, India

Conventional Plant Breeding to Modern Biotechnological Approaches in Crop Improvement

1

Javed Akhatar, Harjeevan Kaur, and Hitesh Kumar

Abstract

In an era of climate change along with increasing world population, the food security can be ensured by developing climate resilient crop varieties having good nutritional quality. Improved crop varieties can be developed through importation of desirable features which encompass the discrimination of available variability through introduction and selection as primitive method. Conventional breeding methods take more time in developing and delivering crop varieties. In the last two decades, modern biotechnological tools have been developed which led to the better understanding of the genetics of traits and are able to assist conventional breeding to release new cultivars in a shorter span of time. Thus, integrating modern biotechnological tools into the conventional breeding offers new opportunities to breed crop cultivars with enhanced quality, quantity, and tolerance to abiotic and biotic stresses. Besides, locally adapted, several landraces can be developed through plant domestication, farmer's selection, and self- and cross-pollinated methods. In this chapter, conventional breeding methods, modern and advanced biotechnological tools, viz., plant tissue culture, genetic transformation, hybrids generation, TILLING, RNAi, genome editing, and nano-biotechnology along with their importance will be discussed which would aid in transferring knowledge from scientist to farmer fields.

J. Akhatar · H. Kaur
Punjab Agricultural University, Ludhiana, Punjab, India

H. Kumar (✉)
Banda University of Agriculture and Technology, Banda, Uttar Pradesh, India

Kamaluddin et al. (eds.), *Technologies in Plant Biotechnology and Breeding of Field Crops*, https://doi.org/10.1007/978-981-16-5767-2_1

Keywords

Plant breeding · Plant tissue culture · Micro-propagation · Speed breeding · Stress resistance

1.1 Introduction

Due to the increasing human population and declining area under cultivation, crop improvement through plant breeding and biotechnological tools are one of the major approaches to meet out the present and future food demand. In past, domestication, introduction and selection of crop varieties led to the increased production and productivity in several agricultural crops (Chahal and Gosal 2002). The movement of new crops and varieties from its origin to different regions where it can adopt with desirable characteristics played a significant role to augment food demand in every part of the globe. Conventional breeding methods were followed for creation and discrimination of variability and development of new cultivars with superior yield and quality traits (Allard 1960; Fehr 1987; Acquaah 2012). The agricultural crop improvement methods are being continuously practiced from thousands of years to introduce the desired traits for the development of nutritionally enriched and high yielding food crops (Poehlman and Sleper 1999) (Fig. 1.1). The twentieth century is evident that crop production developed through conventional plant breeding methods has played an important role in ensuring food security and production (Tester and Langridge 2010; Shiferaw et al. 2013). However, with a rapid increase of world's population and challenges of climate change, these approaches have become insufficient for plant genome enhancement and development of nutritionally enriched crops, within a short period of time. Due to continuous demand for plant-based products and high annual crop yield, integration of conventional methods with modern biotechnological tools has provided an extensive range of innovation and achievements in plant breeding (Varshney et al. 2006; Li et al. 2018).

In the era of advanced biotechnological tools and deciphering the molecular mechanism of the desirable traits, it becomes possible to integrate modern breeding methods with the conventional approaches for the development of elite cultivars. Phenotypic selection made by crop breeder involves focused desirable trait selection. Ever since 1990s, molecular markers and modern plant breeding approaches were practiced for the selection of superior breeding lines based on the genetic makeup of the plant. Several conventional and modern breeding methods combined with genome studies can be used to enhance the accuracy of breeding practices within a

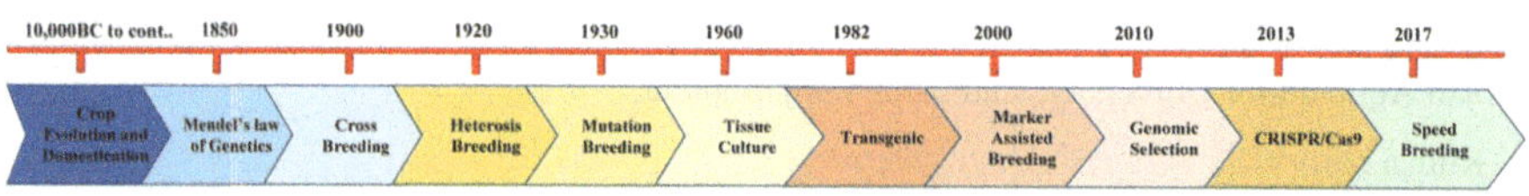

Fig. 1.1 Chronological development of crop evolution, breeding methods, and modern breeding tools and improving crops

short time span (Doust and Diao 2017). There are numerous examples to show that biotechnology plays a crucial role in understanding crop genetics and improving of food crops. The industrial use of many crops has been met through various biotechnological approaches. The ever-increasing demand for food has exerted pressure on the agriculture sector to speed up the breeding process. There is a requirement for highly productive resistant varieties with minimum inputs. One potential solution for these demands is the integration of biotechnological tools with breeding programs (Gosal et al. 2010; Beddington et al. 2012). Gene discovery has been accelerated through genomics, proteomics, and other biotechnological approaches. Identification of novel genes and introducing those genes into the desired species for plant breeding programs is of immense importance to crop improvement. To increase crop yield, quality improvement, resistance to insect pests, crop adaptability to adverse climate conditions, harvesting quality are the major target in the crop improvement program. The main objective of this chapter is to present the role of different plant breeding and biotechnological approaches—plant tissue culture, genetic transformation, hybrids generation, TILLING, RNAi, genome editing, and nano-biotechnology have been discussed which represents the enhanced methods of crop improvement.

1.2 Definition of Plant Breeding and Conventional Breeding

Plant breeding is a purposeful effort by humans to change certain traits of crop plants (Acquaah 2012). The process of evolution is also often linkened to plant breeding (Zohary 1988). The artificial process (orchestrated by humans) of selection is a relatively quick process rather than the evolution of crop spp. (Gepts 2002). The crop adaptability is increased due to crop evolution, whereas the aim of the plant breeder is specific toward population improvement and predetermined goal (Borojevic 1990).

An organized and thought-out method for developing high-quality cultivars with high predictability is now being practiced in plant breeding instead of the traditional approach of crossing the best with the best and hoping for the best. Major category of plant breeding approaches can be divided into two types, i.e., conventional and unconventional breeding programs, which keep changing with advanced methodology and technology (Chahal and Gosal 2002; Acquaah 2012). Conventional breeding methods employed in the development of new cultivars, which is averse to the newer, more advanced, and sometimes revolutionary tools of molecular plant breeding (Jain and Kharkwal 2004). By molecular breeding, sometimes manipulation of the genes in a plant may desecrate due to natural biological barriers, perhaps the ability to transfer desirable genes from alien species (Acquaah 2012).

1.3 Origins of Conventional Plant Breeding

Since the humans started to feed themselves, cultivation started with whatever plants they found edible. More than 10,000 years ago, farmers have been changing the genetic architecture of the available crops and selected the best-looking plants that grow well and used them for the upcoming season. This was the phase to start the era of domestication of plants (Gepts 2002). Then, the selection process is started with the availability of natural biological diversity and gradually increases the utility of desirable traits of the plants. Domesticated plants become more valuable and advantageous to the farmer concerning wild progenitors (Zohary 1988). Selection is the ancient and most widely used method in plant breeding for crop improvement. It consists mostly of phenotypes with a mental picture of the focus traits. Plant breeder discriminates, identifies, and selects the desirable plants among the available variability. Crop improvement techniques by the farmers were followed by the seed selection of best-looking and superior plants from their crop and using to next year's crop. The development of seed by this technique is called landraces (Allard 1960). Dramatic changes were observed in plant species as compared to their wild relatives when the selection process was faster with plant features like, crop duration, high productivity, resistance to insect and pest, seed quality, etc. In modern plant breeding, a rich source of variability is provided by the landraces (Harlan 1975).

1.3.1 Conventional Plant Breeding Forward with Science

Once the knowledge abounds, methodology advances, and science of genetics becomes better understood, plant breeders are gradually dependent on science and reduce the selection process (Chahal and Gosal 2002). However, conventional breeding still relies to some extent on instinct, judgment, and skill of the breeder (breeder's eye), in co-occurrence with plant breeder knowledge (Allard 1960). With the advancement of plant breeding science in the twentieth century, plant breeders created new and improved cultivars for different crops, increasing their efficiency.

The conventional plant breeding process has improved with technique over time, creating a productive framework for the enhancement of crop performance. During this process, the plant breeder decides which parents are to be used for pollination and which are to produce advanced plants. In plant breeding, creating a very large population is a benefit, unlike animal breeding. As a consequence, only a small proportion of the plants are selected for advance breeding program and the vast majority are discarded with undesirable characteristics. This process is applied during many stages to select few individuals from the large population.

1.4 Breeding Methods Based on Observed Variation

1.4.1 Origin of Crops-Plant Domestication

In modern cultivated plants, genetic variability tends to be decrease than in their wild relatives because of the *founder effect* in crop domestication (Breseghello and Coelho 2013). Wild progenitors of crop species were domesticated primarily before 10,000 years ago into the major crops worldwide (Doebley et al. 2006). The domestication process by ancient humans led to selection of rare mutants for adaptation of cultivation, resulting in most of the crop variations in the populations of the cultivated forms. So that many desirable genes related to insect and pest resistance were eliminated from the cultivated gene pool (Zamir 2001).

In the course of the development of crop domestication or breeding program, primitive landraces developed intermediate domestication levels, whereas in modern breeding, cultivars or varieties were developed, and that was the beginning of ideotype breeding. The ideotype concept was evolved to maximize the genetic yield potential and enhancement of germplasm (Donald 1968). Through the domestication and breeding steps within time, germplasm pools were extensively changed and move forward to the narrowing genetic diversity (Tanksley and McCouch 1997). In some examples, broader phenotypic diversity was observed by domestication which is exemplified by the shape, size, and color in tomato (Paran and Van Der Knaap 2007; Rodrigues 2011). In the case of maize, cultivars have developed with narrower phenotypic diversity as compared to their less domesticated landraces (Goodman and Brown 1988; Troyer 1999). Domestication and breeding programs demonstrate the historically useful and parallel domestication syndrome in various crop species. The most challenging task in modern plant breeding is the incorporation of desirable genes into the modern cultivars with the help of applications of molecular tools in the recent breeding program.

1.4.2 Landraces: Inherent Farmer Selection

Landraces are the region-specific popularize population and have been cultivated for many generations, which is resistant to biotic and abiotic stresses, seed handling, and developed on eating habitat. They are continually changing dynamic genetic entities either by the natural or artificial directional selection process and other mechanical impurities. Landraces are keeping their identity by stabilizing selection and also lead to the slow adaptation to environmental changes in a given region. In other cases, landraces quickly change their genetic architecture by the introduction into the different regions or due to the outcrossing with nearly cultivated original landraces. Although modern plant breeding methods are used to develop modern cultivars, landraces can be developed when farmers plant modern cultivars to create landraces. Several key characteristics of landraces have been identified (Zeven 1998): (a) high genetic diversity within populations with limited variation between individuals, (b) greater adaptability in different climate conditions with resistance to insects,

(c) more valuable edible parts picked by local populations, and (d) stable yield yields under normal climate conditions.

A farmer's selection is integral to shaping the varieties to a particular area and environment, based on local preference, well-serving and production being consumed locally. However, unintentional selection by farmer's for a desirable trait may lead to undesirable changes in landraces due to genetic correlations between them. The selection of landraces is based on edible parts such as panicles, ears, spikes in cereals, presenting the low harvest index and they are normally tall plants and prone to lodging. Landraces are one of the most beneficial genetic resources for breeding programs and used for germplasm conservation under "ex situ" conditions. The majority of germplasm conversion regulators encourage local commodities by financially incentivizing them to keep growing their landraces in their "in situ" conservation. This type of conservation will also take advantage of scientific communities for modern crop breeding. In terms of landrace breeding, it refers to initiatives that allow developing varieties that evolved through natural selection in the local cropping system, with or without mass selection. They are developed by using closely related crossing parents in a breeding program.

1.5 Self- and Cross-Pollinated Cultivars

Self-pollinated cultivars are developed either by a single plant seed or a mixture of plant seeds (Briggs and Knowles 1967). The two most common methods for breeding self-pollinated species are hybridization (crossing of pants) and selection (Acquaah 2012). A single plant or a mixture can be made into a final product. At present, several types of cultivar development methods are being used, such as pedigree, single seed descent, pure line, bulk, backcross, etc. Mass selection, synthetic, composite, and recurring selection methods can be used for a plant mixture.

1.5.1 Mass Selection

It is one of the methods that we can use for both self-pollinated and crossbred crop species with different genetic consequences (Allard 1960). It is also a population improvement method that increases the average performance of the base population by increasing the gene frequencies after the selection of desirable genes. This can use to maintain genetic purity, selection for lasting resistance, and wide adaptation to an area of production. When the heritability of the traits of interest is high, the mass selection is more effective (Brown and Caligari 2008).

1.5.2 Pure Line Selection

The varieties of pure lineages are genetically homozygous and homogeneous (Allard 1960). The uniformity is the main feature of such cultivars for the trait of interest with a narrow genetics base. Selection is followed; from a variable population to repeated self pollination until there is no segregation in the progeny (Poehlman and Sleper 1999). This was proposed by Louis de Vilmorin (1856) in self-pollinated crops, though the use of his principles in the nineteenth century by certain farmers (Allard 1999). Until now, homogeneity has prevailed and has been considered into breeding and agriculture, after the realization of this paradigm. In self-pollinated species, landraces are mainly of a mixture of pure lines, due to the low frequency (heterozygous individuals) of cross-pollination. In this case, the crossing of selected single plants develops progenies with higher productivity than their original landraces. However, these superior progenies have some disadvantages: they are less stable than the base population in terms of stress resistance and short-term adaptation under growing conditions. Pedigree, selection, bulk population breeding, single-seed-descent, backcross breeding, multiline breeding, cultivar blends and composites methods in the self-pollinated crop for crop enhancement.

1.6 Plant Breeding Methods Controlled by Mating

The ideotype breeding has been designed to improve gain for quantitative traits, especially in terms of productivity. It is a simple and straightforward method for complex traits by changing other simpler traits that are positively correlated with each other (Donald 1968). The main advantage of this method is, once the underlined hypothesis proves, that could encourage a significant gain for the product, even with a small or smart breeding program. Ideotype breeding can change the variation to the current plant type, beyond the boundaries of elite germplasm. Once the ideotype is assembled with the desired trait and inserted into an elite genetic pool, these lines can be transferred to the advanced breeding program.

The population breeding, a method designed to enhance overall phenotypic performance by increasing the frequency of favourable alleles which is dominant trait of interest in intermating population. Mass selection, is one of the simplest method of population breeding in cross-pollinated species, has developed a cultivar directly as enhanced population. Increasing the value of the working population from source lines is the main objective in modern population breeding improvement programs. Over-performing individuals are selected from average quality improved lines; with variations within those lines to be preserved. These outperforming lines can be used as release cultivars or parents of hybrids in case of self and cross-pollinated species, respectively.

In hybrid cultivars, which are measuring by heterosis, with superiority of individuals over their inbred parents (Shull 1948). Hybrid vigour, deteriorates abruptly because of inbreeding, indicating the presence of heterozygous loci. For this reason, maize breeding programs are currently focused on competitive F_1

hybrids development. In the plant kingdom, heterosis is more prominent in cross-pollinated species rather than self-pollinated ones. With increasing parental genetics distance, heterosis increase significantly (Springer and Stupar 2007).

The major challenges in hybrid breeding are: (a) extend of heterosis exhibited in crop plant species, (b) efficient pollination control mechanism, (c) seed rate of the crop plant and (d) efficient hybrid seed production methods to reduce the cost of seed. In Brassica, a high level of heterosis was found in spring and winter type of *B. napus* for seed yield, up to 40% heterosis has been reported in summer rape for yield and 60–70% for winter form (Lefort-Buson et al. 1982; Sernyk and Stefansson 1983; Erickson et al. 1986). Hybrids have a great advantage irrespective of business purpose, despite using mechanical emasculation ("detasselling" in maize) to avoid self-pollination in high yield for seed production. Due to the segregation of thousands of genes in genetically heterogeneous hybrid seeds, resulting in wide variations in agronomic traits, crop architecture, maturity duration, hereby reducing seed quality and overall yield if they planted in the next generation. So that, farmers have to buy fresh seed every year for the production of hybrid seed.

1.7 Speed Breeding: A Time Saving Tool

Researcher from the University of Queensland has coined the new term "speed breeding" as a method to accelerate the breeding speed in wheat crop. Now, it is has been developed in many crops (Watson et al. 2018; Ghosh et al. 2018). It does not require any specific equipment or technique like doubled haploids (DH), for the production of homozygous lines (Slama-Ayed et al. 2019). In a speed breeding program, using the optimum light intensity, temperature, and controlled day-length, increase the photosynthesis rate which is directly stimulates flowering and shortens to harvesting time (Watson et al. 2018; Chiurugwi et al. 2019). The wavelength and intensity of light are directly proportional to the regulation of flowering in plants. Early and late flowering genotypes were developed in peas, chickpeas, lupins, and faba beans under different light spectrum (Croser et al. 2016). Positive correlation between red:infrared proportion was observed in these species (R:FR) (Moe and Heins 1990; Ribalta et al. 2017). Reduces stem elongation and increases lateral branching by utilizing light with high R:FR and light with low R:FR, enhance stem elongation but reduces flowering and lateral branching. Speed breeding under controlled condition in greenhouse strategies increase the generation cycles by extended photoperiod (Lionneton et al. 2004; Ochatt and Sangwan 2008; Yao et al. 2020). Until now, speed breeding has introduced in some crop species, i.e., Wheat (*Triticum aestivum* L.), Chickpea (*Cicer arietinum* L.), sunflower (*Helianthus annuus*), pepper (*Capsicum annum*), radish (*Raphanus sativus*), and Amaranth (*Amaranthus spp.*) by extending photoperiods to shorten the crop duration (Stetter et al. 2016; Ghosh et al. 2018; Chiurugwi et al. 2019). International Crop Research Institute for the Semi-Arid Tropics (ICRISAT) has also worked on crops such as pigeon pea, sorghum, and millet for developing a protocol for short-day plants. Another success story was reported in speed breeding of peanuts (*Arachis hypogaea*). Induction of rapid crop development and early flowering is one of the main goal by improving protocol and required conditions (Chiurugwi et al. 2019). In conclusion, at present, cutting-edge genomics, modern plant breeding practicies, and innovative agronomic strategies have contributed to the development of improved cultivars that have led to remarkable agriculture outcomes (Ghosh et al. 2018).

1.8 Biotechnological Approaches

1.8.1 Plant Tissue Culture (PTC)

Plant tissue culture is an in vitro method of culturing cells, tissues, organs, single cells, protoplasts, embryos or whole plants (explants) on nutrient media under aseptic controlled conditions. Due to its high reliability and efficiency, PTC has played a crucial role in agricultural research and crop improvement during the past century. Various tissue culture methods have been used for crop improvement such as micro-propagation, shoot apex culture, germplasm conservation, haploid/diploid culture, embryo rescue technique, somaclonal variation, protoplast fusion, etc.

1.8.1.1 Micro-Propagation

Micro-propagation involves clonal propagation from small plant tissues and ensures rapid mass multiplication of genetically identical plants. Tiny parts of leaves, stems, buds, etc. are cultured on a synthetic medium in suitable containers such as glass jars. Later on, these are transferred onto the soil under greenhouse conditions. It is one of the successful commercial applications of PTC and has developed rapidly during the past five decades. It allows large scale production of plants in a smaller duration of time in the laboratory under controlled conditions. Many crop plants such as potato, sugarcane, sweet potato, yam, taro yam, turmeric, ginger, and cassava are multiplied through vegetative micro-propagation (Ahloowalia 2003). Micro-propagation has various advantages as compared to the traditional propagation methods such as rapid multiplication of true-to-type plants, production of disease-free plants (Brown and Thorpe 1995), and in some cases, increase in yield had been observed (Hussain et al. 2012).

Somaclonal Variation

Inheritable genetic or epigenetic variation is sometimes observed under in vitro conditions in callus cultures, isolated protoplasts or undifferentiated cells as compared to their donor plants and is termed as somaclonal variation (Larkin and Scowcroft 1981). The variation is usually due to chromosomal variation. Novel crop genotypes resistant to biotic and abiotic stresses, high and low pH, various diseases, salinity, herbicides could be obtained through somaclonal variation. Somaclonal variants had been reported in sugarcane (Heinz and Mee 1971), potato (Sharma et al. 2007), bananas (Sahijram et al. 2003), tomato (Bhatia et al. 2005), oats (Molnar et al. 2011), etc. and have improved traits. Notable commercial examples of some clones include Bio-13 variety of medicinal plants *Citronella java* and Super tomatoes (Bhatia 2015).

1.8.1.2 Somatic Embryogenesis

In somatic embryogenesis, somatic cells or tissues (cells other than gametes) are induced to differentiate into embryos (without fertilization) and subsequently develop into whole plants. It is a valuable genetic manipulation tool in crop plants. It was used for the development of resistant varieties in cotton (Sun et al. 2003; Han et al. 2009). In vitro cultures of mature and immature embryos are possible for recovering plants from interspecies and intergeneric crosses.

1.8.1.3 Doubled Haploids

Anther or Pollen culture has been used for doubled haploid production. Under normal conditions, it takes around 8–10 years to have desirable gene combinations in homozygous forms. With haploid culture, homozygous true breeding lines can be obtained instantly. Anthers/pollens from an unopened flower bud are cultured into a nutrient medium to form embryoids or callus (haploid plantlets). These are treated with colchicine to obtain homozygous doubled haploid plants that can be used for field testing and selection (Gosal et al. 2010). From F_1 progeny in rice plants, doubled haploids have been produced in less than one year in a study (Davey 2009). Doubled haploids have been produced in wheat, maize, mustard, rapeseed, tomato, and pigeonpea (Guzmán and Zapata Arias 2000; Croser et al. 2006; Seguí-Simarro and Nuez 2007).

1.8.1.4 Genetic Transformation

It is one of the aspects of tissue culture where gene(s) of interest are transferred to the host plants and transgenic plants are obtained. Genetic transformation has a high potency of crop improvement by modifying its genetics and introducing various agronomically important traits. It can be achieved by vector-mediated indirect or vector-less direct gene transfer methods. *Agrobacterium* based vector-mediated method is widely used for gene transformation. DNA and RNA virus-based vector methods are also available for efficient genetic transformation of target regions. *Agrobacterium tumefaciens* are gram-negative soil born bacterium that causes crown gall disease in host plants. Naturally, they infect the wounded plant by transferring T-DNA located on their Ti (tumor inducing) plasmid which integrates into the plant genome (Tisser and Bourgeois 2001; Gelvin 2003). Using this method, plants were regenerated from shoot apices in Jatropha (Purkayastha et al. 2010) and new grass peas varieties have been developed (Girma 2010). Herbicide tolerance in cotton, canola, maize, and soyabean, virus resistance in papaya and squash, insect resistance in wheat, cotton, rice, potato and maize, abiotic stress tolerance in rice, beta carotene enriched potato, bacterial wilt resistance in banana, etc. had also been achieved (Gelvin 2003).

1.8.1.5 Protoplast Fusion

Protoplasts (cells without cell walls) from two different species or genera are fused to form new living entities and the process is termed protoplast fusion or parasexual hybridization. It is one of the best methods to overcome sexual barriers and pre- or post-fertilization barriers. Moreover, nuclear and organellar genomes from distant species could be combined through protoplast fusion. Fusions of protoplasts could be symmetric where whole genomes from both parents combine or asymmetric where partial genome from a donor is transferred to the recipient (Wang et al. 2013). It had been used for the production of somatic hybrids and cytoplasmic hybrids (Downey and Rimmer 1993). Protoplast fusion can be induced by using chemical (polyethylene glycol (PEG) with sucrose and calcium chloride), mechanical, and electrical methods. This method had been used widely in the polyploid *Brassica* species. In the past, various protoplast hybrids were made between different species such as Brassica and Arabidopsis (Arabido brassica) (Hoffmann and Adachi 1981; Bauer-Weston et al. 1993), Brassica and *Sinapis turgida* (Toriyama et al. 1987), *Eruca sativa* and *B. juncea* (Erussica) (Sikdar et al. 1990), *Diplotaxis muralis* and *B. juncea* (Chatterjee et al. 1988), *B. nigra* and *B. oleracea* (Narasimhulu et al. 1992), *B. juncea* and *B. oleracea* (Lian et al. 2011), *Raphanus sativus* and *B. oleracea* (Yamanaka et al. 1992), common wheat and *Agropyron elongatum* (Xia 2009), seedless citrus hybrids (Grosser and Gmitter 2011), cotton hybrids (Sun et al. 2011), etc.

1.8.2 Germplasm Conservation

Endangered genotypes could be conserved with in vitro cultures. Clones can also be preserved with tissue culture methods to conserve the genetic background of a crop. Moreover, the plant species that do not produce seeds or have recalcitrant seeds could be preserved via in vitro methods for long periods in gene banks. In vitro methods are coupled with cryopreservation.

1.8.3 Genomics and Transcriptomics

Genomes and transcriptomes are sequenced at a mass scale with next-generation sequencing (NGS) technologies. These produce a large amount of genomic information that allows the discovery of new genes and regulators. The molecular basis of the traits can be studied with this information which could be used further in marker-assisted selection for the development of crop varieties. Association mapping is one of the advanced genomic technologies for studying genetics of the desirable traits. Genomic technologies are very useful for studying the genetics of complex traits of multi-gene nature in polyploid crops (such as wheat and Brassica) where more than one type of genome is present. These tools allow the detection of QTLs (quantitative trait loci) and alleles with small effects. Various crops have large and complex genomes and their sequencing is a challenging task. Thus, transcriptome sequencing is an alternative tool to understand expression behavior of the genes. Their is another important DNA sequencing technique - whole genome resequencing that allows the genome-wide discovery of markers such as SNPs (single nucleotide polymorphism) and the construction of high-density genetic maps (Bentley 2006).

1.8.4 TILLING and EcoTILLING

TILLING (target induced local lesions in genomes) allows screening of allelic variants in the target loci by screening mutant and germplasm collections. It is a reverse genetics approach that allows rapid and efficient detection of induced point mutations in mutagenized population. It is one of the powerful tools for crop improvement. In the original method, chemically (for example EMS) induced mutagenesis of the plant population is carried out. The seeds are treated with the mutagen and grown to produce M_1 plants which are self-fertilized to generate the M_2 population. DNA is extracted from leaf tissues of M_2 plants, followed by pooling and amplifying the region of interest. It creates heteroduplexes in the pooled DNA and then denaturing high performance liquid chromatography is performed to detect basepair changes (McCallum et al. 2000). EMS is the most widely used mutagen in TILLING and it produces transitions (G/C: A/T). DNA samples are normalized to avoid any biasing and then pooled together into 96 well microtiter plates, pooled samples are amplified for the target region of interest using fluorescently labeled forward and reverse primers ($5'$ end labeled with IRD700 and IRD800 dye). Heteroduplexes are formed during the denaturation and annealing cycles. An endonuclease CEL1 (isolated from celery) is applied for a short incubation period which recognizes the mismatch and cleaves the DNA at $3'$ end of the mismatch. The digested fragments are separated on a denaturing polyacrylamide gel attached to an LI-COR 4300 DNA analysis system. Homo and hetero duplexes can be detected and the location of the mutation can also be estimated in the DNA analysis system. The detected mutation is later on validated by sequencing of that particular region (Till et al. 2003; Comai and Henikoff 2006; Barkley and Wang 2008). TILLING has been widely used in the model plant *Arabidopsis thaliana* (Till et al. 2003), *Lotus japonicus* (Horst et al. 2007), Pea (Triques et al. 2007), maize (Till et al. 2004), Barley (Caldwell et al. 2004), wheat (Slade et al. 2005), rice (Wu et al. 2005), soybean (Cooper et al. 2008), rapeseed (Wells et al. 2014), etc.

EcoTILLING is a type of TILLING where natural genetic variation in the population is analyzed instead of induced variation. It is useful in species where it is not possible to induce mutations. Naturally occurring SNPs can be quickly screened in a population with this method (Barkley and Wang 2008). This approach had been used in various studies including Arabidopsis (Comai et al. 2004), black cottonwood (Gilchrist et al. 2006), Barley (Mejlhede et al. 2006), melon (Nieto et al. 2007), mung bean (Fery 2002), etc.

1.8.5 Genome Editing

Genome editing is an advanced molecular biology method that allows the targeted alteration of an organism's genome in a precise, robust, and efficient manner. It is used to elucidate the gene functions and thus, contributing to crop improvement. This technique uses sequence specific nucleases that recognize specific DNA sequences and generate double stranded breaks. The plant's endogenous repair mechanism heals this double stranded lesion by homologous recombination or non-homologous end joining. It can lead to gene replacements, insertion or deletions, thereby generating gene knockouts (Gao et al. 2015). Originally, genome editing used zinc-finder nucleases (ZFNs) (Kim et al. 1996) and transcription activator-like effect or nucleases (TALENs) (Christian et al. 2010). Recent advances in genome editing employs the use of CRISPER/Cas (Clustered regularly interspaced short palindromic repeats/CRISPER associated system) technology that offers a simple and efficient method of targeted gene editing (Jinek et al. 2012; Cong et al. 2013). It has been used in various important crops such as wheat, rice, rapeseed, soybean, potato, tomato, cotton, barley, soybean, apple, oranges, watermelon, grapes, etc. (Zhang et al. 2018). Examples of genome editing used in various crop improvement programs include tiller spreading phenotype in rice (Miao et al. 2013), enhanced grain number and size in rice (Li et al. 2016), increased shelf life of soybean oil (Haun et al. 2014), increased oleic acid and decreased polyunsaturated fatty acids in *Camelina sativa* (Jiang et al. 2017), high amylopectin maize (Pioneer 2016), waxy potatoes (Andersson et al. 2017), browning resistant mushrooms (Waltz 2016), fragrant rice (Shan et al. 2015), purple tomatoes (Čermák et al. 2015), etc. Resistant genotypes had also been developed in wheat, rice, tomatoes, grapefruits, and cucumber (Zhou et al. 2015; Nekrasov et al. 2017; Zhang et al. 2017, 2018; Ortigosa et al. 2019; Zaidi et al. 2020). Genome editing has the potential to manipulate multiple genes simultaneously and thus, allowing stacking of genes. With this approach, a trait having complex genetics can also be manipulated.

1.8.6 RNAi

RNA interference (RNAi) is a powerful tool in molecular biology and genetic engineering. RNAi is based on the naturally occurring conserved defense mechanism against double stranded RNA of cellular and viral mRNAs. In this process, small non-coding RNAs (micro RNA and small interfering RNA) interfere with target mRNA translation and thus leads to transcriptional or translational repression and hence suppress the expression of the gene. These small RNAs in association with RISC (RNA-induced silencing complex), Argonaute and other effector proteins lead to the phenomenon of RNAi. RNAi construct is designed in a way that self-complementary sequence is homologous to the target gene and forms a hairpin RNA (Redfern et al. 2013; Wilson and Doudna 2013; Saurabh et al. 2014). RNAi has great potential for crop improvement in the fields such as improving the quality, reducing the toxic substances, providing the resistance against various biotic and abiotic stresses, altering phenotype, in therapeutics, allergen and toxin elimination, and many more (Saurabh et al. 2014). Examples of RNAi for crop improvement include increased shelf life of tomato (Xiong et al. 2005; Meli et al. 2010), seedless watermelons (Varoquaux et al. 2000), restoring fertility in male sterile tobacco plants (Nizampatnam and Dinesh Kumar 2011), biofortification of tomatoes with antioxidants and essential elements (Niggeweg et al. 2004), flavor enhancement in canola seeds (Hüsken et al. 2005), lowering the allergen content in peanut (Dodo et al. 2008).

1.8.7 Nano-Biotechnology

Nano-biotechnology is one of the most promising technologies of the twenty-first century for sustainable agriculture practice. It combines the use of nanotechnology or nano-engineering in biology. It has high potential in crop improvement by increasing production, assuring sustainability, and lowering crop losses. Nanotools (nanobiosensors) helps in the precise, controlled and efficient management of agrochemicals such as fertilizers, pesticides and herbicides (Shang et al. 2019). For the detection of environmental stress and enhancing the plant's potential against diseases, nanosensors have high potential (Kwak et al. 2017). Agrochemicals are usually applied to the crops by spraying which results in ultralow levels reaching the target sites as there are losses by degradation or leaching. Nanotechnology offers controlled delivery techniques by nanoparticles that would release ideal amount of the agrochemical over the period and also, lower the harmful effects of over application (Nair et al. 2010; Pandey 2018; Rashid et al. 2018). Potassium nitrate was capsulated in graphene oxide films that slowly release the fertilizer (Zhang et al. 2014). The use of porous nanomaterials (zeolite, chitosan or clay) reduced the nitrogen loss by regulating the release on demand. They also increase the solubility of phosphate minerals, thus increasing the availability and hence uptake by the plants (Abdel-Aziz et al. 2016; Dwivedi et al. 2016). Nanotubes were used for extended release of active ingredients of pesticides, better control, and minimum environmental effect (Dwivedi et al. 2016). For foreign DNA and chemical deliveries inside the cell to manipulate the target gene(s), nanoparticles, nanocapsules, and nanofibers could be used (Torney et al. 2007). For instance, silica nanoparticles were used to deliver DNA inserts into tobacco and corn plants (Galbraith 2007). In the particle bombardment method of genetic transformation, nanoparticles delivery systems had been used (Vijayakumar et al. 2010). Nanoparticles based delivery had also been used for CRISPER/Cas9 technology (Mout et al. 2017).

References

Abdel-Aziz HMM, Hasaneen MNA, Omer AM (2016) Nano chitosan-NPK fertilizer enhances the growth and productivity of wheat plants grown in sandy soil. Spanish J Agric Res 14:e0902. https://doi.org/10.5424/sjar/2016141-8205

Acquaah G (2012) Principles of plant genetics and breeding, 2nd edn. Wiley

Ahloowalia BS (2003) Tissue culture | Clonal propagation, in vitro. In: Encyclopedia of applied plant sciences. Elsevier, pp 1360–1364

Allard RW (1960) Principles of plant breeding. Graham Toms Fine & Rare Books

Allard RW (1999) History of plant population genetics. Annu Rev Genet 33:1–27

Andersson M, Turesson H, Nicolia A et al (2017) Efficient targeted multiallelic mutagenesis in tetraploid potato (Solanum tuberosum) by transient CRISPR-Cas9 expression in protoplasts. Plant Cell Rep 36:117–128. https://doi.org/10.1007/s00299-016-2062-3

Barkley N, Wang M (2008) Application of TILLING and EcoTILLING as reverse genetic approaches to elucidate the function of genes in plants and animals. Curr Genomics 9:212–226. https://doi.org/10.2174/138920208784533656

Bauer-Weston B, Keller W, Webb J, Gleddie S (1993) Production and characterization of asymmetric somatic hybrids between Arabidopsis thaliana and Brassica napus. Theor Appl Genet 86: 150–158. https://doi.org/10.1007/BF00222073

Beddington JR, Asaduzzaman M, Clark ME et al (2012) The role for scientists in tackling food insecurity and climate change. Agric Food Secur 1:1–9. https://doi.org/10.1186/2048-7010-1-10

Bentley DR (2006) Whole-genome re-sequencing. Curr Opin Genet Dev 16:545–552. https://doi.org/10.1016/j.gde.2006.10.009

Bhatia S (2015) Application of plant biotechnology. Mod Appl Plant Biotechnol Pharm Sci:157–207. https://doi.org/10.1016/B978-0-12-802221-4.00005-4

Bhatia P, Krauss SL, Ashwath N et al (2005) Genetic analysis of cotyledon derived regenerants of tomato using AFLP markers. Curr Sci 88(2):280–284

Borojevic S (1990) Principles and methods of plant breeding. Elsevier

Breseghello F, Coelho ASG (2013) Traditional and modern plant breeding methods with examples in rice (Oryza sativa L.). J Agric Food Chem 35:8277–8286

Briggs F, Knowles P (1967) Introduction to plant breeding. New York, Reinhold Pub. Corp.

Brown J, Caligari PDS (2008) An introduction to plant breeding. Blackwell Publishing Ltd

Brown DCW, Thorpe TA (1995) Crop improvement through tissue culture. World J Microbiol Biotechnol 11:409–415. https://doi.org/10.1007/BF00364616

Caldwell DG, McCallum N, Shaw P et al (2004) A structured mutant population for forward and reverse genetics in Barley (Hordeum vulgare L.). Plant J 40:143–150. https://doi.org/10.1111/j.1365-313X.2004.02190.x

Čermák T, Baltes NJ, Čegan R et al (2015) High-frequency, precise modification of the tomato genome. Genome Biol 16:232. https://doi.org/10.1186/s13059-015-0796-9

Chahal GS, Gosal SS (2002) Principles and procedures of plant breeding: biotechnological and conventional approaches. Narosa Publishing House, New Delhi, Vol. 21, 64-89

Chatterjee G, Sikdar SR, Das S, Sen SK (1988) Intergeneric somatic hybrid production through protoplast fusion between Brassica juncea and Diplotaxis muralis. Theor Appl Genet 76:915–922. https://doi.org/10.1007/BF00273681

Chiurugwi T, Kemp S, Powell W, Hickey LT (2019) Speed breeding orphan crops. Theor Appl Genet 132:607–616

Christian M, Cermak T, Doyle EL et al (2010) Targeting DNA double-strand breaks with TAL effector nucleases. Genetics 186:757–761. https://doi.org/10.1534/genetics.110.120717

Comai L, Henikoff S (2006) TILLING: practical single-nucleotide mutation discovery. Plant J 45(4):684–694

Comai L, Young K, Till BJ et al (2004) Efficient discovery of DNA polymorphisms in natural populations by Ecotilling. Plant J 37:778–786. https://doi.org/10.1111/j.0960-7412.2003.01999.x

Cong L, Ran FA, Cox D et al (2013) Multiplex genome engineering using CRISPR/Cas systems. Science 339:819–823. https://doi.org/10.1126/science.1231143

Cooper JL, Till BJ, Laport RG et al (2008) TILLING to detect induced mutations in soybean. BMC Plant Biol 8:9. https://doi.org/10.1186/1471-2229-8-9

Croser JS, Lülsdorf MM, Davies PA et al (2006) Toward doubled haploid production in the fabaceae: Progress, constraints, and opportunities. Crit Rev Plant Sci 25:139–157

Croser JS, Pazos-Navarro M, Bennett RG et al (2016) Time to flowering of temperate pulses in vivo and generation turnover in vivo–in vitro of narrow-leaf lupin accelerated by low red to far-red ratio and high intensity in the far-red region. Plant Cell Tissue Organ Cult 127:591–599. https://doi.org/10.1007/s11240-016-1092-4

Davey MR (2009) Advances in haploid production in higher plants. Ann Bot 104. https://doi.org/10.1093/aob/mcp248

Dodo HW, Konan KN, Chen FC et al (2008) Alleviating peanut allergy using genetic engineering: the silencing of the immunodominant allergen Ara h 2 leads to its significant reduction and a decrease in peanut allergenicity. Plant Biotechnol J 6:135–145. https://doi.org/10.1111/j.1467-7652.2007.00292.x

Doebley JF, Gaut BS, Smith BD (2006) The molecular genetics of crop domestication. Cell 127: 1309–1321

Donald CM (1968) The breeding of crop ideotypes. Euphytica 17:385–403. https://doi.org/10.1007/BF00056241

Doust A, Diao X (2017) Genetics and genomics of setaria. Springer, Cham

Downey RK, Rimmer SR (1993) Agronomic improvement in oilseed brassicas. In: Advances in agronomy. Academic Press, pp 1–66

Dwivedi S, Saquib Q, Al-Khedhairy AA, Musarrat J (2016) Understanding the role of nanomaterials in agriculture. In: Microbial inoculants in sustainable agricultural productivity, vol 2. Functional Applications

Erickson L, Grant I, Beversdorf W (1986) Cytoplasmic male sterility in rapeseed (Brassica napus L.) - 1. Restriction patterns of chloroplast and mitochondrial DNA. Theor Appl Genet 72:145–150. https://doi.org/10.1007/BF00266985

Fehr W (1987) Principles of cultivar development: crop species. Agron Books

Fery RL (2002) New opportunities in Vigna. In: Trends in new crops and new uses. ASHS Press, Alexandria, VA

Galbraith DW (2007) Nanobiotechnology: silica breaks through in plants. Nat Nanotechnol 2:272–273

Gao J, Wallis JG, Browse J (2015) Mutations in the prokaryotic pathway rescue the fatty acid biosynthesis1 mutant in the cold. Plant Physiol 169:442–452. https://doi.org/10.1104/pp.15.00931

Gelvin SB (2003) Agrobacterium-mediated plant transformation: the biology behind the "gene-jockeying" tool. Microbiol Mol Biol Rev 67:16–37. https://doi.org/10.1128/MMBR.67.1.16-37.2003

Gepts P (2002) A comparison between crop domestication, classical plant breeding, and genetic engineering. Crop Sci 42:1780–1790

Ghosh S, Watson A, Gonzalez-Navarro OE et al (2018) Speed breeding in growth chambers and glasshouses for crop breeding and model plant research. Nat Protoc 13:2944–2963. https://doi.org/10.1038/s41596-018-0072-z

Gilchrist EJ, Haughn GW, Ying CC et al (2006) Use of Ecotilling as an efficient SNP discovery tool to survey genetic variation in wild populations of Populus trichocarpa. Mol Ecol 15:1367–1378. https://doi.org/10.1111/j.1365-294X.2006.02885.x

Girma D (2010) Ethiopian grass pea (Lathyrus sativus L.) started the genomics era. In: Transient genetic transformation of grass pea. Lambert Academic Publishing, Koeln

Goodman M, Brown W (1988) Races of corn. 33–79
Gosal SS, Wani SH, Kang MS (2010) Biotechnology and crop improvement. J Crop Improv 24: 153–217. https://doi.org/10.1080/15427520903584555
Grosser JW, Gmitter FG (2011) Protoplast fusion for production of tetraploids and triploids: applications for scion and rootstock breeding in citrus. Plant Cell Tissue Organ Cult 104:343–357. https://doi.org/10.1007/s11240-010-9823-4
Guzmán M, Zapata Arias FJ (2000) Increasing anther culture efficiency in rice (Oryza sativa L.) using anthers from ratooned plants. Plant Sci 151:107–114. https://doi.org/10.1016/S0168-9452(99)00204-6
Han GY, Wang X, Zhang G, Ma Z (2009) Somatic embryogenesis and plant regeneration of recalcitrant cottons (Gossypium hirsutum). African J Biotechnol 8:432–437
Harlan JR (1975) Our vanishing genetic resources. Science 188:618–621. https://doi.org/10.1126/science.188.4188.617
Haun W, Coffman A, Clasen BM et al (2014) Improved soybean oil quality by targeted mutagenesis of the fatty acid desaturase 2 gene family. Plant Biotechnol J 12:934–940. https://doi.org/10.1111/pbi.12201
Heinz DJ, Mee GWP (1971) Morphologic, cytogenetic, and enzymatic variation in saccharum species hybrid clones derived from callus tissue. Am J Bot 58:257–262. https://doi.org/10.1002/j.1537-2197.1971.tb09971.x
Hoffmann F, Adachi T (1981) "Arabidobrassica": chromosomal recombination and morphogenesis in asymmetric intergeneric hybrid cells. Planta 153:586–593. https://doi.org/10.1007/BF00385546
Horst I, Welham T, Kelly S et al (2007) Tilling mutants of Lotus japonicus reveal that nitrogen assimilation and fixation can occur in the absence of nodule-enhanced sucrose synthase. Plant Physiol 144:806–820. https://doi.org/10.1104/pp.107.097063
Hüsken A, Baumert A, Milkowski C et al (2005) Resveratrol glucoside (Piceid) synthesis in seeds of transgenic oilseed rape (Brassica napus L.). Theor Appl Genet 111:1553–1562. https://doi.org/10.1007/s00122-005-0085-1
Hussain A, Qarshi IA, Nazir H, Ullah I (2012) Plant tissue culture: current status and opportunities. In: Leva A, Rinaldi LMR (eds) Recent advances in plant in vitro culture. InTech
Jain HK, Kharkwal MC (eds) (2004) Plant breeding. Springer, Dordrecht
Jiang WZ, Henry IM, Lynagh PG et al (2017) Significant enhancement of fatty acid composition in seeds of the allohexaploid, Camelina sativa, using CRISPR/Cas9 gene editing. Plant Biotechnol J 15:648–657. https://doi.org/10.1111/pbi.12663
Jinek M, Chylinski K, Fonfara I et al (2012) A programmable dual-RNA-guided DNA endonuclease in adaptive bacterial immunity. Science (80-) 337:816–821. https://doi.org/10.1126/science.1225829
Kim YG, Cha J, Chandrasegaran S (1996) Hybrid restriction enzymes: zinc finger fusions to Fok I cleavage domain. Proc Natl Acad Sci 93:1156–1160. https://doi.org/10.1073/pnas.93.3.1156
Kwak SY, Wong MH, Lew TTS et al (2017) Nanosensor technology applied to living plant systems. Annu Rev Anal Chem 10:113–140
Larkin PJ, Scowcroft WR (1981) Somaclonal variation — a novel source of variability from cell cultures for plant improvement. Theor Appl Genet 60:197–214. https://doi.org/10.1007/BF02342540
Lefort-Buson M, Dattée Y, Lavoisier O, et al (1982) Genetic study of some agronomic characters in winter oilseed rape (Brassica napus L.) I.-Heterosis Genetic study of some agronomic characters in winter oilseed rape (Brassica napus L.)
Li M, Li X, Zhou Z et al (2016) Reassessment of the four yield-related genes Gn1a, DEP1, GS3, and IPA1 in rice using a CRISPR/Cas9 system. Front Plant Sci 7:377. https://doi.org/10.3389/fpls.2016.00377
Li H, Rasheed A, Hickey LT, He Z (2018) Fast-forwarding genetic gain. Trends Plant Sci 23:184–186

Lian YJ, Lin GZ, Zhao XM, Lim HT (2011) Production and genetic characterization of somatic hybrids between leaf mustard (Brassica juncea) and broccoli (Brassica oleracea). Vitr Cell Dev Biol Plant 47:289–296. https://doi.org/10.1007/s11627-011-9355-6

Lionneton E, Aubert G, Ochatt S, Merah O (2004) Genetic analysis of agronomic and quality traits in mustard (Brassica juncea). Theor Appl Genet 109:792–799. https://doi.org/10.1007/s00122-004-1682-0

McCallum CM, Comai L, Greene EA, Henikoff S (2000) Targeted screening for induced mutations. Nat Biotechnol 18:455–457. https://doi.org/10.1038/74542

Mejlhede N, Kyjovska Z, Backes G et al (2006) EcoTILLING for the identification of allelic variation in the powdery mildew resistance genes mlo and Mla of barley. Plant Breed 125:461–467. https://doi.org/10.1111/j.1439-0523.2006.01226.x

Meli VS, Ghosh S, Prabha TN et al (2010) Enhancement of fruit shelf life by suppressing N-glycan processing enzymes. Proc Natl Acad Sci U S A 107:2413–2418. https://doi.org/10.1073/pnas.0909329107

Miao J, Guo D, Zhang J et al (2013) Targeted mutagenesis in rice using CRISPR-Cas system. Cell Res 23:1233–1236. https://doi.org/10.1038/cr.2013.123

Moe R, Heins R (1990) Control of plant morphogenesis and flowering by light quality and temperature. Acta Hortic:81–90. https://doi.org/10.17660/actahortic.1990.272.11

Molnar SJ, Tinker NA, Kaeppler HF, Rines HW (2011) Molecular genetics of quality in oats. In: Oats: chemistry and technology, 2nd edn. Elsevier, pp 51–75

Mout R, Ray M, Yesilbag Tonga G et al (2017) Direct cytosolic delivery of CRISPR/Cas9-ribonucleoprotein for efficient gene editing. ACS Nano 11:2452–2458. https://doi.org/10.1021/acsnano.6b07600

Nair R, Varghese SH, Nair BG et al (2010) Nanoparticulate material delivery to plants. Plant Sci 179(3):154–163

Narasimhulu SB, Kirti PB, Prakash S, Chopra VL (1992) Resynthesis of Brassica carinata by protoplast fusion and recovery of a novel cytoplasmic hybrid. Plant Cell Rep 11:428–432. https://doi.org/10.1007/BF00234376

Nekrasov V, Wang C, Win J et al (2017) Rapid generation of a transgene-free powdery mildew resistant tomato by genome deletion OPEN. Sci Rep 7:482. https://doi.org/10.1038/s41598-017-00578-x

Nieto C, Piron F, Dalmais M et al (2007) EcoTILLING for the identification of allelic variants of melon elF4E, a factor that controls virus susceptibility. BMC Plant Biol 7:34. https://doi.org/10.1186/1471-2229-7-34

Niggeweg R, Michael AJ, Martin C (2004) Engineering plants with increased levels of the antioxidant chlorogenic acid. Nat Biotechnol 22:746–754. https://doi.org/10.1038/nbt966

Nizampatnam NR, Dinesh Kumar V (2011) Intron hairpin and transitive RNAi mediated silencing of orfH522 transcripts restores male fertility in transgenic male sterile tobacco plants expressing orfH522. Plant Mol Biol 76:557–573. https://doi.org/10.1007/s11103-011-9789-6

Ochatt SJ, Sangwan RS (2008) In vitro shortening of generation time in Arabidopsis thaliana. Plant Cell Tissue Organ Cult 93:133–137. https://doi.org/10.1007/s11240-008-9351-7

Ortigosa A, Gimenez-Ibanez S, Leonhardt N, Solano R (2019) Design of a bacterial speck resistant tomato by CRISPR/Cas9-mediated editing of SlJAZ2. Plant Biotechnol J 17:665–673. https://doi.org/10.1111/pbi.13006

Pandey G (2018) Challenges and future prospects of Agri-nanotechnology for sustainable agriculture in India. Environ Technol Innov 11:299–307

Paran I, Van Der Knaap E (2007) Genetic and molecular regulation of fruit and plant domestication traits in tomato and pepper. J Exp Bot 58:3841–3852

Pioneer (2016) DuPont pioneer announces intentions to commercialize first CRISPR-Cas product. JOHNSTON, IOWA (PRWEB). http://www.prweb.com/releases/dupont-pioneer-seed/crispr-cas-corn/prweb13349828.htm

Poehlman JM, Sleper D (1999) Breeding field crops. Iowa State University Press

Purkayastha J, Sugla T, Paul A et al (2010) Efficient in vitro plant regeneration from shoot apices and gene transfer by particle bombardment in Jatropha curcas. Biol Plant 54:13–20. https://doi.org/10.1007/s10535-010-0003-5

Rashid SA, Othman RNIR, Hussein MZ (2018) Synthesis, technology and applications of carbon nanomaterials. Elsevier

Redfern AD, Colley SM, Beveridge DJ et al (2013) RNA-induced silencing complex (RISC) proteins PACT, TRBP, and Dicer are SRA binding nuclear receptor coregulators. Proc Natl Acad Sci U S A 110:6536–6541. https://doi.org/10.1073/pnas.1301620110

Ribalta FM, Pazos-Navarro M, Nelson K et al (2017) Precocious floral initiation and identification of exact timing of embryo physiological maturity facilitate germination of immature seeds to truncate the lifecycle of pea. Plant Growth Regul 81:345–353. https://doi.org/10.1007/s10725-016-0211-x

Rodrigues ASL (2011) Improving coarse species distribution data for conservation planning in biodiversity-rich, data-poor, regions: no easy shortcuts. Anim Conserv 14:108–110

Sahijram L, Soneji JR, Bollamma KT (2003) Analyzing somaclonal variation in micropropagated bananas (Musa spp.). Vitr Cell Dev Biol Plant 39:551–556. https://doi.org/10.1079/IVP2003467

Saurabh S, Vidyarthi AS, Prasad D (2014) RNA interference: concept to reality in crop improvement. Planta 239:543–564. https://doi.org/10.1007/s00425-013-2019-5

Seguí-Simarro JM, Nuez F (2007) Embryogenesis induction, callogenesis, and plant regeneration by in vitro culture of tomato isolated microspores and whole anthers. J Exp Bot 58:1119–1132. https://doi.org/10.1093/jxb/erl271

Sernyk JL, Stefansson BR (1983) Heterosis in summer rape (Brassica napus L.). Can J Plant Sci 63: 407–413. https://doi.org/10.4141/cjps83-046

Shan Q, Zhang Y, Chen K et al (2015) Creation of fragrant rice by targeted knockout of the OsBADH2 gene using TALEN technology. Plant Biotechnol J 13:791–800. https://doi.org/10.1111/pbi.12312

Shang Y, Kamrul Hasan M, Ahammed GJ et al (2019) Applications of nanotechnology in plant growth and crop protection: a review. Molecules 24(14):255

Sharma SK, Bryan GJ, Winfield MO, Millam S (2007) Stability of potato (Solanum tuberosum L.) plants regenerated via somatic embryos, axillary bud proliferated shoots, microtubers and true potato seeds: a comparative phenotypic, cytogenetic and molecular assessment. Planta 226: 1449–1458. https://doi.org/10.1007/s00425-007-0583-2

Shiferaw B, Smale M, Braun HJ et al (2013) Crops that feed the world 10. Past successes and future challenges to the role played by wheat in global food security. Food Secur 5:291–317. https://doi.org/10.1007/s12571-013-0263-y

Shull GH (1948) What is "heterosis" ? Genetics 33:439–446

Sikdar SR, Chatterjee G, Das S, Sen SK (1990) "Erussica", the intergeneric fertile somatic hybrid developed through protoplast fusion between Eruca sativa Lam. and Brassica juncea (L.) Czern. Theor Appl Genet 79:561–567. https://doi.org/10.1007/BF00226168

Slade AJ, Fuerstenberg SI, Loeffler D et al (2005) A reverse genetic, nontransgenic approach to wheat crop improvement by TILLING. Nat Biotechnol 23:75–81. https://doi.org/10.1038/nbt1043

Slama-Ayed O, Bouhaouel I, Ayed S et al (2019) Efficiency of three haplomethods in durum wheat (Triticum turgidum subsp. durum desf.): isolated microspore culture, gynogenesis and wheat × maize crosses. Czech J Genet Plant Breed 55:101–109. https://doi.org/10.17221/188/2017-CJGPB

Springer NM, Stupar RM (2007) Allelic variation and heterosis in maize: how do two halves make more than a whole? Genome Res 17:264–275

Stetter MG, Zeitler L, Steinhaus A et al (2016) Crossing methods and cultivation conditions for rapid production of segregating populations in three grain amaranth species. Front Plant Sci 7: 816. https://doi.org/10.3389/fpls.2016.00816

Sun Y, Zhang X, Jin S et al (2003) Somatic embryogenesis and plant regeneration in wild cotton (Gossypium klotzschianum). Plant Cell Tissue Organ Cult 75:247–253. https://doi.org/10.1023/A:1025895631247

Sun Y, Liu S, Wang Y et al (2011) An interspecific somatic hybrid between upland cotton (G. hirsutum L. cv. ZDM-3) and wild diploid cotton (G. klotzschianum a.). Plant Cell Tissue Organ Cult 106:425–433. https://doi.org/10.1007/s11240-011-9939-1

Tanksley SD, McCouch SR (1997) Seed banks and molecular maps: unlocking genetic potential from the wild. Science (80-) 277:1063–1066. https://doi.org/10.1126/science.277.5329.1063

Tester M, Langridge P (2010) Breeding technologies to increase crop production in a changing world. Science (80-) 327:818–822

Till BJ, Reynolds SH, Greene EA et al (2003) Large-scale discovery of induced point mutations with high-throughput TILLING. Genome Res 13:524–530. https://doi.org/10.1101/gr.977903

Till BJ, Reynolds SH, Weil C et al (2004) Discovery of induced point mutations in maize genes by TILLING. BMC Plant Biol 4:12. https://doi.org/10.1186/1471-2229-4-12

Tisser A, Bourgeois P (2001) Reverse genetics in plants. Curr Genomics 2:269–284

Toriyama K, Kameya T, Hinata K (1987) Selection of a universal hybridizer in Sinapis turgida Del. and regeneration of plantlets from somatic hybrids with Brassica species. Planta 170:308–313. https://doi.org/10.1007/BF00395021

Torney F, Trewyn BG, Lin VSY, Wang K (2007) Mesoporous silica nanoparticles deliver DNA and chemicals into plants. Nat Nanotechnol 2:295–300. https://doi.org/10.1038/nnano.2007.108

Triques K, Sturbois B, Gallais S et al (2007) Characterization of Arabidopsis thaliana mismatch specific endonucleases: application to mutation discovery by TILLING in pea. Plant J 51:1116–1125. https://doi.org/10.1111/j.1365-313X.2007.03201.x

Troyer AF (1999) Background of U.S. Hybrid Corn Crop Sci 39:601–626. https://doi.org/10.2135/CROPSCI1999.0011183X003900020001XA

Varoquaux F, Blanvillain R, Delseny M, Gallois P (2000) Less is better: new approaches for seedless fruit production. Trends Biotechnol 18(6):233–242

Varshney RK, Hoisington DA, Tyagi AK (2006) Advances in cereal genomics and applications in crop breeding. Trends Biotechnol 24:490–499

Vilmorin LD (1856) Note sur la création d'une nouvelle race de betteraves à sucre. – Considérations sur l'hérédité dans les végétaux. Comptes Rendus de l'Académie des Sciences 43:113 n. 1

Vijayakumar PS, Abhilash OU, Khan BM, Prasad BLV (2010) Nanogold-loaded sharp-edged carbon bullets as plant-gene carriers. Adv Funct Mater 20:2416–2423. https://doi.org/10.1002/adfm.200901883

Waltz E (2016) Gene-edited CRISPR mushroom escapes US regulation. Nature 532(7599):293

Wang J, Jiang J, Wang Y (2013) Protoplast fusion for crop improvement and breeding in China. Plant Cell Tissue Organ Cult 112:131–142. https://doi.org/10.1007/s11240-012-0221-y

Watson A, Ghosh S, Williams MJ et al (2018) Speed breeding is a powerful tool to accelerate crop research and breeding. Nat Plants 4:23–29. https://doi.org/10.1038/s41477-017-0083-8

Wells R, Trick M, Soumpourou E et al (2014) The control of seed oil polyunsaturate content in the polyploid crop species Brassica napus. Mol Breed 33:349–362. https://doi.org/10.1007/s11032-013-9954-5

Wilson RC, Doudna JA (2013) Molecular mechanisms of RNA interference. Annu Rev Biophys 42:217–239. https://doi.org/10.1146/annurev-biophys-083012-130404

Wu JL, Wu C, Lei C et al (2005) Chemical- and irradiation-induced mutants of indica rice IR64 for forward and reverse genetics. Plant Mol Biol 59:85–97

Xia G (2009) Progress of chromosome engineering mediated by asymmetric somatic hybridization. J Genet Genomics 36:547–556. https://doi.org/10.1016/S1673-8527(08)60146-0

Xiong AS, Yao QH, Peng RH et al (2005) Different effects on ACC oxidase gene silencing triggered by RNA interference in transgenic tomato. Plant Cell Rep 23:639–646. https://doi.org/10.1007/s00299-004-0887-7

Yamanaka H, Kuglnuki Y, Kanno T, Nishio T (1992) Efficient production of somatic hybrids between Raphanus sativus and Brassica oleracea. Japanese J Breed 42:329–339. https://doi.org/10.1270/jsbbs1951.42.329

Yao M, Guan M, Zhang Z et al (2020) GWAS and co-expression network combination uncovers multigenes with close linkage effects on the oleic acid content accumulation in Brassica napus. BMC Genomics. https://doi.org/10.1186/s12864-020-6711-0

Zaidi SS-E-A, Mahas A, Vanderschuren H, Mahfouz MM (2020) Engineering crops of the future: CRISPR approaches to develop climate-resilient and disease-resistant plants. Genome Biol 21:1–19. https://doi.org/10.1186/s13059-020-02204-y

Zamir D (2001) Improving plant breeding with exotic genetic libraries. Nat Rev Genet 2:983–989

Zeven AC (1998) Landraces: a review of definitions and classifications. Euphytica 104:127–139. https://doi.org/10.1023/A:1018683119237

Zhang M, Gao B, Chen J et al (2014) Slow-release fertilizer encapsulated by graphene oxide films. Chem Eng J 255:107–113. https://doi.org/10.1016/j.cej.2014.06.023

Zhang Y, Bai Y, Wu G et al (2017) Simultaneous modification of three homoeologs of TaEDR1 by genome editing enhances powdery mildew resistance in wheat. Plant J 91:714–724. https://doi.org/10.1111/tpj.13599

Zhang Y, Massel K, Godwin ID, Gao C (2018) Applications and potential of genome editing in crop improvement. Genome Biol 19:210. https://doi.org/10.1186/s13059-018-1586-y

Zhou J, Peng Z, Long J et al (2015) Gene targeting by the TAL effector PthXo2 reveals cryptic resistance gene for bacterial blight of rice. Plant J 82:632–643. https://doi.org/10.1111/tpj.12838

Zohary D (1988) Domestication of plants in the old world: the origin and spread of cultivated plants in West Asia, Europe, and the Nile Valley. Clarendon Press

Marker-Assisted Selection for Value Addition in Crop Plants

2

Kamaluddin, Preeti Sonkar, Vijay Sharma, Hitesh Kumar, Mukul Kumar, H. S. Negi, Usha Kiran, M. Z. Abdin, and A. K. Choubey

Abstract

Modern-day varieties of different agricultural crops are the result of long-time selection and domestication of wild plants by human race. To feed the growing population and to overcome the problem of malnutrition, there is a challenge before the breeder to increase production of nutrient-rich vegetables, cereals and fruits for achieving food and nutritional security. Marker-assisted selection is one of the most important modern molecular tools of genetic improvement of crops that select plant having desirable genotype efficiently at early stage and thus reduce time of development of improved cultivars. It is a technology that selects gene responsible for expression of particular traits such as disease resistance, quality traits (aroma in basmati rice, flavour in vegetables and fruits) and abiotic stress tolerance using linked marker and is helpful in developing improved cultivars with high yield and enhanced nutritional quality of cereals, fruits and vegetables. Marker-assisted selection is also helpful in identification of cultivars,

Kamaluddin (✉) · P. Sonkar · V. Sharma · H. Kumar · M. Kumar
Department of Genetics and Plant Breeding, College of Agriculture, Banda University of Agriculture and Technology, Banda, Uttar Pradesh, India

H. S. Negi
Department of Plant Pathology, College of Agriculture, Banda University of Agriculture and Technology, Banda, Uttar Pradesh, India

U. Kiran
Vanercia Institute of Technical Education, Gautam Buddh Nagar, Uttar Pradesh, India

M. Z. Abdin
Department of Biotechnology, School of Chemical and Life Sciences, Jamia Hamdard, India

A. K. Choubey
Department of Soil Science and Agriculture Chemistry, College of Agriculture, Banda University of Agriculture and Technology, Banda, Uttar Pradesh, India

Kamaluddin et al. (eds.), *Technologies in Plant Biotechnology and Breeding of Field Crops*, https://doi.org/10.1007/978-981-16-5767-2_2

assessment of genetic diversity and purity, selection of parents, study of heterosis, etc. MAS has been categorized into five broad areas, viz., marker-assisted evaluation of breeding material, gene pyramiding, early generation selection, marker-assisted backcrossing and combined MAS. Marker-assisted selection is helpful in value addition of varieties by incorporating gene responsible for aroma, flavour, enhanced content of micronutrients, etc. from different sources to the elite varieties.

Keywords

Marker-assisted evaluation · Marker-assisted selection · Heterosis · Quantitative trait locus · Elite variety

2.1 Introduction

The present-day agricultural crops are the result of long-time selection and domestication of wild plants by human race. Nevertheless, this resulted in the action of elimination, acquisition and retention of certain traits, which proved to be beneficial (Ranjisha et al. 2020). The world grew in both population and economic development. As industrialization and urbanization is fast, lifespan and survival rate of the people improved, and the world's population become nearly triple from about 2.5 billion in 1961 to 7.2 billion in 2016 in these 56 years (Tandzi et al. 2020). It was also estimated that the human population will be around 10 billion in 2050. Besides, almost 800 million people all over the world and mostly in developing countries are malnourished. Apart from this, about 2 billion people globally are suffering from hidden hunger which is mainly due to meagre intake of essential micronutrients in their daily diet.

Although Green Revolution has increased yield of field crops significantly, the improved varieties lack in important micro nutrients. To feed this ever-growing population and to overcome the problem of malnutrition in human population, there is need of hour and important challenge before the breeder to increase production of nutrient-rich vegetables, cereals, and fruits for achieving food and nutritional security. To achieve this goal, there is need to improve infrastructure, judicious usage of pesticides and fertilizers, disease diagnostics and use of high-throughput technologies of crop improvement to boost agricultural productivity. Marker-assisted selection may be one of the best high-throughput technologies of molecular breeding that speed up breeding of varieties with high yield and good nutritional quality.

Marker-assisted selection (MAS) is a strategy to select a gene or genes responsible for expression of a particular trait like disease resistance, quality traits (aroma in basmati rice, flavour in vegetables), heat stress, drought stress, salinity stress tolerance, etc. (Prabhu et al. 2009). It involves selection of those plants which possess genomic regions that control expression of the traits of interest through use of molecular markers (Das et al. 2017). It is also useful in identification of cultivars, assessment of genetic diversity and purity, selection of parents and study of heterosis, also helpful to improve yield and nutritional quality of the crops and subsequently may help in bridging the gap between demand and supply of food for the continuously growing population (Gouda et al. 2020).

Many researchers reported various types of molecular tools including MAS that can be reasonable and environmental friendly approach for the improvement of biotic and abiotic stress-resistant crops with enhanced quality of vegetables and fruits (Das et al. 2017).

2.2 Marker-Assisted Selection (MAS)

The important steps in marker-assisted selection are selection of desirable parents, development of breeding population, isolation of DNA from each plant, scoring of molecular marker and correlation of molecular marker with molecular traits (Fig. 2.1).

However, MAS has an advantage in early generation selections by eliminating undesirable gene combinations which lack essential disease resistance genes. Barr et al. reported that this is fantasy for public sector breeders, as MAS can only

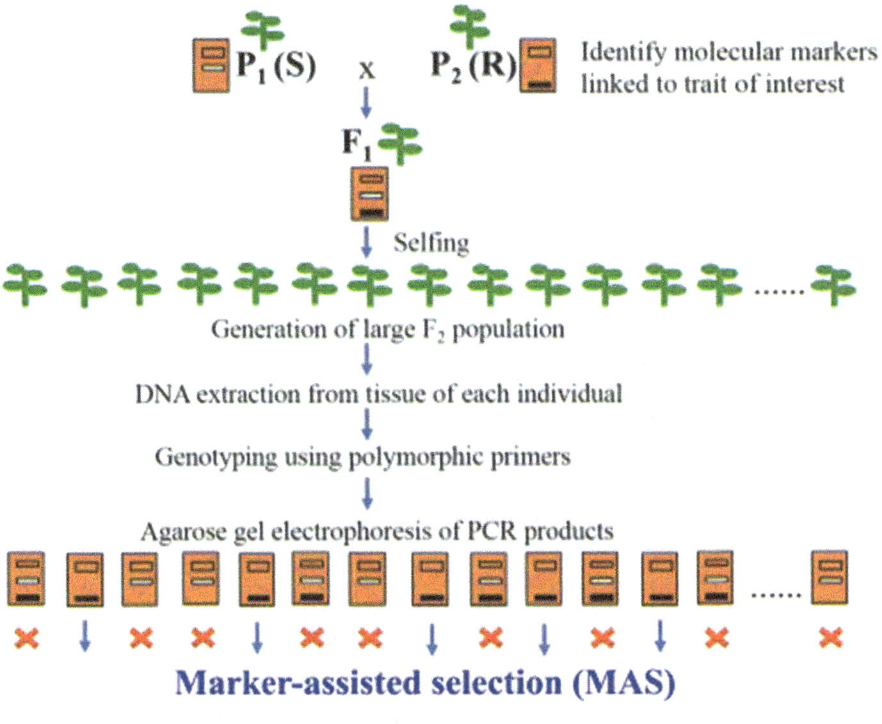

Fig. 2.1 Schematic representation of marker-assisted selection

be used in early generation screening for very important material. Markers are also used to select parents. The achievable genetic gain through marker-assisted recurrent selection (MARS) is higher than achievable through marker-assisted backcrossing (MABC). Collard and Mackill (2008) classified MAS in these broad areas, viz., marker-assisted evaluation of breeding material, gene pyramiding, early generation selection, marker-assisted backcrossing and combined MAS.

2.3 Marker-Assisted Evaluation of Breeding Material

This approach included cultivar identity, assessment of genetic diversity and parent selection and confirmation of hybrids. This technique has been done based on visual selection and analysing data based on morphological characteristics. Cultivar identity is also known as assessment of purity, i.e., seed of different strains is mixed due to handling of large numbers, and markers can be used to confirm the true identity of plants as well as maintenance of high levels of genetic purity (Yashitola et al. 2002).

2.3.1 Assessment of Genetic Diversity and Parental Selection

Breeding programmes depend on high level of genetic diversity for achieving progress from selection (Reif et al. 2005). Marker-assisted selection is useful in selection of those plants which possess desirable allele using linked marker.

2.3.2 Study of Heterosis

DNA markers have been used for hybrid crop to identify more heterotic parents and useful to classify them in to different heterotic groups (Reif et al. 2003). Upon crossing, parents under different heterotic groups show more heterosis than the other.

2.3.3 Identification of Genomic Regions under Selection

Marker-assisted selection is an important tool to identify specific gene or genes or haplotypes' shifts in allele frequencies within genome (Steele et al. 2004). This may enable breeder to devise appropriate breeding strategies to breed improved cultivars.

2.4 Marker-Assisted Backcrossing (MABC)

Marker-assisted backcross breeding is an important application of MAS which is used in plant breeding to transfer desirable traits from a donor plant into an elite genotype (recurrent parent) using linked molecular marker (Zhou et al. 2003b). In repeated crossings, the original cross is backcrossed with the recurrent parent until most of the genes stemming from the donor are eliminated (Ragimekula et al. 2013; Frisch et al. 2005). MABC is efficient if single allele is to be transferred in all background. In MABC, DNA markers can be scored as a dominant or codominant trait prior to flowering and facilitate the backcrossing programme as saving time and saving resources if phenotyping is difficult (Fig. 2.2).

MABC is an artistic tool for development of improved varieties and is simplest form of MAS where goal is to incorporate a major gene from agronomically inferior source into desired cultivar. Nevertheless, it is advantageous as selection can be carried out at seedling stage and is simpler than phenotypic screening which can save time, resources and efforts. Three general levels of marker-assisted backcrossing (Holland 2004) are given below.

2.4.1 Foreground Selection

In the first level, markers can be used in combination with or to replace screening for the target gene or quantitative trait locus (QTL), known as foreground selection (Holland 2004). It is an approach under MAS and is performed in backcross population to track the target alleles and selected plants having the desire allele of donor parent at target locus using tightly linked markers. This target locus is present in a heterozygous state (one is donor allele and one recurrent parent allele).

2.4.2 Recombinant Selection

The second level involves selection of BC progeny with the target gene and recombination events between the target locus and flanking marker, and hence, it is also called as recombination selection (Salina et al. 2003). The purpose of

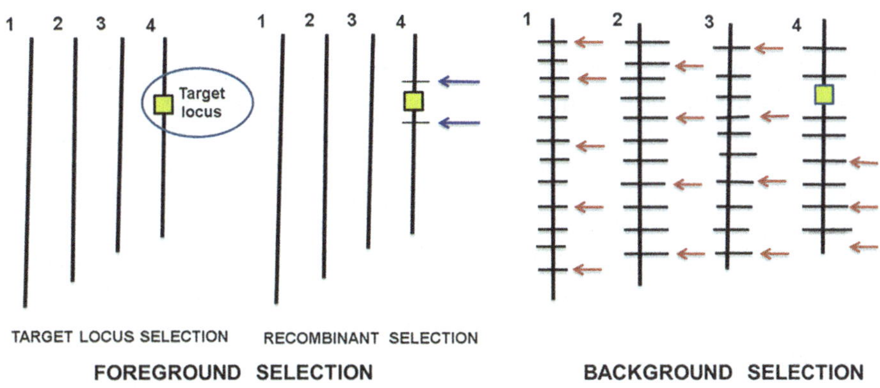

Fig. 2.2 Principles of marker-assisted backcross breeding

recombinant selection is to reduce the size of the donor chromosome segment containing the target locus. Here, fragment size of donor slowly decreases.

2.4.3 Background Selection

Background selection involves selection of backcross progeny with the greatest proportion of recurrent from recurrent parent genome, using markers that are unlinked to target locus (Frisch et al. 1999). Ribaut and Ragot (2007) improved drought adaptation in maize using MABC. Different foreground and background methods were compared and suggested that over 99% recovery of the recipient QTL alleles for the background traits were achieved after three generations of backcrossing. Some important examples of transfer of important traits from donor to recipient parents using marker-assisted backcross breeding are enlisted in Table 2.1.

2.5 Marker-Assisted Recurrent Selection (MARS)/Population Improvement

The improvement of complex traits via phenotypic recurrent selection is generally possible. On the other hand, long selection cycles impose restrictions on the practicability of this breeding method. With the use of markers, recurrent selection can be accelerated considerably, and several selection cycles are possible within 1 year, accumulating favourable QTL alleles in the breeding population (Ragimekula et al. 2013).

2.6 Marker-Assisted Gene Pyramiding

Gene pyramiding is a way to combine into a single genotype, a series of target genes identified in different parents. Servin et al. (2000) reported to find out the best gene-pyramiding scheme in which there are eight targets in three generations less than the reference method while requiring few genotyping. Here, it's assumed that recombination fractions between the loci are known and want to derive the 'ideal' genotype (called ideotype), i.e., homozygous. Gene-pyramiding scheme is divided into two parts: one is pedigree and second is fixation steps (aimed at fixing the target gene into a homozygous state) as well as derived the ideotype from the root genotype.

Rice cultivars ASD 16 and ADT 43 were improved through introgression of blast (*Pi 54*), sheath blight (*qSBR7-1*, *qSBR11-1* and *qSBR11-2*) and bacterial blight (*xa5*, *xa13*, and *Xa21*) resistance genes/QTLs using marker-assisted backcross breeding (Ramalingam et al. 2020) (Table 2.2).

Table 2.1 Marker-Assisted Backcross breeding for different traits in crop plants

Crops	Trait(s)	Gene/QTLs	Foreground selection	Background selection	Reference
Barley	Barley yellow dwarf virus	*Yd2*	STS	Not performed	Jefferies et al. (2003)
	Leaf rust	*Rphq6*	AFLP	AFLP	Van Berloo et al. (2001)
	Stripe rust	QTLs on 4H and 5H	RFLP	Not performed	Toojinda et al. (1998)
	Yield	QTLs on 2HL and 3HL	RFLP	RFLP	Schmierer et al. (2004)
Maize	Corn borer resistance	QTLs on chromosomes 7, 9 and 10	RFLP	RFLP	Wilcox et al. (2002)
	Earliness and yield	QTLs on chromosomes 5, 8 and 10.	RFLP	RFLP	
Rice	Bacterial blight	*Xa21*	STS	RFLP	Chen et al. (2000)
	Bacterial blight	*Xa21*	STS	RFLP	Chen et al. (2001)
	Bacterial blight	*xa5*, *Xa13*, *Xa21*	STS and CAPS	Not performed	
	Bacterial blight and quality	*Xa13*, *xa21*	STS and SSR	AFLP	Joseph et al. (2004)
	Blast	*Pi1*	SSR	ISSR	Liu et al. (2003)
	Deep roots	QTLs on chromosomes 1, 2, 7 and 9	RFLP and SSR	SSR	Shen et al. (2001)
	Quality	Waxy	RFLP	AFLP	Zhou et al. (2003a)
	Root traits and aroma	QTLs on chromosomes 2, 7, 8, 9 and 11	RFLP and SSR	RFLP and SSR	
	Submergence tolerance	*Sub1* QTLs	Phenotyping and SSR	SSR	Mackill et al. (2006)
	Submergence tolerance, disease resistance and quality	*Subchr9* QTL, *Xa21*, *Bph* and blast QTLs and quality loci	SSR and STS	Not performed	
Wheat	Powdery mildew	*22Pm* genes	Phenotyping	AFLP	

Table 2.2 Marker-assisted gene pyramiding for important traits in crop plants

S. No.	Crops	Trait(s)	Gene from parent1	Genes from parent2	Selection stage	Marker	Reference
1.	Barley	Barley yellow mosaic virus	*Rmy1*	*Rym5*	F_2	RFLP, CAPS	Okada et al. (2004)
2.	Barley	Barley yellow mosaic virus	*Rym4, rym9, rym11*	*Rym4, rym9, rym11*	F_1	RAPD, SSR	Werner et al. (2005)
3.	Barley	Stripe rust	Rspx, Rspx	QTLs 4,7 QTLs5	F_1	SSR	Castro et al. (2003)
4.	Rice	Bacterial blight	Xa5, xa13	Xa4, Xa21	F_2	RFLP, STS	Huang et al. (1997)
5.	Rice	Bacterial blight, yellow stem borer, sheath blight	Xa21, Bt	RC7 chitinase gene, Bt	F_2	STS	Datta et al. (2002)
6	Rice	Blast disease	Pi1, Piz-5	Pil, Pita	F_2	RFLP, STS	Hittalmani et al. (2000)
7	Rice	Brown plant hopper	Bph1	Bph2	F_4	STS	Sharma et al. (2004)
8	Rice	Insect resistance and bacterial blight	Xa21	Bt	F_2	STS	Jiang et al. (2004)
9	Wheat	Powdery mildew	Pm2	Pm4a	F_2	RFLP	Liu et al. (2000)

2.7 Early Generation Marker-Assisted Selection

In the plant breeding programme, MAS is a great advantage for early generations because undesirable gene combinations can be eliminated. Early generation called as single large-scale MAS increases efficiency of MAS due to increasing probability of recombination between the marker and QTL.

2.8 Combined Marker-Assisted Selection

When phenotypic screening can be successively combined with MAS, this technique is more efficient to single phenotypic screening. In some situations, there is low level of recombination between a marker and QTL, and markers flanking the QTL are used.

2.9 Application of Marker-Assisted Selection for Value Addition in Crops

MAS is very efficient, effective and rapid method for transferring desirable genes like resistance genes for biotic and abiotic stresses and genes responsible for quality traits in crop plants. It is useful in gene pyramiding for disease and insect resistance and is being used for transfer of male sterility and photo period insensitivity into cultivated genotypes from different sources. However, MAS is being used for improvement of quality characters in different crops such as for protein quality in maize, fatty acid (linolenic acid) content in soybean and storage quality in vegetables and fruit crops. Nevertheless, MAS can be successfully used for transferring desirable transgene (such as *Bt* gene) from one cultivar to another. MAS is useful in genetic improvement of tree species also where fruiting takes very long time (say 20 years) because for application of phenotypic selection, we have to wait for such a long time.

2.10 MAS for Disease and Insect Pest Resistance

Marker-assisted selection (MAS) is analysed by marker to select for particular DNA segments, i.e., genetically linked and resistance to disease. The impact of DNA marker-assisted selection on breeding disease-resistant cultivar has been impressive as reflected by number of studies. In lieu of restriction fragment length polymorphism (RFLP), PCR-based molecular tools in the form of STS and SSR have been developed to implement genotypic selection. Two important diseases, BLB and blast, are being developed through successful pyramiding of multiple disease resistance genes. A number of molecular markers, such as STS and SSR, linked to various BLB and blast genes of rice and their primer sequences are listed in database.

Among several STS markers generated from AFLP fragments linked to the rice bacterial blight resistance gene *Xa7*, *M5* was found to be cosegregating with gene of interest. Gu et al. saturated the *Xa 27* (t) genomic region with markers derived from the markers derived from genomic sequence of *Oryza sativa* cv. Nipponbare and developed markers, viz., M631, M1230 and M449, cosegregate with the gene of interest. SSR and STS markers linked to BB resistance genes, viz., *xa5*, *xal3* and *xa21,* have been identified and developed in rice. PCR-based markers and mapping for blast resistance genes have been developed and analysed by molecular profile. Bacterial blight resistance of two elite restorer line '6078' and 'Minghui63' improved by incorporating *Xa21* from 'IRBB21' through MAS (Chen et al. 2000).

Medaner et al. incorporated wheat rust resistance genes *Lr34* and *Yr36*, the eyespot resistance genes *rym4* and *rym5* to barley yellow mosaic viruses, *mlo* to barley powdery mildew and two QTL for resistance *Fhb*1 and *Qfhs.ifa-54* to Fusarium head blight in wheat and barley, respectively, using MAS. Risk et al. (2013) observed that barley lines showed enhanced resistance against barley leaf rust and powdery mildew after incorporation of resistance genes against these diseases applying MAS.

Das and Rao (2015) reported successful incorporation of gene/QTLs to confer resistance to blast (*Pi2* and *Pi9*) and gall midge (*GM1* and *GM4*) in a rice variety CRMAS2621-7-1 using marker-assisted selection. Six pyramided wheat lines with two disease resistance genes and high-quality genes were obtained as well as many improved agronomical traits (Zhang et al. 2020). Jamaloddin et al. (2020) developed two rice lines, viz., TH-625-159 and TH-625-491, possessing four genes that exhibited a high level of resistance to bacterial blight and blast diseases using gene-specific marker (pTA248 (*Xa21*), *xa13prom* (*xa13*), *Pi54MAS* (*Pi54*) and RM224 (*Pi1*)).

Dixit et al. (2020) emphasized the threat of biotic stresses on yield of crops, i.e., blast (Pi9), bacterial leaf blight (*Xa4, Xa5, Xa13* and *Xa21*), brown plant hopper (*Bph3* and *Bph17*) and gall midge (*GM4* and *GM8*) and suggested that these could be improved through marker-assisted forward breeding (MAFB) approach. They also analysed 3 ILs, viz., IL1 (Pi9 + Xa4 + Xa5 + Xa21 + Bph17 + Gm8 + 9DTY$_{1.1}$ + 9DTY$_{3.1}$), that showed resistance against the above diseases both in field and glasshouse conditions.

Hanson et al. (2016) developed F_7 multiple disease-resistant tomato lines that showed resistant to tomato yellow leaf curl disease, late blight, bacterial wilt, grey leaf spot and tobacco mosaic virus using marker-assisted selection. Some examples of lines/stocks carrying disease resistance gene in their genetic background using marker-assisted selection are given in Table 2.3.

2.11 MAS for Insect Pest Resistance

Notorious insect pests mainly caused damage to agricultural crops that have been estimated around 14% of total agricultural crop production including 83% in rice, 52% in wheat, 59% in maize, 74% in potato, 58% in soybean and 84% in cotton, respectively. Economic injuries due to pesticide uses and control measures of biotic and abiotic factors have been valuated at 10 billion and 2 billion USD, respectively. Smith (2021) advocated that development of insect-resistant varieties is the best way to minimize yield losses and to feed the world population. Among the mapped insect-resistant genes, considerable progress has been made in development of PCR-based markers for gall midge resistance genes. Since rice breeders evaluate their segregating population in endemic areas where the pest occurrence is severe in particular parts of the year, it is very time consuming and labour intensive to breed variety. The brown plant hopper (BPH) is one of the most serious insect pests of rice. Genetic analysis of BPH based on molecular marker was resistance of B5, a highly resistant line (Huang et al. 2001). Kammar and Nitin (2019) reported that location and identification genes conferring resistance are significant for breeding insect-

Table 2.3 Marker-assisted selection for disease resistance in different crops

S. No.	Crops	Resistance gene/disease	Reference
1.	Rice	Resistance to BLB (*Xa4*, *Xa5*, *Xa13* and *Xa21*)	Chukwu et al. (2020)
		Bacterial blight-resistant gene	Hsu et al. (2020).
		Xa5, Xa13, Xa21	Ramalingam et al. (2020)
		Developed stem rot resistance	Luo et al. (2020a, b)
		QTL for stem rot resistance	Dodia et al. (2019)
2.	Peanut	Bacterial wilt resistance	Luo et al. (2020a, b)
3.	Wheat	Yellow rust	
		QTL for leaf rust resistance	Beukert et al. (2020)
		QTL *QLr2135* for leaf rust	Zhang et al. (2021)
4.	Grain legume	*Fusarium* wilt resistance in major grain legumes	Jha et al. (2020)

Table 2.4 Marker-assisted selection for insect pest resistance in different cops

S. No.	Traits	Reference
1	Sorghum shoot fly resistance in sorghum	Folkertsma et al. (2003)
2	Brown plant hopper resistance in rice	Huang et al. (2001)
3	Green bug tolerance in sorghums	Nagaraj et al. (2005)
4	Resistance to insects in sorghums	Sharma et al. (1992)
5	Insect resistance in plant	Yencho et al. (2000)
6	Russian wheat aphid (RWA) resistance	
8	Insect-resistant crop plants are best way to feed the world population	Smith (2021)

resistant varieties. These technologies along with other approaches can help to reduce such yield losses from insect pests, ultimately leading to increased agricultural produce. Important insect pest resistances are listed below n Table 2.4.

2.12 MAS for Nutritional Quality

Researches are in progress for enhanced nutrition in fruit, vegetables and agricultural crops. A number of genotypes for different fruit trees with enhanced nutritional quality were developed either by increasing the content of beneficial compounds like phenolic compounds, vitamins and carotenoids or by decreasing the levels of specific anti nutrients.

Advance breeding lines of rice with high grain zinc content were developed (Swamy et al. 2016). Descalsota-Empleo et al. (2019) carried out work on Zn biofortification in rice and identified QTL *OsZFP252* which is responsible for high zinc content.

Table 2.5 Marker-assisted selection for improved nutritional quality in different cops

Crops	Traits	Gene/markers/QTL	Reference
Rice	Grain zinc	QTLs (OsHMA9, OsMAPK6, OsNRAMP7, OsMADS13, OsZFP252)	Descalsota-Empleo et al. (2019)
	Grain iron and zinc	QTL (OsZIP6)	Calayugan et al. (2020)
	Grain Zn	QTLs (qzpr.1.1, qzpr.11.1, qZBR.2.1, qzPR.2.1)	Suman et al. (2021)
	Grain Zn and yield	QTLs (LOC_Os03g47980, LOC_Os07g47950, LOC_Os07g48050, rMQTL7.1, OsNAS3)	Babu et al. (2020)
	Protein content	RI SBZ1 gene in rice FMO2EST1	Sood et al. (2016)
Wheat	Protein content	Xucw108	Vishwakarma et al. (2016)
	Protein content	Xuhw89	Hussein et al. (2014)
	Gluten strength	QTLs (QGlu.spa-1A, QGlu.spa-3A.2)	Ruan et al. (2020)
	Protein content	QTLs (QGpc.spa-3A.3, QGpc.spa-7A)	
	Protein content	Six QTLs	
Finger millet	Calcium content	M2, M6, M11, M16, M26, M27, M36, M45, M65, calcium exchangers, calcium ATPase and calcium sensors like calmodulin of cereals	Kumar et al. (2015)
Maize	Tryptophan content	27 KDa-Zein gene of *opaque2*, modifiers maize *OM5*, *FM8*	Babu et al. (2014)
	Pro-vitamin A	*crtRB13′TE, crtRB1-5′TE-2, crtRB1-3′TE-1, crtRB1-5′TE-2, crtRB1-3′TE-1, LCYE, PSY1*, and *crtRb1* for	Babu et al. (2013), Liu et al. (2015), Yan et al. (2010)
Watermelon	Lycopene content and flesh colour	*Claa005011, Claa005012, LCYB*	Wang et al. (2019)
	Red flesh colour	*CILCYB*	Zhang et al. (2020)
	High lycopene content red flesh	SNP	
	Fruit growth and ripening	*PPR* genes	
	Fruit shape	*Cla011257*	Dou et al. (2018)

Calayugan et al. (2020) developed a variety of rice which is nutritionally important. Suman et al. (2021) identified RILs for high grain Zn content in rice. Developed and characterized GR2E Golden rice introgression lines for enhanced vitamin A using marker-assisted backcross breeding. Raza et al. (2020) worked out on genetic diversity in rice based on SSR markers for grain Fe and Zn (Table 2.5).

2.13 Future Prospects of Marker-Assisted Selection

Different types of stresses caused severe yield losses throughout the world. Thus, there is a need to use most effective and reliable method for management of stresses and enhancement of host resistance through an economical and environment-friendly approach.

The MAS is the most frequently used tool for selection of QTLs from breeding lines and introgression of genes from donor to recipient parent. It is also applied in germplasm evaluation and use, as with assistance of marker can be identified trait of crop. MAS could greatly assist plant breeders in development of improved variety.

References

Babu R, Rojas NP, Gao S et al (2013) Validation of the effects of molecular marker polymorphisms in LcyE and CrtRB1 on provitamin a concentrations for 26 tropical maize populations. Theor Appl Genet 126:389–399. https://doi.org/10.1007/s00122-012-1987-3

Babu BK, Agrawal P, Pandey D et al (2014) Association mapping of agro-morphological characters among the global collection of finger millet genotypes using genomic SSR markers. Mol Biol Rep 41:5287–5529. https://doi.org/10.1007/s11033-014-3400-6

Babu PM, Neeraja CN, Rathod S et al (2020) Stable SNP allele associations with high grain zinc content in polished rice (*Oryza sativa* L.) identified based on ddRAD sequencing. Front Genet 11:763. https://doi.org/10.3389/fgene.2020.00763

Beukert U, Thorwarth P, Zhao Y et al (2020) Comparing the potential of marker-assisted selection and genomic prediction for improving rust resistance in hybrid wheat. Front Plant Sci 28(11): 594–113. https://doi.org/10.3389/fpls.2020.594113

Calayugan MIC, Formantes AK, Amparado A et al (2020) Genetic analysis of agronomic traits and grain iron and zinc concentrations in a doubled haploid population of rice (*Oryza sativa* L.). Sci Rep 10(1):22–83. https://doi.org/10.1038/s41598-020-59184-z

Castro AJ, Capettini F, Corey AE et al (2003) Mapping and pyramiding of qualitative and quantitative resistance to stripe rust in barley. Theor Appl Genet 107(5):922–930. https://doi.org/10.1007/s00122-003-1329-6

Chen S, Lin XH, Xu CG et al (2000) Improvement of bacterial blight resistance of 'Minghui 63, an elite restorer line of hybrid rice, by molecular marker assisted selection. Crop Sci 40:239–244

Chen S, Xu CG, Lin XH et al (2001) Improving bacterial blight resistance of '6078, an elite restorer line of hybrid rice, by molecular marker-assisted selection. Plant Breed 120:133–137. https://doi.org/10.1046/j.1439-0523.2001.00559

Chukwu SC, Rafii MY, Ramlee SI et al (2020) Recovery of recurrent parent genome in a marker-assisted backcrossing against rice blast and blight infections using functional markers and SSRs. Plants (Basel) 9(11):1411. https://doi.org/10.3390/plants9111411

Collard BCY, Mackill DJ (2008) Marker assisted selection: an approach for precision plant breeding in the twenty-first century. Philos Trans R Soc Lond Ser B Biol Sci 363(1491): 557–572. https://doi.org/10.1098/rstb.2007.2170

Das G, Rao GJ (2015) Molecular marker assisted gene stacking for biotic and abiotic stress resistance genes in an elite rice cultivar. Front Plant Sci 30:6–698. https://doi.org/10.3389/fpls.2015.00698

Das G, Patra JK, Bae KH et al (2017) Insight into MAS: a molecular tool for development of stress resistant and quality of rice through gene stacking. Front Plant Sci 13:8–985

Datta K, Baisakh N, Thet KM et al (2002) Pyramiding transgenes for multiple resistance in rice against bacterial blight, yellow stem borer and sheath blight. Theor Appl Genet 106:1–8

Descalsota-Empleo GI, Noraziyah AAS, Navea IP et al (2019) Genetic dissection of grain nutritional traits and leaf blight resistance in rice. Genes (Basel) 8(10):1–30. https://doi.org/10.3390/genes10010030

Dixit S, Singh UM, Singh AK, Alam S, Venkateshwarlu C, Nachimuthu VV, Yadav S, Abbai R, Selvaraj R, Devi MN, Ramayya PJ, Badri J, Ram T, Lakshmidevi G, Lrk JV, Padma Kumari AP, Laha GS, Prasad MS, Seetalam M, Singh VK, Kumar A et al (2020) Marker assisted forward breeding to combine multiple biotic-abiotic stress resistance/tolerance in rice. Rice (NY) 29(13):1–29

Dodia SM, Joshi B, Gangurde SS et al (2019) Genotyping-by-sequencing based genetic mapping reveals large number of epistatic interactions for stem rot resistance in groundnut. Theor Appl Genet 132(4):1001–1016. https://doi.org/10.1007/s00122-018-3255-7

Dou J, Zhao S, Lu X et al (2018) Genetic mapping reveals a candidate gene (ClFS1) for fruit shape in watermelon (*Citrullus lanatus* L.). Theor Appl Genet 131(4):947–958. https://doi.org/10.1007/s00122-018-3050-5

Folkertsma RT, Sajjanar GM, Reddy BVS, Sharma HC, Hash CT et al (2003) Genetic mapping of QTL associated with sorghum shoot fly (*Atherigona soccata*) resistance in sorghum (*Sorghum bicolour*). In: Final abstracts guide, plant and animal genome XI, p 42

Frisch M, Bohn M, Melchinger AE et al (1999) Minimum sample size and optimal positioning of flanking markers in marker-assisted backcrossing for transfer of a target gene. Crop Sci 39: 967–975

Frisch M, Melchinger AE et al (2005) Selection theory for marker-assisted backcrossing. Genetics 170(2):909–917. https://doi.org/10.1534/genetics.104.035451

Gouda G, Gupta MK, Donde R, Mohapatra T, Vadde R, Behera L et al (2020) Marker-assisted selection for grain number and yield-related traits of rice (*Oryza sativa* L.). Physiol Mol Biol Plants 26(5):885–898. https://doi.org/10.1007/s12298-020-00773-7

Hanson P, Lua SF, Wanga JF, Chena W, Kenyona L, Tana CW, Tee KL, Wanga YY, Hsu YC, Schafleitner R, Ledesmaa D, Yang RY (2016) Conventional and molecular marker-assisted selection and pyramiding of genes for multiple disease resistance in tomato Peter. Scient Hort 20:346–354

Hittalmani S, Parco A, Mew TV, Zeigler RS, Huang N et al (2000) Fine mapping and DNA marker-assisted pyramiding of the three major genes for blast resistance in rice. Theor Appl Genet 100: 1121–1128. https://doi.org/10.1007/s001220051395

Holland JB (2004) Implementation of molecular markers for quantitative traits in breeding programs—challenges and opportunities. In: Proceedings of the 4th International Crop Science Congress, Brisbane

Hsu YC, Chiu CH, Yap R, Tseng YC, Wu YP et al (2020) Pyramiding bacterial blight resistance genes in tainung82 for broad-spectrum resistance using marker-assisted selection. Int J Mol Sci 21(4):1281. https://doi.org/10.3390/ijms21041281

Huang N, Angeles E, Domingo J et al (1997) Pyramiding of bacterial blight resistance genes in rice: marker-assisted selection using RFLP and PCR. Theor Appl Genet 95:313–320. https://doi.org/10.1007/s001220050565

Huang Z, He G, Shu L, Li X, Zhang Q et al (2001) Identification and mapping of two brown plant hopper resistance genes in rice. Theor Applied Genet 102(67):929–934

Hussein HA, Ebtissam Hussein HA, El-Sayed OE, Al-Ansary AMF, Khatab SA, Rabie Sally AA et al (2014) Inter specific crosses and marker assisted selection for improving the nutritional value of Egyptian wheat cultivars. Inter J Agric Res 9:119–135

Jamaloddin M, Durga Rani CV, Swathi G, Anuradha C, Vanisri S, Rajan CPD et al (2020) Marker assisted gene pyramiding (MAGP) for bacterial blight and blast resistance into mega rice variety "Tellahamsa". PLoS One 15(6):e0234088. https://doi.org/10.1371/journal.pone.0234088

Jefferies SP, King BJ, Barr AR, Warner P, Logue SJ, Langridge P et al (2003) Marker-assisted backcross introgression of the Yd2 gene conferring resistance to barley yellow dwarf virus in barley. Plant Breed 122:52–56. https://doi.org/10.1046/j.1439-0523.2003.00752.x

Jha UC, Bohra A, Pandey S, Parida SK et al (2020) Breeding, genetics, and genomics approaches for improving *Fusarium* wilt resistance in major grain legumes. Front Genet 23(11):1001. https://doi.org/10.3389/fgene.2020.01001

Jiang GH, Xu CG, Tu JM, Li XH, He YQ, Zhang QF et al (2004) Pyramiding of insect- and disease resistance genes into an elite indica, cytoplasm male sterile restorer line of rice, 'Minghui 63'. Plant Breed 123:112–116. https://doi.org/10.1046/j.1439-0523.2003.00917.x

Joseph M, Gopalakrishnan S, Sharma RK, Singh VP, Singh AK, Singh NK, Mohapatra T et al (2004) Combining bacterial blight resistance and Basmati quality characteristics by phenotypic and molecular marker assisted selection in rice. Mol Breed 13:377–387

Kammar V and Nitin KS (2019) Molecular marker-assisted selection of plant genes for insect resistance. In: Kumar Chakravarthy A, Selvanarayanan V(eds) Experimental techniques in host-plant resistance. Springer, Singapore 981(13):2652–2327

Kumar A, Yadav S, Panwar P, Gaur VS, Sood S et al (2015) Identification of anchored simple sequence repeat markers associated with calcium content in finger millet (*Eleusine coracana*). Proc Natl Acad Sci India B Biol Sci 85:311–317. https://doi.org/10.1007/s40011-013-0296-1

Liu J, Liu D, Tao W, Li W, Wang S, Chen P, Cheng S, Gao D et al (2000) Molecular marker-facilitated pyramiding of different genes for powdery mildew resistance in wheat. Plant Breed 119:21–24. https://doi.org/10.1046/j.1439-0523.2000.00431

Liu SP, Li X, Wang CY, Li XH, He YQ et al (2003) Improvement of resistance to rice blast in Zhenshan 97 by molecular marker-aided selection. Acta Bot Sin 45:1346–1350

Liu L, Jeffers D, Zhang Y, Ding M, Chen W, Kang MS et al (2015) Introgression of the crtRB1 gene into quality protein maize inbred lines using molecular markers. Mol Breed 35:154. https://doi.org/10.1007/s11032-015-0349-7

Luo H, Pandey MK, Zhi Y, Zhang H, Xu S, Guo J, Wu B, Chen H, Ren X, Zhou X, Chen Y, Chen W, Huang L, Liu N, Sudini HK, Varshney RK, Lei Y, Liao B, Jiang H et al (2020a) Discovery of two novel and adjacent QTLs on chromosome B02 controlling resistance against bacterial wilt in peanut variety Zhonghua 6. Theor Appl Genet 133(4):1133–1148. https://doi.org/10.1007/s00122-020-03537-9

Luo Z, Cui R, Chavarro C, Tseng YC, Zhou H, Peng Z, Chu Y, Yang X, Lopez Y, Tillman B, Dufault N, Brenneman T, Isleib TG, Holbrook C, Ozias-Akins P, Wang J et al (2020b) Mapping quantitative trait loci (QTLs) and estimating the epistasis controlling stem rot resistance in cultivated peanut (Arachis hypogaea). Theor Appl Genet 133(4):1201–1212. https://doi.org/10.1007/s00122-020-03542-y

Mackill DJ, Collard BCY, Neeraja CN, Maghirang RR, Heuer S, Ismail AM et al (2006) QTLs in Marker-assisted selection in plant breeding B. C. Y. Collard & D. J. Mackill 569 Phil. Trans. R. Soc. B (2008) rice breeding: examples for abiotic stresses. In: Fifth International Rice Genetics Symposium, Manila, Philippines. International Rice Research Institute, Los Baños, The Philippines

Nagaraj N, Reese JC, Tunistra MR, Smith CM, Amand PT, Kirkham MB, Kofoid KD, Campbell LR, Gerald EW et al (2005) Molecular mapping of sorghum genes expressing tolerance to damage by green bug (Homoptera: Ahidae). J Econ Entomol 98(2):595–602

Okada Y, Kanatani R, Arai S, Ito K et al (2004) Interaction between barley yellow mosaic disease-resistance genes rym1 and rym5, in the response to BaYMV strains. Breed Sci 54:319–325

Prabhu AS, Filippi MC, Silva GB, Silva-Lobo VL, Morais OP et al (2009) An unprecedented outbreak of rice blast on a newly released cultivar BRS Colosso in Brazil. In: Wang GL, Valent B (eds) Advances in genetics, genomics and control of rice blast. Springer, Dordrecht, pp 257–267

Ragimekula N, Varadarajula NN, Mallapuram SP, Govardhan G, Reddy RK, Kondreddy HR et al (2013) Marker assisted selection in disease resistance breeding. J Plant Breed Genet 01(02): 90–109

Ramalingam J, Raveendra C, Savitha P, Vidya V, Chaithra TL, Velprabakaran S, Saraswathi R, Ramanathan A, Arumugam Pillai MP, Arumugachamy S, Vanniarajan C et al (2020) Gene

pyramiding for achieving enhanced resistance to bacterial blight, blast, and sheath blight diseases in rice. Front Plant Sci 19(11):591–457. https://doi.org/10.3389/fpls.2020.591457
Ranjisha KR, Surendran K, Nair RA, Pillai PP et al (2020) Bioprospecting of biodiversity for improvement of agronomic traits in plants. Trans Tech Based Val Addit Plant Biotechnol. https://doi.org/10.1016/B978-0-12-818632-9.00001-0
Raza Q, Riaz A, Saher H, Bibi A, Raza MA, Ali SS, Sabar M et al (2020) Grain Fe and Zn contents linked SSR markers based genetic diversity in rice. PLoS One 15(9):239–739. https://doi.org/10.1371/journal.pone.0239739
Reif JC, Melchinger AE, Xia XC, Warburton ML, Hoisington DA, Vasal SK, Beck D, Bohn M, Frisch M et al (2003) Use of SSRs for establishing heterotic groups in subtropical maize. Theor Appl Genet 107:947–957. https://doi.org/10.1007/s00122-003-1333-x
Reif JC, Hamrit S, Heckenberger M, Schipprack W, Maurer HP, Bohn M, Melchinger AE et al (2005) Trends in genetic diversity among European maize cultivars and their parental components during the past 50 years. Theor Appl Genet 111:838–845. https://doi.org/10.1007/s00122-005-0004-5
Ribaut JM, Ragot M (2007) Marker-assisted selection to improve drought adaptation in maize: the backcross approach, perspectives, limitations, and alternatives. J Exp Bot 58(2):351–360. https://doi.org/10.1093/jxb/erl214
Risk JM, Selter LL, Chauhan H, Krattinger SG, Kumlehn J, Hensel G, Viccars LA, Richardson TM, Buesing G, Troller A, Lagudah ES, Keller B et al (2013) The wheat Lr34 gene provides resistance against multiple fungal pathogens in barley. Plant Biotechnol J 11(7):847–854
Ruan Y, Yu B, Knox RE, Singh AK, De PR, Cuthbert R, Zhang W, Piche I, Gao P, Sharpe A, Fobert P et al (2020) High density mapping of quantitative trait loci conferring gluten strength in Canadian durum wheat. Front Plant Sci 11:170. https://doi.org/10.3389/fpls.2020.00170
Salina E, Dobrovolskaya O, Efremova T, Leonova I, Roder MS et al (2003) Microsatellite monitoring of recombination around the Vrn-B1 locus of wheat during early backcross breeding. Plant Breed 122:116–119. https://doi.org/10.1046/j.1439-0523.2003.00817.x
Schmierer DA, Kandemir N, Kudrna DA, Jones BL, Ullrich SE, Kleinhofs A et al (2004) Molecular marker-assisted selection for enhanced yield in malting barley. Mol Breed 14:463–473. https://doi.org/10.1007/s11032-004-0903-1
Servin B, Martin CO, Mezard M, Hospital F et al (2000) Toward a theory of marker-assisted gene pyramiding. Genetics 168:1513–1523
Sharma HC, Taneja SL, Leuschner K and Nwanze KF et al (1992) Techniques to screen sorghums for resistance to insects. Inform. Andhra Pradesh: International Crops Research Institute for the Semi-Arid Tropics. bulletin no 32, p 48
Sharma PN, Torii A, Takumi S, Mori N, Nakamura C et al (2004) Marker-assisted pyramiding of brown plant hopper (Nilaparvata lugens Stål) resistance genes Bph1 and Bph2 on rice chromosome 12. Hereditas 140:61–69. https://doi.org/10.1111/j.1601-5223.2004.01726.x
Shen L, Courtois B, McNally KL, Robin S, Li Z et al (2001) Evaluation of near-isogenic lines of rice introgressed with QTLs for root depth through marker-aided selection. Theor Appl Genet 103:75–83. https://doi.org/10.1007/s001220100538
Smith CM (2021) Conventional breeding of insect-resistant crop plants: still the best way to feed the world population. Curr Opin Insect Sci 45:7–13
Sood S, Kumar A, Kalyana BB, Gaur SV, Pandey D, Kant L, Pattnayak A et al (2016) Gene discovery and advances in finger millet [*Eleusine coracana*(L.) *Gaertn.*] genomics-an important nutri-cereal of future. Front Plant Sci 7:1634. https://doi.org/10.3389/fpls.2016.01634
Steele KA, Edwards G, Zhu J, Witcombe JR et al (2004) Marker-evaluated selection in rice: shifts in allele frequency among bulks selected in contrasting agricultural environments identify genomic regions of importance to rice adaptation and breeding. Theor Appl Genet 109:1247–1260. https://doi.org/10.1007/s00122-004-1732-7
Suman K, Neeraja CN, Madhubabu P, Rathod S, Bej S, Jadhav KP, Kumar JA, Chaitanya U, Pawar SC, Rani SH, Subbarao LV, Voleti SR et al (2021) Identification of promising RILs for high grain zinc through genotype × environment analysis and stable grain zinc QTL using SSRs and

SNPs in rice (*Oryza sativa* L.). Front Plant Sci 18(12):587–482. https://doi.org/10.3389/fpls.2021.587482

Swamy BPM, Rahman MA, Inabangan-Asilo MA, Amparado A, Manito C, Chadha-Mohanty P, Reinke R, Slamet-Loedin IH et al (2016) Advances in breeding for high grain zinc in rice. Rice (N Y) 9(1):49. https://doi.org/10.1186/s12284-016-0122-5

Tandzi NL, Mutengwa SC et al (2020) Factor affecting yield of crops. https://doi.org/10.5772/intechopen.90672

Toojinda T, Baird E, Booth A, Broers L, Hayes P, Powell W, Thomas W, Vivar H, Young G et al (1998) Introgression of quantitative trait loci (QTLs) determining stripe rust resistance in barley: an example of marker-assisted line development. Theor Appl Genet 96:123–131. https://doi.org/10.1007/s001220050718

Van Berloo BR, Aalbers H, Werkman A, Niks RE et al (2001) Resistance QTL confirmed through development of QTL–NILs for barley leaf rust resistance. Mol Breed 8:187–195. https://doi.org/10.1023/A:101372200856

Vishwakarma MK, Arun B, Mishra VK et al (2016) Marker-assisted improvement of grain protein content and grain weight in Indian bread wheat. Euphytica 208:313–321. https://doi.org/10.1007/s10681-015-1598-6

Wang C, Qiao A, Fang X, Sun L, Gao P, Davis AR, Liu S, Luan F (2019) Fine mapping of lycopene content and flesh color related gene and development of molecular marker-assisted selection for flesh color in watermelon (Citrullus lanatus). Front Plant Sci 10:1240

Werner K, Friedt W, Ordon F et al (2005) Strategies for pyramiding resistance genes against the barley yellow mosaic virus complex (BaMMV, BaYMV, BaYMV-2). Mol Breed 16:45–55. https://doi.org/10.1007/s11032-005-3445-2

Wilcox MC, Khairallah MM, Bergvinson D, Crossa J, Deutsch JA, Emeades GO, González-de-León D, Jiang C, Jewell DC, Mihm JA, Williams WP, Hoisington D et al (2002) Selection for resistance to southwestern corn borer using marker-assisted selection and conventional backcrossing. Crop Sci 42:1516–1528

Yan J, Kandianis CB, Harjes CE, Bai L, Kim E-H, Yang X et al (2010) Rare genetic variation at Zea mays crtRB1 increases β-carotene in maize grain. Nat Genet 42:322–327. https://doi.org/10.1038/ng.551

Yashitola J, Thirumurugan T, Sundaram RM, Naseerullah MK, Ramesha MS, Sarma NP, Sonti RV et al (2002) Assessment of purity of rice hybrids using microsatellite and STS markers. Crop Sci 42:1369–1373

Yencho GC, Cohen MB, Byrne PF et al (2000) Applications of tagging and mapping insect resistance loci in plants. Ann Rev Entoml 45:393–422

Zhang H, San ML, Jang SG, Lee JH, Kim NE, Lee AR, Park SY, Cao FY, Chin JH, Kwon SW et al (2020) Genome-wide association study of root system development at seedling stage in rice. Genes (Basel) 25(11):12–1395. https://doi.org/10.3390/genes11121395

Zhang P, Yan X, Gebrewahid TW, Zhou Y, Yang E, Xia X, He Z, Li Z, Liu D et al (2021) Genome-wide association mapping of leaf rust and stripe rust resistance in wheat accessions using the 90K SNP array. Theor Appl Genet. https://doi.org/10.1007/s00122-021-03769-3

Zhou PH, Tan YF, He YQ, Xu CG, Zhang Q et al (2003a) Simultaneous improvement for four quality traits of Zhenshan 97, an elite parent of hybrid rice, by molecular marker-assisted selection. Theor Appl Genet 106:326–331

Zhou YC, Wu WR, Qi JM et al (2003b) Approximate estimation of minimal sample size required for marker-assisted backcross breeding. Yi Chuan Xue Bao 30(7):625–630

Application of Gene Mining and Editing Technologies for Agricultural Research and Breeding

3

Lee-Ann Niekerk, Mogamat Fahiem Carelse, Olalekan Bakare, Ashwil Klein, Arun Gokul, and Marshall Keyster

Abstract

The current world population is rapidly approaching the eight billion mark, and the current projected annual increase of 1.1% will take the population to approximately ten billion people in the year 2050. Therefore, maintaining and increasing our food production systems to feed the ever-growing population is a major concern. Abiotic and biotic stresses have increased in onset, duration, and intensity, and this could be attributed to more severe cycles of climate change. These stresses will severely hamper the future output and yield from crop plant-based agricultural systems all over the world. To improve yield, genetic engineering technologies have become an attractive tool for rapidly improving plant tolerance to abiotic and biotic stresses. For this purpose, proper bioinformatic

L.-A. Niekerk · M. F. Carelse · O. Bakare
Environmental Biotechnology Laboratory, Department of Biotechnology, University of the Western Cape, Bellville, South Africa
e-mail: 3255882@myuwc.ac.za; 3341863@myuwc.ac.za; 3779970@myuwc.ac.za

A. Klein
Plant Omics Laboratory, Department of Biotechnology, University of the Western Cape, Bellville, South Africa
e-mail: aklein@uwc.ac.za

A. Gokul
Department of Plant Sciences, University of Free State, Phuthaditjhaba, South Africa
e-mail: gokula@ufs.ac.za

M. Keyster (✉)
Environmental Biotechnology Laboratory, Department of Biotechnology, University of the Western Cape, Bellville, South Africa

DST-NRF Centre of Excellence in Food Security, University of the Western Cape, Bellville, South Africa
e-mail: mkeyster@uwc.ac.za

Kamaluddin et al. (eds.), *Technologies in Plant Biotechnology and Breeding of Field Crops*, https://doi.org/10.1007/978-981-16-5767-2_3

gene mining tools have laid the foundation for selecting and identifying genes for downstream editing. This chapter describes some of the gene mining tools that led to selected genes for engineering plant tolerance to abiotic and biotic stresses in the four major food crop plant families.

Keywords

Abiotic and biotic stresses · Genome-wide association studies · Quantitative trait loci (QTL) · CRISPR/Cas9 system · Drought tolerance

3.1 Introduction

The human population is increasing at a very rapid rate and will require optimal food for nutrition and health. Agricultural innovations will play a major role in providing enough food for an overpopulated human race. To keep up with the increasing population numbers and to sustain food needs and requirements, the growing population will have to double annual food production on very limited land supply. Therefore, the agricultural sector will be under severe pressure from increasing urbanization and industrialization which often overlaps with fertile lands (Gokul et al. 2020). With the fertile land rapidly depleting, the human population will have to find innovative means to optimally grow and produce food and feed crops in order to sustain plant and meat-based diets. An estimated 67% of the world's population relies primarily on plant-based diets (Pimentel and Pimentel 2003); therefore, Gokul et al. 2020 suggested that most of the fertile land should be made available for crop plant production in the future. In addition to limited fertile land, current environmental changes brought about by climate change will increase in severity over time. Therefore, limited fertile land and increases in the severity and time of climate change will result in the ever-growing population outweighing food production in the near future (Gokul et al. 2020).

Crop plant breeding efforts can only benefit from the rapid increase in genomic data for gene mining as well as the growing genome editing toolbox. These technological advances are key to the improvement of existing crops as well as the domestication of new crops (Batley and Edwards 2016). The lower sequencing cost of DNA has contributed drastically to the assembly and annotation of plant genomes. In addition, RNA sequencing and gene prediction algorithms have facilitated the annotation of many crop plant genomes, and the improvement of bioinformatic pipelines has contributed to improved assemblies from scaffold-only-based draft genomes to high-level chromosome reference assemblies. Genome-wide association studies (GWAS) and quantitative trait loci (QTL) analyses are rapidly identifying huge numbers of candidate regions that are linked to major agronomic traits that can be used in breeding programs (Furbank and Tester 2011).

The advances in bioinformatic gene mining pipelines have led to major advances in genome editing. These gene editing approaches make plants more tolerant to abiotic stresses and provide a rapid and sustainable solution for increased food and feed crop production (Bhatnagar-Mathur et al. 2008). Various strategies and technologies have been developed for efficient and effective transferring of candidate genes from multiple organisms and into plants cells (Wani et al. 2016; Khanna and Deo 2016). In addition, technologies have been developed to precisely edit sequenced genes at the molecular level in order to control gene expression and protein function. However, not many publications focus on providing an overview of the successes of improving plant growth and development in the major crop plant families (*Poaceae*, *Fabaceae*, *Brassicaceae,* and *Solanaceae*) using gene editing as a tool. Therefore, this chapter highlights few of the successful studies which used gene mining techniques to edit plant genes in order to improve plant growth and increase the content of important plant biomolecules for human consumption and health.

3.2 Gene Mining and Editing in Poaceae

Oryza sativa is an important food crop and the main staple food for more than half of the world's population. *O. sativa* is categorized as a salt-sensitive crop since it is originally grown in freshwater marshes and swamps (Kumar et al. 2013). Throughout the past decade, many salted-related genes have been successfully cloned to improve *O. sativa* tolerance to biotic and abiotic stress (Table 3.1). OsRR22 is one of the candidate genes which encode a 696-amino acid B-type response regulator transcription factor involved in cytokinin metabolism and signal transduction. Studies have shown that a loss of function of the OsRR22 gene leads to significantly increased salt tolerance (Takagi et al. 2015). A study by Zhang et al. (2019a) proposed using the CRISPR/Cas9 system to mutate the OsRR22 gene in an effort to increase salinity tolerance in *O. sativa*. Due to the detrimental effects salt has on *O. sativa,* it has become one of the most important breeding goals (Qin et al. 2020). A BLAST search using the NCBI online tool was used to confirm the target

Table 3.1 Gene editing approaches in *O. sativa*

Targeted Gene	Genome editing strategy	Molecular function	Reference
BADH2	CRISPR/ CAS9	Improved yield, quality: Enhanced fragrance	Shao et al. (2017)
OsSWEET13	TALENs	Biotic stress tolerance: Enhanced resistance to bacterial blight	Blanvillain-Baufumé et al. (2017)
Os SAPK2	CRISPR/ CAS9	Abiotic stress tolerance: Drought tolerance	Lou et al. (2017)
OsPDS, *OsSBEIIb*	Base editing	Nutritional improvement	Li et al. (2017)
Gn1a, *DEP1*, *GS3*, and *IPA1*	CRISPR/ CAS9	Improvement of grain number, panicle architecture, grain size, and plant architecture	Li et al. (2016)

sequences in *O. sativa*. The appropriate gene identified for the purpose of this study was *O. sativa* response regulator 22 (OsRR22). The *O. sativa* response regulator 22 encodes a regulator protein that acts as a transcription factor which regulates genes which are involved in osmotic response and ion transport (Sun et al. 2019). The elite japonica *O. sativa* cultivar WPB106 was used in this study and was a product of breeding Huan9/Huxiangjing//Huhan3/Huhan11 in the laboratory. Zhang et al. (2019a) used the CRISPR/Cas9 system and the Cas9-OsRR22-gRNA expression vector (Fig. 3.1). The Cas9-OsRR22-gRNA-expressing vector was constructed following the previously described protocol of Ma et al. (2015). The vector was inserted into *Agrobacterium tumefaciens,* and Agrobacterium-mediated transformation was performed on WPB106. The mutant lines WPB106-cas-1 and WPB106-cas-2 were observed to have a decrease in fresh weight of 10% and 2%, respectively, when treated with 0.75% NaCl. However, the wild type (WT) performed far worse with a decrease in fresh weight of 50% when treated with 0.75% NaCl. The height of the *O. sativa* plants was also affected by the NaCl treatments. The 0.75% NaCl treatment caused a decrease of 32%, 20%, and 18% in plant height of the WT, WPB106-cas-1, and WPB106-cas-2, respectively. It should be noted that under normal growth conditions, no significant difference was observed in the growth characteristics (number of tiller per plants, number of grains per panicle, and yield per plant) between the WT and T2 mutant lines.

Zea mays is a major nutrient source for both human and animal feed. It is one of the most cultivated crops worldwide and a crucial commodity crop for world food security (Pechanova and Pechan 2017). Drought stress is a major factor affecting *Z. mays* development in many areas. Therefore, the development of drought-tolerant *Z. mays* is of utmost important; however, the physiological processes and metabolic networks underlying drought tolerance are complicated. Nevertheless, the phytohormone ethylene is known to play an important role in regulating plant response to abiotic stress, including water deficits and high temperature (Hays et al. 2007; Kawakami et al. 2010). ARGOS genes, and more specifically the ARGOS8 gene, are negative regulators of ethylene responses in *Z. mays*. ARGOS8 proteins physically interact with the ethylene receptor signaling complex, subsequently modulating ethylene perception and the early stages of the ethylene signal transduction, and this leads to enhance drought tolerance when overexpressed in transgenic *Z. mays* plants (Guo et al. 2014; Shi et al. 2015, 2016). Shi et al. (2017) reported that modifying single native genes to change expression patterns can increase *Z. mays* grain yield under drought stress conditions. In this study, CRISPR/Cas-enabled advanced breeding technology was employed to generate novel variants of ARGOS8. This included a two-step procedure for ARGOS8 editing, firstly, duplication of the GOS2 promoter and, secondly, translocation to the ARGOS8 locus. Their results showed that the two genome-edited variants ARGOS8-v1 and ARGOS8-v2 had enhanced ARGOS8 mRNA expression in all the tissues compared to that of the WT. Furthermore, a field study showed the ARGOS8 variants increased grain yield by five bushels per acre under flowering stress conditions and had no yield loss under well-watered conditions. Their results demonstrate that modifying single

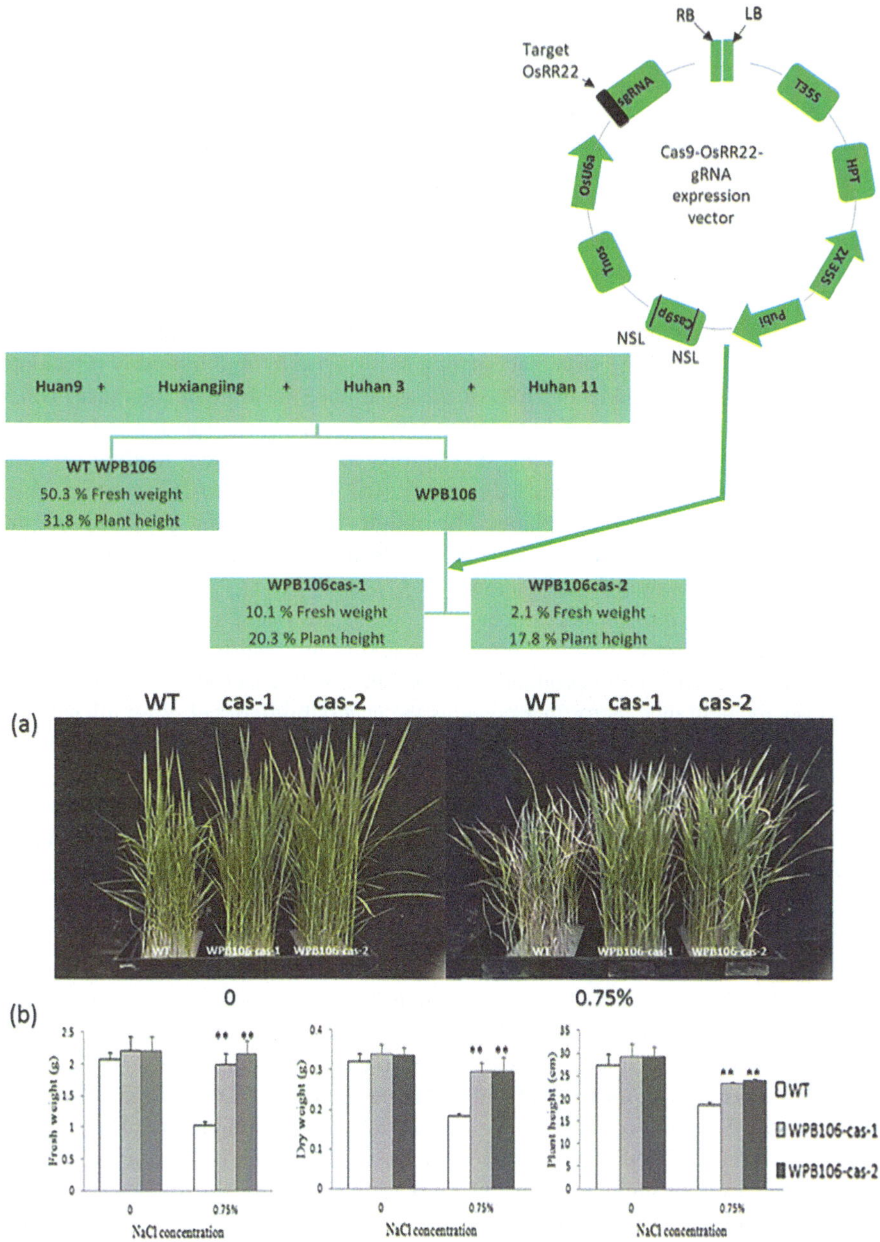

Fig. 3.1 A modified diagram representing the workflow and results of introducing the Cas9-OsRR22-gRNA expression vector into *O. sativa* plants. Zhang et al. (2019a) bred WPB106 line from four parents (Huan9, Huxiangjing, Huhan3, and Huhan11). Using CRISPR/Cas9 technology, they generated two mutants (WPB106cas-1 and WPB106cas-2). (**a**) and (**b**) represent the results they obtained from treating the plants to control conditions or 0.75% NaCl conditions. Both mutant lines represented higher tolerance to NaCl stress, WPB106cas-1 line higher than both the control and WPB106cas-2

native genes to change expression patterns can increase *Z. mays* grain yield under drought stress conditions.

A study to map the regulatory variants controlling the gene expression in the drought response and tolerance in *Z. mays* was performed by Liu et al. (2020). The authors performed RNA-seq analysis followed by sequencing reads, quality control, and mapping. The sequences were then subjected to gene expression quantification and clustering analysis. Genome-wide association studies were performed to identify important expression quantitative trait loci (eQTL). Mendelian randomization analysis was then performed to identify the relationship between the gene expression level and the phenotype of the plants. The gene Zm00001d051554 (abh2) was identified as a candidate for further study. The abh2 gene has been shown to be involved in promoting the stomatal opening in response to high humidity (Okamoto et al. 2009). To understand the role of the gene abh2 in drought tolerance, abh2-CRISPR lines were created. *Z. mays* immature embryos were transformed with the knockout vector constructs using an Agrobacterium-mediated method. It was observed that the CRISPR lines exhibited a 30% increase in ABA content when compared to the WT *Z. mays* plants. The authors suggested that by disrupting the abh2 gene, a higher level of ABA may be sustained by the *Z. mays* plants which may result in an increase in drought tolerance.

Saccharum officinarum is commercially one of the world's most efficient feedstock due to its superior biomass production and accumulation of sucrose in stems which can be used for the production of bioethanol (Byrt et al. 2011). However, lignin is a recalcitrance factor that prevents the accessibility of cellulose microfibrils by cellulase enzyme as well as adsorbs hydrolytic enzymes; thereby, it inhibits the release of cell wall-bound sugars for biofuel production from lignocellulosic biomass (Chen and Dixon 2007; Weng et al. 2008). Studies have shown that RNAi suppression of the lignin biosynthetic gene caffeic acid *O*-methyltransferase (COMT) has been demonstrated to improve bioethanol production from lignocellulosic biomass by catalyzing *O*-methylation of 5-hydroxyconiferaldehyde and 5-hydroxyconiferyl alcohol, thereby diverting metabolic flux to the formation of the syringyl (S) lignin monomer. In a study by Kannan et al. (2018), a conserved region of COMT was targeted with a single transcription activator-like effector nuclease (TALEN) pair for multiallelic mutagenesis to modify lignin biosynthesis in sugarcane. A single TALEN pair designed to target the highly conserved domain of COMT was able to mutate 98% of the more than 100 COMT copies/alleles as revealed by Sanger sequencing of cloned long PCR amplicons. Field-grown TALEN-mediated COMT mutants showed up to 20% lignin reduction and significantly decreased syringyl to guaiacyl ratio resulting in an up to 44% improved saccharification efficiency. Furthermore, no significant difference in biomass production and agronomic performance compared to the original *S. officinarum* cultivar (WT) was observed.

As the fourth most important cereal crop, *Hordeum vulgare* has a particularly high level of Vitamin E (classified into two categories, tocopherols and tocotrienols). Vitamin E is a lipid-soluble antioxidant and an essential nutrient for human health. Many human diseases such as cardiovascular disease and certain cancers are associated with insufficient vitamin E intake (Vardi et al. 2013). Therefore, to meet the demand for human consumption, it is necessary to modify the level and composition of vitamin E in food crops via genome engineering and precision plant breeding. HGGT and HPT, two committed enzymes in the vitamin E biosynthesis pathway, are the main targets for metabolic engineering of vitamin E. In a study by Zeng et al. (2020), targeted knockout mutations of HvHPT and HvHGGT in barley were created with CRISPR/Cas9-enabled genome editing. The content of tocochromanol isomers in transgene-free homozygous HvHPT and HvHGGT mutants was measured with high-performance liquid chromatography (HPLC). Mutagenesis efficiency among T0-regenerated plantlets was 50% to 65% as a result of two simultaneously expressed guide RNAs targeting each gene; most of the mutations were stably inherited by the next generation. The transgene-free homozygous mutants of HvHPT and HvHGGT exhibited decreased grain size and weight, and the HvHGGT mutation led to a shrunken phenotype and significantly lower total starch content in grains. HPLC analysis revealed that targeted mutation of HvHPT significantly reduced the content of both tocopherols and tocotrienols, whereas mutations in HvHGGT completely blocked tocotrienol biosynthesis in *H. vulgare* grains. Our results functionally validated that HvHGGT is the only committed gene for the production of tocotrienols, whereas HvHPT is partly responsible for tocopherol biosynthesis in *H. vulgare*.

Zhang et al. (2019b) used base editing techniques to improve herbicide tolerance in *Triticum aestivum*. Weeds are one of the major threats to world food production. In an effort to control weeds, herbicides such as Imidazolinone are used but may have a negative effect on the agricultural crops as well. The gene Acetolactate synthase (ALS) was identified to be appropriate for the purpose of this study. Acetolactate synthase was identified as one of the first common enzymes in the biosynthetic pathway of the branched amino acids such as valine, leucine, and isoleucine (Ogawa et al. 2008). The ALS gene was observed to be the target site for more than 50 commercial herbicides from five structurally different chemical classes (Whaley et al. 2007). Due to the function of the gene, the genetic modification of the gene would lead to improved herbicide tolerance to a wide array of herbicides. The authors use the Cytidine base editing system to target the ALS gene. The expression vector construct was delivered into immature embryos of *T. aestivum* cultivars Kenong199 and Kenong9204 using particle bombardment. To test the herbicide resistance, the authors exposed the plants to 1, 3, and 9 times the field recommendation of the herbicide Mesosulfuron. The plants containing the mutations grew better at 1, 3, and 9 times the field recommendation when compared to the WT which showed stunting of growth as well as visible injury at concentration as low as 1 and 3 times field recommendation.

3.3 Gene Mining and Editing in Fabaceae

In the beginning of the genome editing technology era, only single genes were edited, and no approach focused on targeting multiple genes (Zhang et al. 2020). A bottleneck was thus introduced when investigating relationships among several related genes and to investigate the function of gene families. Thus, an alteration of a genome editing technology had to be discovered. Therefore, Zhang et al. (2020) developed a multiplex mutation system which was designed to integrate multiple sgRNA and a Cas9 expression cassette into a single binary vector. The most detrimental disease to *Glycine max* plants is caused by the Soy bean mosaic virus (SMV) (Zhang et al. 2020). This disease leads to low *G. max* crop yield and reduces the quality of the *G. max* crops. Isoflavones are an essential cluster of secondary metabolites that serves an important role in plant disease resistance and an important role in plant–environment interactions (Cheng et al. 2015). Isoflavone synthase (IF) is an important enzyme in the biosynthesis of isoflavone; however, the competitive relationship between IF, Flavanone-3-hydroxylase (F3H), and Flavone synthase II (FNSII) in the phenylpropanoid pathway reduces the synthesis of isoflavones (Liu et al. 2002). All these enzymes target the same substrate Naringenin (flavone). The interaction between the three enzymes with the substrate, naringenin, is shown in Fig. 3.2a. Thus, an approach was required to control this event in which only IF binds to naringenin (Zhang et al. 2020). Two studies conducted by Jiang et al. (2014) and Jiang et al. (2010) noted that when GmF3H-1 and GmFNSII-2 were silenced either separately or simultaneously in *G. max* hairy roots by RNA interference technology, an increase in isoflavone content was observed. Thus, Zhang et al. (2020) wanted to implement a multiplex system to improve isoflavone content in *G. max*. Jiang et al. (2014) characterized two FNSII genes (GmFNSII-1 and GmFNSII-2) from the *G. max* cultivar "Hefeng 47" using the cDNA PCR amplification to clone it into the pMD18-T vector and then sequenced the fragments, they used the BLAST program to do a homology search, and multiple sequence alignments were performed with the Clustal W software and the Genedoc software. Then, in the new study, Zhang et al. (2020) exploited the CRISPR/Cas9 technology to design a multiplex gene-editing system which could simultaneously target GmF3H1, GmF3H2, and GmFNSII-1 in the *G. max* hairy roots (Fig. 3.2b). This multiplex technology was employed to metabolically engineer *G. max* isoflavone content. A vector constructed via the step-by-step assembly strategy of Du et al. (2016) was employed to design and construct the multiplex sgRNA modules. The promoters used in this system were amplified from *G. max* cultivar "Williams 84" (GmUbi3 and GmU3-19g-1). A single carrier plasmid was generated using the step-by-step strategy, containing a binary vector and two scaffold vectors (Zhang et al. 2020; Du et al. 2016). The target sequences of the sgRNA (for GmF3H1, GmF3H2, and GmFNSII-1) were identified by using web-based tools CRISPR-P (Lei et al. 2014) and CRISPR-PLANT (Xie et al. 2014). The vector targeting all three sites was then termed pGmUbi-Cas9-4XsgR. Agrobacterium-mediated transformation was conducted to transform the pGmUbi-Cas9-4XsgR vector into hairy roots of *G. max* plants. RNA was isolated from leaves or seeds, and metabolic analysis

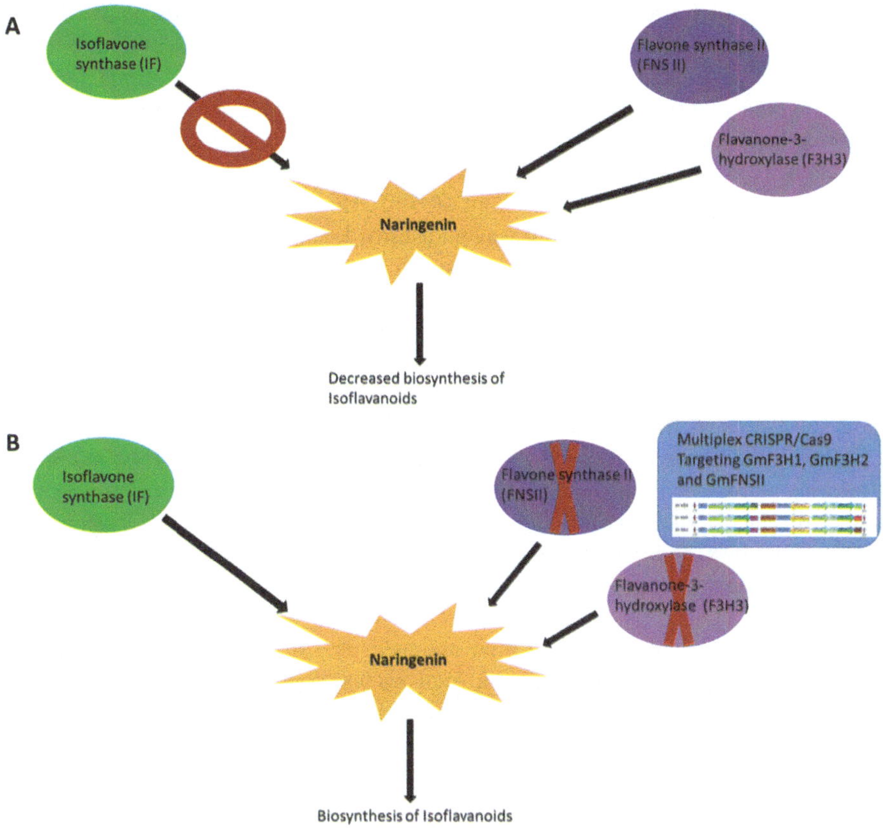

Fig. 3.2 Diagram depicting the interactions of enzymes from multiple pathways on a single substrate. Isoflavone synthase interacting with naringenin is the first step in the Isoflavone biosynthesis pathway; however, the interaction of other enzymes from other pathways results in the decrease in isoflavone synthesis. (**a**) Illustrates how Flavone synthase II (FNS II) and Flavanone-3-hysroxylase (F3H3) interact with the substrate naringenin, inhibiting normal binding of IF to naringenin. (**b**) Illustrates how the introduction of a multiplex CRISPR/Cas9 system simultaneously inactivates both FNS II and F3H3, allowing the interaction of IF and naringenin to occur for the proper synthesis of isoflavonoid

was conducted via liquid chromatography, and the isoflavone content, seed protein content, and oil concentration were determined. The T3 generation leaves were used for SMV resistance tests, in which Zhang et al. (2020) determined the SMV virus coat protein using Enzyme-Linked immunosorbent assay kit (ELISA). In the triple mutant, the expression of F3H genes did not significantly change however, some of the derivatives significantly decreased. Similar results were observed when looking at FNS gene expression, in which a comparison to the control showed that the expression did not change, but a significant decrease was observed in some of the derivatives of FNS (Zhang et al. 2020). The results show that downstream material flow was reduced after the knockout of the GmF3H and GmFNS genes. In the mutants, an increase of 12 folds was observed in naringenin when compared to the control plants. Compared to WT, the T3 generation homozygous triple mutants showed about two times more leaf isoflavone content, and a one-third reduction in Soy bean mosaic virus coat protein content was exhibited in the knockout lines, after the plants were infected with the strain SC7 (Zhang et al. 2020). Zhang et al. (2020) thus concluded that due to the knockout of GmF3H and GmFNS genes, the interaction of Isoflavone with naringenin was undisturbed, and the synthesis of isoflavone was improved, ultimately enhancing resistance of *G. max* leaves to SMV.

All plants have particular shortfalls to various stress conditions, and in *G. max*, photoperiod sensitivity is one of those shortfalls (Cai et al. 2018). *G. max* are sensitive to seasonal changes in day length, and this sensitivity limits the geographical range and cultivation of this legume crop (Cai et al. 2018). The photoperiods have an effect on the flowering pathways. The *Arabidopsis thaliana* FLOWER LOCUS T (FT) gene encodes for florigen, and FT has an essential function in the flowering pathways. FT proteins are transported through the phloem and function as a long-distance signal that induces floral initiation in Arabidopsis (Corbesier et al. 2007). It has been shown that some homologous genes of FT have similar functions in *G. max* plants (Kong et al. 2010). One of those genes is GmFT2a which Kong et al. (2010) obtained by amplifying the entire coding region of the FT homologs, and the transcribed regions were then assigned by comparison of the nucleotide sequence with the BLASTN algorithm. Analysis was conducted on the transcripts of GmFT2a, which revealed that the expression of GmFT2a is regulated by the photoperiod and is associated with flowering induction and maintenance (Sun et al. 2011). Thus, taking all of these findings into account, the study by Cai et al. (2018) was aimed at examining the photoperiod response in flowering induction, by means of using the CRISPR/Cas9 system to specifically induce target mutagenesis of the GmFT2a gene in *G. max*. GmFT2a is a *G. max* ortholog of FT and thus plays an important role in floral initiation in *G. max*. Cai et al. (2018) used ViewSolid Biotech to synthesize the sequences. The GmFT2a sequence and other information were downloaded from the Phytozome website. CRISPR/Cas9 expression vector was generated by using the CRISPR/Cas9-mediated genome-editing tool to knockout the *G. max* exogenous gene GmFT2a, and this was called the GmFT2a-CRISPR/Cas9 vector (Cai et al. 2018). Three target sites were used in this study targeting GmFT2a and sgRNA (GmFT2a-SP1, GmFT2a-SP2, and GmFT2a-SP3), and the complexes were designed using CRISPR-P web tool, and the CRISPR/Cas9 expression vector was transformed into *G. max* plants using an Agrobacterium strain via electrophoresis. Cai et al. (2018) recorded that under short-day conditions, the T2 ft2a mutants represented no floral buds when the WT were flowering already at this point, and when the ft2a mutant did start flowering, the WT plants had produced pods already. Under long-day conditions, the T2 ft2a mutant represented no floral buds when the WT was flowering, and when the ft2a came into the flowering stage, the WT flowers were just falling off and the initial pods were being formed. Thus, T2 ft2a mutants depicted a delayed flowering phenotype under both short- and long-day conditions. Three target sites in the first exon of GmFT2a and mutations were induced in the coding regions to inactivate GmFT2a. When comparing flowering time, the homozygous ft2a mutants showed late-flowering phenotypes when compared to the control under natural, short-day, and long-day conditions. Ft2a mutants under any conditions had significantly lower GmFT2 transcript levels in the T2 generation than the controls plants. T1 generation of homozygous-null alleles of GmFT2a caused a frameshift mutation in a 1 bp insertion or short deletion, and this mutant exhibited late flowering under normal conditions (summer). The targeted mutations were stably inherited and maintained consistent mutation types from the T1 to T2 generation. Homozygous T2 ft2a mutants exhibited (T2 generation) late flowering under both long-day and short-day conditions.

Arachis hypogaea falls in the top five highly important oilseed crops globally. *A. hypogaea* crops are very important because approximately two-third of the global production is used as edible oil in the daily consumption of humans (Wen et al. 2018). In addition to this, the fatty acid composition of *A. hypogaea* is essential for the shelf-life of *A. hypogaea* products and provides heat stability. The protein fatty acid desaturase 2 (FAD2) converts monosaturated fatty acid (oleic acid) to polyunsaturated fatty acid (PUFA), linoleic acid (Janila et al. 2016; Chi et al. 2011). The concern about this conversion is that in literature it is documented that high levels of PUFA in oil lead to greater chances of oxidation resulting in distasteful flavors and reduce the shelf-life of the products. Therefore, a good strategy to increase shelf-life of *A. hypogaea* products is to increase the oleic acid content and to reduce the activity of FAD genes. This was accomplished in a study by Wen et al. (2018) where the authors utilized the TALENs technology to create target mutations in the conserved coding sequences of *A. hypogaea* AhFAD2 gene. TALENs were designed to attach and concurrently cleave specific DNA sequences in the FAD2A and FAD2B genes, thus inactivating (by knockout) all the alleles of the ahFAD2A and ahFAD2B genes in *A. hypogaea* plants (Wen et al. 2018). To design TALENs, Wen et al. (2018) used the software proposed by Doyle et al. (2012) called the TAL-Effector-Nucleotide Targeter (TALE-NT) 2.0 program, in which DNA sequences were input into the program to identify TAL effector-binding sites. To construct the TALEN vector, Wen et al. (2018) used the FastTALETM TALEN Assembly Kit (SIDANSAI) to perform a one-step ligation assay, and the final construct was subcloned into a modified binary vector pCAMBIA 1301, and then by Agrobacterium-mediated transformation, the construct was introduced into the *A. hypogaea* plants via direct embryogenesis (Tiwari et al. 2015). Wen et al. (2018) collected seeds from the T2 generation of transgenic *A. hypogaea* plants and used the seeds for total fatty acid profiling and protein concentration analysis. Following the Soxhlet extraction protocol, the total oil content of *A. hypogaea* plants was calculated, and the profile analysis of fatty acids in *A. hypogaea* was performed using gas chromatography. The TALEN-mediated target mutagenesis allowed for the successful inactivation of the ahFAD2 genes which resulted in an accumulation of oleic acid. An increase of 0.5–2 fold was recorded in oleic acid in comparison to the nontransgenic controls. The oleic acid content was increased with a maximum value as high as 80%, and the oil characteristics were altered by reducing linoleic acid to 7% in the seeds of transgenic *A. hypogaea* plants when compared to the controls. Therefore, it was concluded that the TALEN-induced FAD2 mutants can be used to efficiently improve the quality of edible *A. hypogaea* oil. In consideration of other agronomical traits (plant height, mature pod weight), there was no significant difference observed between the transgenic *A. hypogaea* and nontransgenic control *A. hypogaea* plants. Wen et al. (2018) concluded that the TALEN-mediated targeted mutagenesis of the FAD2 gene displays an improvement over the traditional breeding methods mainly because the genome editing process can be greatly accelerated in *A. hypogaea* plants.

Medicago sativa is a valuable forage crop plant which is frequently cultivated in regions subjected to drought stress and water shortages (Arshad et al. 2017). Due to exposure to frequent stress conditions, Arshad et al. (2017) sought to improve *M. sativa* plants' response to drought by focusing on the role and editing of microRNA156 (miRNA156) and more specifically in silencing the target gene of miRNA156, called the Squamosa Promoter Binding Protein-like 13 (SPL13). Aung et al. (2015) assembled an overexpression line identified as miRNA156OE. The miRBase was employed to obtain the miR156 precursor sequence, and Aung et al. (2015) used this sequence in a BLAST search against *M. sativa* and *Medicago truncatula* sequences on the NCBI database. The miR156 overexpression gene was cloned into the pENTR/D-TOPO entry vector and then transformed into *Agrobacterium tumefaciens*. Finally, the vector was transformed into *M. sativa* plants via *Agrobacterium tumefaciens*-mediated transformation (Aung et al. 2015). Then, Arshad et al. (2017) amplified the SPL13 fragment from *M. sativa* cDNA. The primer designs were based on *M. truncatula,* and the SPL13-RNAi construct was made using the Gateway system (Invitrogen). The construct was cloned into pENTR/D-TOPO entry vector, and after validation of the sequence of SPL13, the fragment was recombined into pHELLSGATE12 destination vector by the LR clonase reaction and transformed into *M. sativa* via *Agrobacterium tumefaciens*-mediated transformation. An empty vector was used for control plants (Arshad et al. 2017). For drought experiments, four genotypes were set up [WT, miRNA156OE genotypes, the SPL-13 RNAi genotypes, and the empty vector control (EV)]. Arshad et al. (2017) documented that miRNA156 was indeed induced by drought stress as the expression levels of miRNA156 in the leaf and root tissues were more than that of the WT plants grown under control and drought stress. RT-qPCR was conducted and illustrated that there was a significant downregulation of SPL13 in the RNAi genotypes when compared to the EV plants. Survival rate of the miR156OE genotype plants was 71%, whereas the controls were 46% after rewatering. miR156OE retains about 4–8% more water than WT after 120 and 150 min. The water loss was decreased in both overexpressed lines and RNAi-silenced SPL13 plants when compared to the WT and EV (Arshad et al. 2017). The SPL13-RNAi plants exhibited denser thicker and longer phenotypes than the EV control plants. In the plants with reduced expression of miR156-targeted SPL13, reduced water loss, enhanced stromatal conductance, chlorophyll content, and photosynthetic assimilation were observed. MiRNA156 targets a number of SPL13 genes for silencing by transcript cleavage in *M. sativa*. Arshad et al. (2017) observed induced expression of SPL13 in miRNA156OE genotypes relative to WT under both control and drought stress.

Thus, it appears that reduced SPL13 expression may be a mechanism by which *M. sativa* tolerates drought. Lower expression of SPL13 in the miR156OE leaves and roots may indicate that SPL13 negatively regulates drought tolerance. Therefore, Arshad et al. (2017) concluded that miR156 improves drought tolerance in *M. sativa* at least by silencing SPL13 genes.

3.4 Gene Mining and Editing in Brassicaceae

The methods to efficiently edit plant genomes have rapidly improved in recent years, which have led to the implementation of these techniques to improve breeding and agriculture research. However, stable integration of these genome editing tools and their prolonged expression may lead to mutations in off-target sites (Cox et al. 2015). To overcome the previous mentioned limitations, preassembled ribonucleoprotein complexes (RNPs) or endogenous tRNA-processing systems may be used with the Cas9 system (Ma et al. 2019; Murovec et al. 2018).

Murovec et al. (2018) proposed a protocol for DNA-free genome editing of different species of Brassica. To identify appropriate endogenous genes to be used in this study, the authors used the Brassica database as well as the NCBI GenBank, and ultimately, FRIGIDA (FRI) and Phytoene desaturase (PDS) genes were selected. The DNA sequence for FRI was originally isolated from *Brassica oleracea* L. var. capitate L accession (Bra035723) (Sun et al. 2013), and the DNA sequence for PDS was identified from *Brassica napus* accession (HM989807). The FRI gene is required in the vernalization pathways and affects the flowering time of plants (Yi et al. 2018). The PDS gene has shown implications on fruit ripening in certain plants (Naing et al. 2019). The modification of the FRI gene expression in plants would have an effect on tolerance to cold temperatures. The plants which were used as subjects for genome editing in the Murovec et al. (2018) study were Cabbage (*B. oleracea* var. capitate f. alba) and Chinese cabbage (*Brassica rapa* subsp. Pekinensis). The authors chose to use a CRISPR/Cas9 Ribonucleoprotein complex to edit the genome of the two plants to obtain stable modification. Leaf tissues from *B. oleracea* and *B. rapa* were used to isolate the protoplast of the plant followed by the transfection of the protoplast. This technique shows promise as it may enable the development of plant phenotype editing without the need of transgenesis.

Zaman et al. (2019) used genome editing technology to understand what role the JAGGED genes play in *B. napus* and its implication on seed shattering tolerance. Homologs of the JAG genes in *B. napus* were identified by blasting an *A. thaliana* genome sequence (At1G68480.1) using the online NCBI software followed by a phylogenetic tree being constructed using an online tool (http://www.phylogeny.fr/). Five JAG homologs were identified in *B. napus,* namely, BnaA02g13870D, BnaC06g30050D, BnaA07g27150D, BnaC02g18270D, and BnaA08g24290D. The JAG genes are activated in immature organs and are involved in organ development and growth. Further analysis has shown that it may also be involved in increased proliferation, cell enlargement, and cell size homeostasis (Schiessl et al. 2014). The plant that was used as the subject of this study was *B. napus* var.

Zhongshuang. The CRISPR/Cas9-mediated multiplex genome editing was used for the purpose of this experiment. The transformation vectors were inserted into *A. tumefaciens* and introduced into the hypocotyl of *B. napus* plants. The knockout of the five JAG genes resulted in ovary development without any distinct valves, replum, septum, and valve margins. A decrease in pod length of 51% was observed when comparing the JAG mutant to the WT. The number of seeds in the JAG mutants and the WT were 2 and 22, respectively, which amounted to a 90% reduction in pod filling.

A study attempting to modify the fatty acid profile of *B. napus* though gene editing was performed by Huang et al. (2020). The authors used a bioinformatics approach to identify the expression profiles of different BnaFAD2 genes. The NCBI sequence read archive was used to obtain the candidate gene. The FAD gene encodes an enzyme called the fatty acid desaturase which is responsible for catalyzing desaturation of oleic acid (Okuzaki et al. 2018). By identifying important genes in fatty acid synthesis such as BnaFAD2, these genes can be manipulated to produce oils rich in oleic acid as it is more desirable (Dar et al. 2017). The *B. napus* cultivar J9707 was modified for the purpose of the study. The CRISPR/Cas9 system was used to produce the modification of BnaFAD2. The *A. tumefaciens* method was used to introduce the constructed vector into the hypocotyl of the plants. The mutants produced seeds which had an alteration in oleic acid content which ranged from 73% to 82%.

Yang et al. (2018) used precise gene editing technology to knockout the CLAVATA genes in plants in an effort to increase yield. The CLAVATA genes (CLV3) were identified as suitable candidates for gene editing. The CLV genes have been documented to play a role in the regulation of stem cells in the shoot apical meristem (Fletcher 2018). The alteration of the CLV genes has shown to increase the number of floral organs in plants such as *Z. mays* and *O. sativa* leading to increases in yields (Basu et al. 2019; Fletcher 2018). The *B. napus* J9707 line was transformed with binary vectors using the *A. tumefaciens*-hypocotyl method. The CRISPR/Cas9 system was used in this study due to its greater precision over other mutagenic techniques available. The authors observed an increase in growth parameters such as number of seeds per silique (NSS), 1000 seed weight (TSW), and seed weight per silique (SW) in the CLV double mutant. An increase of more than 44% and 27% was observed for the NSS and TSW parameters, respectively. This study clearly shows the value of gene editing technology as a tool to improve the current agricultural research and breeding programs to enhance commodity crop yields.

The use of targeted genome editing opens new possibilities for crop improvement. A study by Sun et al. (2018) investigated the use of genome editing to enhance pathogen tolerance in *B. napus* L. The genes AtWRKY11 and AtWRKY70 were first observed in Arabidopsis and were involved in Jasmonic and Salicylic acid resistance to pathogens (Sun et al. 2019; Pandey and Somssich 2009). The homologs of the aforementioned genes in *B. napus* L were BnWRKY11 and BnWRKY70, which were selected for further study. The CRISPR/Cas9 system was used in conjunction with binary vectors as well as the specific sgRNA. The *B. napus* J9712 line was transformed by making an incision in the hypocotyl of germinated seedlings and floating them on infection medium. The authors observed that the BnWRKY70 mutants were more resistant to the pathogenic fungus *Sclerotinia sclerotiorum* than their nonmutated counterparts as smaller lesion areas were observed when infected (Fig. 3.3). This study once again shows that genome editing tools allow for not only abiotic stress tolerance improvement but can also be used to improve biotic stress tolerance.

3.5 Gene Mining and Editing in Solanaceae

Solanum lycopersicum belongs to the Solanaceae family and has become popular because of its economic importance which includes nutritional values, protection against diseases, and antioxidant properties which have been conferred by the presence of certain bioactive compounds such as lycopene and beta-carotene, vitamin C, flavonoids, and hydroxycinnamic acid derivatives (Del Carmen Orozco-Mosqueda et al. 2019). However, understanding the *S. lycopersicum* immune mechanisms is imperative to reduce the losses caused by pathogens (Liu et al. 2019). These approaches include targeting the conserved microbial-associated molecular patterns (MAMPs) (Mélida et al. 2020) or strain-specific effectors by pattern recognition receptors (PRRs) or resistance (R) proteins triggering MTI (MAMP-triggered immunity) (Teixeira et al. 2019) and ETI (effector-triggered immunity) (Ray et al. 2019). To achieve this target, the *S. lycopersicum* can activate various pathways which include signal transduction pathways (Adhikari et al. 2019), novel protein interactions (Liu et al. 2019), and coordinated alterations in gene expression (Quinet et al. 2019). Gene microarray data are a gene expression data which are used either to detect differentially expressed genes when comparing two states or to build gene clusters with close gene expression profile (Filiz et al. 2019). Multivariate and data mining techniques have also been used to detect genes with similar expression pattern in *S. lycopersicum* on a time course where the most influentially resistant gene was identified with special consideration to time measurement (Ait Issad et al. 2019).

S. lycopersicum species have a reservoir of essential genes used for breeding programs both at genetic and genomic levels with a diploid genetics (24 somatic chromosomes) and a small genome size (950 Mb per haploid nucleus) (Roth et al. 2019). These gene mining techniques would have three main goals in the Solanaceae (Table 3.2): (1) improvement of a desired trait, (2) reduced expression of an unwanted trait, and (3) discovery of a novel trait (Rothan et al. 2019). Some qualities have been established that make *S. lycopersicum* amenable to genetic analysis which includes self-pollination, short generation time, simple reproduction through seed and vegetative reproduction, and cross-compatibility to many WTs. The advantage of these on the mutants includes high expression levels of recombinant proteins, absence of epigenetic effects (gene silencing), improved biosafety due to transgenes and recombinant products, failure of pollen dispersal in the environment, and expression of multiple transgenes in the prokaryotic-like operon (Muñoz-Sanz et al. 2020). The lycopene β-cyclase genes from eubacterium *Erwinian herbicola*

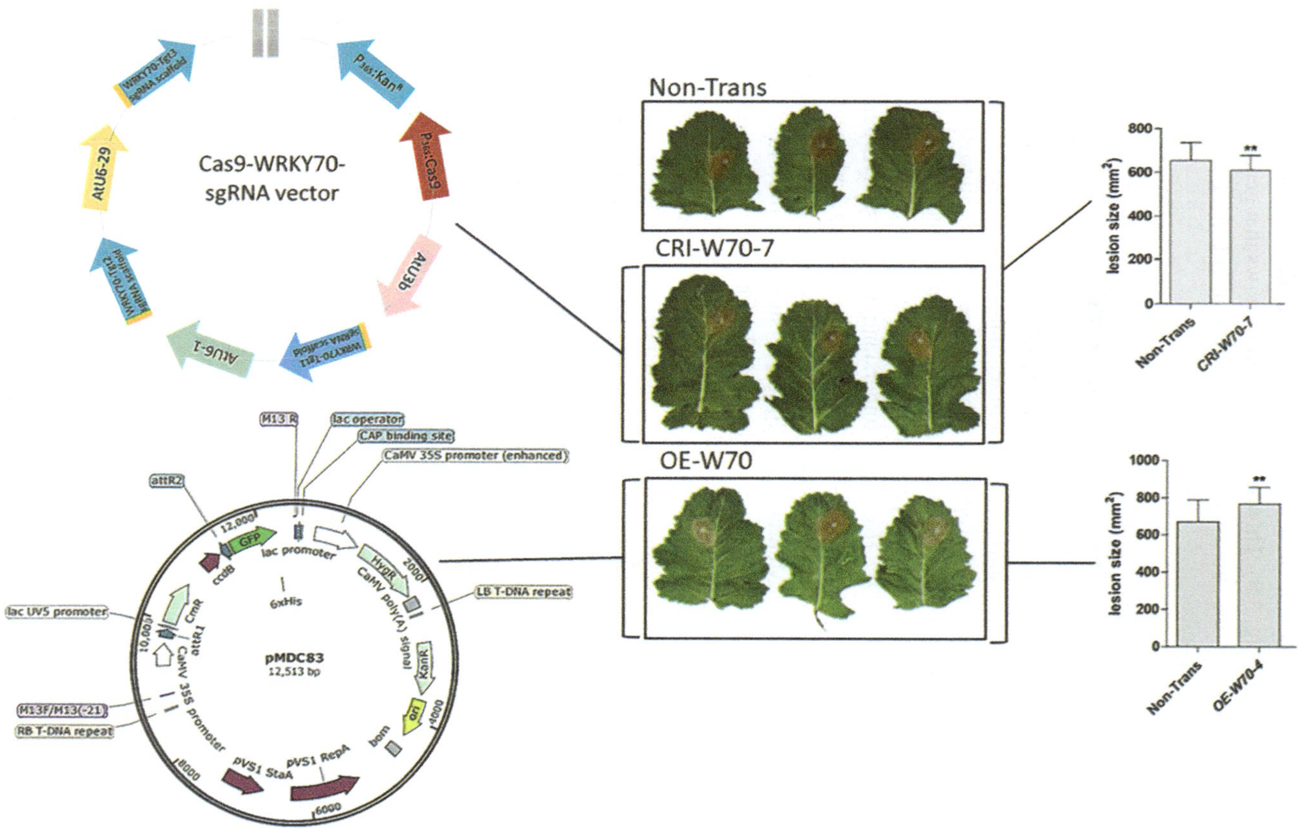

Fig. 3.3 A modified diagram representing the work conducted by Sun et al. (2018) on *Brassica napus*. The image illustrates the two plant lines generated: one incorporating CRISPR/Cas9 technology to knockout of BnWRKY70 (CRI-W70-7) and the other overexpressing the BnWRKY70 gene (OE-W70). On the left is the two vector constructs generated. The leaf images indicate the lesions caused by the *S. sclerotiorum* infection, and the graphs on the right are the graphical representations of lesion sizes of each plant line. Nontrans line represented the control plants in the study

Table 3.2 Novel genes from gene mining techniques used in Solanaceae

Identified genes	Gene function	Roles in plant processes	Editing technology	Reference
Fw2.2	Size	Increased size	Dosage series of single gene	Gonda et al. (2019)
Arf8 and IAA9	Parthenocarpy	Induced parthenocarpy	Single TF	Batista-Silva et al. (2019), Serrani et al. (2008)
Lin5	Soluble solids content	Reduced sugars accumulation	Single biosynthetic key gene	Brekke et al. (2019)
GalLDH, GME	Ascorbic acid content	Increase/ decreased fruit ascorbic content	Single biosynthetic key gene	Liao et al. (2019)
GCHI and/or ADCS	Ascorbic acid content	Increased fruit folate	Genes targeting consecutive biosynthetic steps	Robinson et al. (2019)
CHI	Flavonoid content	Increased fruit peel flavonol	Single biosynthetic key gene	Zhao et al. (2019)
MYB12	Flavonoid content	Accumulation of flavonols	Single TF	Chen et al. (2019)
Del/Ros1	Flavonoid content	High levels of anthocyanins	Multiple TF	Blando et al. (2019)

and *Narcissus pseudonarcissus* have been introduced into the *S. lycopersicum* plastid genome for enhanced carotenoid synthesis, essential pigments of the photosynthetic apparatus, and an indispensable component of the human diet, in order to overproduce lycopene-to-provitamin A conversion at a rate greater than 50% total carotenoid accumulation (Yarra 2020).

Overexpression of either LeAADC1A or LeAADC2 which encodes for phenylalanine decarboxylases involved in the synthesis of 2-phenylethanol (a volatile compound that is a major contributor to flavor in many foods) from phenylalanine resulted in fruits with tenfold emission increase of the pathway products: 2-phenylacetaldehyde, 2-phenylethanol, and 1-ntro, 2-phenylethane, while significant reduction of these emission volatile products was observed using antisense reduction of LeAADC2 (Tieman et al. 2006). Modification of fruit-softening and total firmness traits in transgenic *S. lycopersicum* plants has been successfully carried out by engineering single enzymatic steps in polygalacturonase, pectin methylesterase, expansin, and β-galactosidase controlling genes in cell wall-associated pathways (Romero and Rose 2019). A negative regulator of cell division, gene fw2.2, that is responsible for approximately 30% of the difference of the *S. lycopersicum* fruit size, has been characterized in *S. lycopersicum* using genetic transformation for fruit weight modulation using a dosage series without affecting the fruit size in pericarp and placenta tissues (Nesbitt and Tanksley 2001). Expression of thaumatin, a biologically active, sweet tasting, and flavor-enhancing protein from *Thaumatococcus daniellii* Benth, has been carried out in transgenic

S. lycopersicum using metabolic engineering procedures for enhanced sweetness with a specific aftertaste (Bartoszewski et al. 2003).

Solanum tuberosum has a long and successful history of modified varieties after breeding, and this makes the crop plant a subject of genetic resource development using gene mining approaches such as large expressed sequence tag (EST) data collections which are rich source for the generation of single-nucleotide polymorphism and simple sequence repeat, microarrays, and virus-induced gene silencing (VIGS) (Feingold et al. 2005). Other gene mining and editing approaches used in *S. tuberosum* include the use of mutagenesis (Andersson et al. 2017), somatic hybridization (Tiwari et al. 2018), direct gene transfer through transgenics (Upadhyaya et al. 2020), Digital Gene Expression (DGE) tag profiling technology to characterize a *S. tuberosum* clone (Lemke et al. 2020), and genome (gene) editing using TALEN and CRISPR/Cas9 (Sevestre et al. 2020). To achieve this target, *S. tuberosum* can activate various pathways which include signal transduction pathways, novel protein interactions, and coordinated alterations in gene expression (Li et al. 2020). The key areas for improvement include increased utilization of the crop plant with low Glycaemic index, addition to healthy diets to address consumer demands, development of tolerant species to drought, heat, high carbon-(iv) oxide levels which can maintain yield and quality, and development of species that are pest and disease resistant (Tiwari et al. 2020).

The expressions of CONSTANS (CO) that encodes a nuclear protein similar to zinc finger transcription factors, and FLOWERING LOCUS T (FT), a homolog of the floral inductor referred to as SP6A, which are responsible for the photoperiodic regulation of *S. tuberosum* tuberization, are a promising target for breeding of heat-tolerant *S. tuberosum* (Tiwari et al. 2020). Along with these, two mobile RNA molecules, miR172 and StBEL5, have been reported to elicit a long transmissible signals that are related to day-length control of tuber formations. Such photoperiodic control of tuberization can be abolished in *S. tuberosum* ssp. andigena plants by reducing phytochrome B through antisense-mediated suppression for tuberization under short day, long day, and short day supplemented with a night break conditions (Dutt et al. 2017). Two full-length potato cDNAs, POTM1-1 and POTM1-2, have been identified that encode 250 identical amino acids (aa) including 56 aa of the MADS-box domain and 53 aa of the K-box domain, which function as transcription factors (Kang and Hannapel 1995). These transcripts of the cDNAs reside in the vegetative organs, prompting the notion that these novel MADS genes are involved in vegetative organ development.

3.6 Conclusion

This chapter highlights some of the studies which achieved successes with transgenic technologies in the four major crop plant families. In most cases, we highlight the gene mining tools which led to the selection of a candidate gene. We also highlighted the gene editing tool, and we observed that most studies are now using the CRISPR tool for precision gene editing in plants. Some of the successes are discussed with regard to improved growth in general or improved growth under biotic and abiotic stresses. In addition, many of the gene editing studies, in literature, focused on the regulation of important human nutrition and human health biomolecules. Still, not all crop plants can be efficiently transformed, and this remains a major drawback of gene editing and breeding programs.

Acknowledgments Miss. Niekerk, Mr. Carelse, and Dr. Bakare were financially supported by the National Research Foundation (NRF) of South Africa (Grant numbers: 111871 and 113857). Prof. Keyster's research was financially supported by the NRF (Grant numbers: 116346 and 109083), DST-NRF Centre of Excellence in Food Security (Project ID: 170202), GrainSA (GB0200066), and the University of the Western Cape (UWC). Dr. Klein research was financially supported by the NRF (Grant numbers: 107023 and 115280), GrainSA (GB0200065), and UWC. Dr. Gokul's research was financially supported by the NRF (Grant number: 129493) and the University of Free State (UFS). Therefore, the authors would like to express sincere gratitude toward these funding bodies.

References

Adhikari B, Adhikari M, Ghimire B et al (2019) Cold atmospheric plasma-activated water irrigation induces defense hormone and gene expression in tomato seedlings. Sci Rep 9(1):1–15. https://doi.org/10.1038/s41598-019-52646-z

Ait Issad H, Aoudjit R, Rodrigues JJPC (2019) A comprehensive review of Data Mining techniques in smart agriculture. Eng Agric Environ Food 12(4):511–525. https://doi.org/10.1016/j.eaef.2019.11.003

Andersson M, Turesson H, Nicolia A et al (2017) Efficient targeted multiallelic mutagenesis in tetraploid potato (Solanum tuberosum) by transient CRISPR-Cas9 expression in protoplasts. Plant Cell Rep 36(1):117–128. https://doi.org/10.1007/s00299-016-2062-3

Arshad M, Feyissa BA, Amyot L et al (2017) MicroRNA156 improves drought stress tolerance in alfalfa (*Medicago sativa*) by silencing SPL13. Plant Sci 258:122–136. https://doi.org/10.1016/j.plantsci.2017.01.018

Aung B, Gruber MY, Amyot L et al (2015) MicroRNA156 as a promising tool for alfalfa improvement. Plant Biotechnol J 13(6):779–790. https://doi.org/10.1111/pbi.12308

Bartoszewski G, Niedziela A, Szwacka M et al (2003) Modification of tomato taste in transgenic plants carrying a thaumatin gene from Thaumatococcus daniellii Benth. Plant Breed 122(4): 347–351. https://doi.org/10.1046/j.1439-0523.2003.00864.x

Basu U, Narnoliya L, Srivastava R et al (2019) CLAVATA signaling pathway genes modulating flowering time and flower number in chickpea. Theor Appl Genet 132(7):2017–2038. https://doi.org/10.1007/s00122-019-03335-y

Batista-Silva W, Medeiros DB, Rodrigues-Salvador A et al (2019) Modulation of auxin signalling through DIAGETROPICA and ENTIRE differentially affects tomato plant growth via changes in photosynthetic and mitochondrial metabolism. Plant Cell Environ 42(2):448–465. https://doi.org/10.1111/pce.13413

Batley J, Edwards D (2016) The application of genomics and bioinformatics to accelerate crop improvement in a changing climate. Curr Opin Plant Biol 30:78–81. https://doi.org/10.1016/j.pbi.2016.02.002

Bhatnagar-Mathur P, Vadez V, Sharma KK (2008) Transgenic approaches for abiotic stress tolerance in plants: retrospect and prospects. Plant Cell Rep 27(3):411–424. https://doi.org/10.1007/s00299-007-0474-9

Blando F, Berland H, Maiorano G et al (2019) Nutraceutical characterization of anthocyanin-rich fruits produced by "sun black" tomato line. Front Nutr 6:133. https://doi.org/10.3389/fnut.2019.00133

Blanvillain-Baufumé S, Reschke M, Solé M et al (2017) Targeted promoter editing for rice resistance to Xanthomonas oryzae pv. oryzae reveals differential activities for SWEET14-inducing TAL effectors. Plant Biotechnol J 15(3):306–317. https://doi.org/10.1111/pbi.12613

Brekke TD, Stroud JA, Shaw DS et al (2019) QTL mapping in salad tomatoes. Euphytica 215(7): 1–12. https://doi.org/10.1007/s10681-019-2440-3

Byrt CS, Grof CPL, Furbank RT (2011) C4 plants as biofuel feedstocks: optimising biomass production and feedstock quality from a lignocellulosic perspective. J Integr Plant Biol 53(2): 120–135. https://doi.org/10.1111/j.1744-7909.2010.01023.x

Cai Y, Chen L, Liu X et al (2018) CRISPR/Cas9-mediated targeted mutagenesis of GmFT2a delays flowering time in soya bean. Plant Biotechnol J 16(1):176–185. https://doi.org/10.1111/pbi.12758

Chen F, Dixon RA (2007) Lignin modification improves fermentable sugar yields for biofuel production. Nat Biotechnol 25(7):759–761. https://doi.org/10.1038/nbt1316

Chen H, Wang JP, Liu H et al (2019) Hierarchical transcription factor and chromatin binding network for wood formation in populus trichocarpa. Plant Cell 31(3):602–626. https://doi.org/10.1105/tpc.18.00620

Cheng Q, Li N, Dong L, Zhang D, Fan S, Jiang L, Wang X, Xu P, Zhang S (2015) Overexpression of soybean isoflavone reductase (GmIFR) enhances resistance to phytophthora sojae in soybean. Front Plant Sci 6:1–11. https://doi.org/10.3389/fpls.2015.01024

Chi X, Yang Q, Pan L et al (2011) Isolation and characterization of fatty acid desaturase genes from peanut (Arachis hypogaea L.). Plant Cell Rep 30(8):1393–1404. https://doi.org/10.1007/s00299-011-1048-4

Corbesier L, Vincent C, Jang S et al (2007) FT protein movement contributes to long-distance signaling in floral induction of Arabidopsis. Science 316(5827):1030–1033. https://doi.org/10.1126/science.1141752

Cox DBT, Platt RJ, Zhang F (2015) Therapeutic genome editing: prospects and challenges. Nat Med 21(2):121–131. https://doi.org/10.1038/nm.3793

Dar AA, Choudhury AR, Kancharla PK et al (2017) The FAD2 gene in plants: occurrence, regulation, and role. Front Plant Sci 8:1789. https://doi.org/10.3389/fpls.2017.01789

Del Carmen Orozco-Mosqueda M, Duan J, DiBernardo M et al (2019) The production of ACC deaminase and trehalose by the plant growth promoting bacterium Pseudomonas sp. UW4 synergistically protect tomato plants against salt stress. Front Microbiol 10:1392. https://doi.org/10.3389/fmicb.2019.01392

Doyle EL, Booher NJ, Standage DS et al (2012) TAL effector-nucleotide targeter (TALE-NT) 2.0: tools for TAL effector design and target prediction. Nucleic Acids Res 40(W1):W117. https://doi.org/10.1093/nar/gks608

Du H, Zeng X, Zhao M et al (2016) Efficient targeted mutagenesis in soybean by TALENs and CRISPR/Cas9. J Biotechnol 217:90–97. https://doi.org/10.1016/j.jbiotec.2015.11.005

Dutt S, Manjul AS, Raigond P et al (2017) Key players associated with tuberization in potato: potential candidates for genetic engineering. Crit Rev Biotechnol 37(7):942–957. https://doi.org/10.1080/07388551.2016.1274876

Feingold S, Lloyd J, Norero N et al (2005) Mapping and characterization of new EST-derived microsatellites for potato (Solanum tuberosum L.). Theor Appl Genet 111(3):456–466. https://doi.org/10.1007/s00122-005-2028-2

Filiz E, Ozyigit II, Saracoglu IA et al (2019) Abiotic stress-induced regulation of antioxidant genes in different Arabidopsis ecotypes: microarray data evaluation. Biotechnol Biotechnol Equip 33(1):128–143. https://doi.org/10.1080/13102818.2018.1556120

Fletcher JC (2018) The CLV-WUS stem cell signaling pathway: a roadmap to crop yield optimization. Plan Theory 7(4). https://doi.org/10.3390/plants7040087

Furbank RT, Tester M (2011) Phenomics - technologies to relieve the phenotyping bottleneck. Trends Plant Sci 16(12):635–644. https://doi.org/10.1016/j.tplants.2011.09.005

Gokul A, Niekerk L-A, Carelse MF et al (2020) Transgenic technology for efficient abiotic stress tolerance in plants. In: Transgenic technology based value addition in plant biotechnology, pp 95–122

Gonda I, Ashrafi H, Lyon DA et al (2019) Sequencing-based bin map construction of a tomato mapping population, facilitating high-resolution quantitative trait loci detection. Plant Genome 12(1):180010. https://doi.org/10.3835/plantgenome2018.02.0010

Guo M, Rupe MA, Wei J et al (2014) Maize ARGOS1 (ZAR1) transgenic alleles increase hybrid maize yield. J Exp Bot 65(1):249–260. https://doi.org/10.1093/jxb/ert370

Hays DB, Do JH, Mason RE et al (2007) Heat stress induced ethylene production in developing wheat grains induces kernel abortion and increased maturation in a susceptible cultivar. Plant Sci 172(6):1113–1123. https://doi.org/10.1016/j.plantsci.2007.03.004

Huang H, Cui T, Zhang L et al (2020) Modifications of fatty acid profile through targeted mutation at BnaFAD2 gene with CRISPR/Cas9-mediated gene editing in Brassica napus. Theor Appl Genet 133:2401–2411. https://doi.org/10.1007/s00122-020-03607-y

Janila P, Pandey MK, Shasidhar Y et al (2016) Molecular breeding for introgression of fatty acid desaturase mutant alleles (ahFAD2A and ahFAD2B) enhances oil quality in high and low oil containing peanut genotypes. Plant Sci 242:203–213. https://doi.org/10.1016/j.plantsci.2015.08.013

Jiang YN, Wang B, Li H et al (2010) Flavonoid production is effectively regulated by RNAi interference of two flavone synthase genes from glycine max. J Plant Biol 53(6):425–432. https://doi.org/10.1007/s12374-010-9132-9

Jiang Y, Hu Y, Wang B et al (2014) Bivalent RNA interference to increase isoflavone biosynthesis in soybean (glycine max). Brazilian Arch Biol Technol 57(2):163–170. https://doi.org/10.1590/S1516-89132013005000018

Kang SG, Hannapel DJ (1995) Nucleotide sequences of novel potato (Solanum tuberosum L.) MADS-box cDNAs and their expression in vegetative organs. Gene 166(2):329–330. https://doi.org/10.1016/0378-1119(95)00593-5

Kannan B, Jung JH, Moxley GW et al (2018) TALEN-mediated targeted mutagenesis of more than 100 COMT copies/alleles in highly polyploid sugarcane improves saccharification efficiency without compromising biomass yield. Plant Biotechnol J 16(4):856–866. https://doi.org/10.1111/pbi.12833

Kawakami EM, Oosterhuis DM, Snider JL (2010) Physiological effects of 1-methylcyclopropene on well-watered and water-stressed cotton plants. J Plant Growth Regul 29(3):280–288. https://doi.org/10.1007/s00344-009-9134-3

Khanna HK, Deo PC (2016) Novel gene transfer technologies. In: Banana: genomics and transgenic approaches for genetic improvement. Springer, Singapore, pp 127–140

Kong F, Liu B, Xia Z et al (2010) Two coordinately regulated homologs of FLOWERING LOCUS T are involved in the control of photoperiodic flowering in Soybean. Plant Physiol 154(3): 1220–1231. https://doi.org/10.1104/pp.110.160796

Kumar K, Kumar M, Kim SR et al (2013) Insights into genomics of salt stress response in rice. Rice 6(1):1–15. https://doi.org/10.1186/1939-8433-6-27

Lei Y, Lu L, Liu HY et al (2014) CRISPR-P: a web tool for synthetic single-guide RNA design of CRISPR-system in plants. Mol Plant 7(9):1494–1496. https://doi.org/10.1093/mp/ssu044

Lemke P, Moerschbacher BM, Singh R (2020) Transcriptome analysis of solanum tuberosum genotype RH89-039-16 in response to chitosan. Front Plant Sci 11:1193. https://doi.org/10.3389/fpls.2020.01193

Li M, Li X, Zhou Z et al (2016) Reassessment of the four yield-related genes Gn1a, DEP1, GS3, and IPA1 in rice using a CRISPR/Cas9 system. Front Plant Sci 7:377. https://doi.org/10.3389/fpls.2016.00377

Li J, Sun Y, Du J et al (2017) Generation of targeted point mutations in rice by a modified CRISPR/Cas9 system. Mol Plant 10(3):526–529. https://doi.org/10.1016/j.molp.2016.12.001

Li Q, Qin Y, Hu X et al (2020) Transcriptome analysis uncovers the gene expression profile of salt-stressed potato (Solanum tuberosum L.). Sci Rep 10(1):1–19. https://doi.org/10.1038/s41598-020-62057-0

Liao G, He Y, Li X et al (2019) Effects of bagging on fruit flavor quality and related gene expression of AsA synthesis in Actinidia eriantha. Sci Hortic 256:108511. https://doi.org/10.1016/j.scienta.2019.05.038

Liu CJ, Blount JW, Steele CL et al (2002) Bottlenecks for metabolic engineering of isoflavone glycoconjugates in Arabidopsis. Proc Natl Acad Sci U S A 99(22):14578–14583. https://doi.org/10.1073/pnas.212522099

Liu C, Chen L, Zhao R et al (2019) Melatonin induces disease resistance to botrytis cinerea in tomato fruit by activating jasmonic acid signaling pathway. J Agric Food Chem 67(22):6116–6124. https://doi.org/10.1021/acs.jafc.9b00058

Liu S, Li C, Wang H et al (2020) Mapping regulatory variants controlling gene expression in drought response and tolerance in maize. Genome Biol 21(1):1–22. https://doi.org/10.1186/s13059-020-02069-1

Lou D, Wang H, Liang G et al (2017) OsSAPK2 confers abscisic acid sensitivity and tolerance to drought stress in rice. Front Plant Sci 8:993. https://doi.org/10.3389/fpls.2017.00993

Ma X, Zhang Q, Zhu Q et al (2015) A robust CRISPR/Cas9 system for convenient, high-efficiency multiplex genome editing in monocot and dicot plants. Mol Plant 8(8):1274–1284. https://doi.org/10.1016/j.molp.2015.04.007

Ma C, Zhu C, Zheng M et al (2019) CRISPR/Cas9-mediated multiple gene editing in Brassica oleracea var. capitata using the endogenous tRNA-processing system. Hortic Res 6(1):1–15. https://doi.org/10.1038/s41438-018-0107-1

Mélida H, Bacete L, Ruprecht C et al (2020) Arabinoxylan-oligosaccharides act as damage associated molecular patterns in plants regulating disease resistance. Front Plant Sci 11:1210. https://doi.org/10.3389/fpls.2020.01210

Muñoz-Sanz JV, Zuriaga E, Cruz-García F et al (2020) Self-(In)compatibility systems: target traits for crop-production, plant breeding, and biotechnology. Front Plant Sci 11:195. https://doi.org/10.3389/fpls.2020.00195

Murovec J, Guček K, Bohanec B et al (2018) DNA-free genome editing of brassica oleracea and B. Rapa protoplasts using CRISPR-cas9 ribonucleoprotein complexes. Front. Plant Sci 871(1594). https://doi.org/10.3389/fpls.2018.01594

Naing AH, Kyu SY, Pe PPW et al (2019) Silencing of the phytoene desaturase (PDS) gene affects the expression of fruit-ripening genes in tomatoes. Plant Methods 15(1):1–10. https://doi.org/10.1186/s13007-019-0491-z

Nesbitt TC, Tanksley SD (2001) fw2. 2 directly affects the size of developing tomato fruit, with secondary effects on fruit number and photosynthate distribution. Plant Physiol 127(2):575–583. https://doi.org/10.1104/pp.010087

Ogawa T, Kawahigashi H, Toki S et al (2008) Efficient transformation of wheat by using a mutated rice acetolactate synthase gene as a selectable marker. Plant Cell Rep 27(8):1325–1331. https://doi.org/10.1007/s00299-008-0553-6

Okamoto M, Tanaka Y, Abrams SR et al (2009) High humidity induces abscisic acid 8-′-hydroxylase in stomata and vasculature to regulate local and systemic abscisic acid responses in Arabidopsis. Plant Physiol 149(2):825–834. https://doi.org/10.1104/pp.108.130823

Okuzaki A, Ogawa T, Koizuka C et al (2018) CRISPR/Cas9-mediated genome editing of the fatty acid desaturase 2 gene in Brassica napus. Plant Physiol Biochem 131:63–69. https://doi.org/10.1016/j.plaphy.2018.04.025

Pandey SP, Somssich IE (2009) The role of WRKY transcription factors in plant immunity. Plant Physiol 150(4):1648–1655. https://doi.org/10.1104/pp.109.138990

Pechanova O, Pechan T (2017) Proteomics as a tool to understand maize biology and to improve maize crop. In: Proteomics in food science: from farm to fork. Elsevier, pp 35–56

Pimentel D, Pimentel M (2003) Sustainability of meat-based and plant-based diets and the environment. Am J Clin Nutr 78:660–663

Qin H, Li Y, Huang R (2020) Advances and challenges in the breeding of salt-tolerant rice. Int J Mol Sci 21(21):1–15. https://doi.org/10.3390/ijms21218385

Quinet M, Angosto T, Yuste-Lisbona FJ et al (2019) Tomato fruit development and metabolism. Front Plant Sci 10:1554. https://doi.org/10.3389/fpls.2019.01554

Ray SK, Macoy DM, Kim WY et al (2019) Role of RIN4 in regulating PAMP-triggered immunity and effector-triggered immunity: current status and future perspectives. Mol Cells 42(7): 503–511. https://doi.org/10.14348/molcells.2019.2433

Robinson BR, Salinas CG, Parra PR et al (2019) Expression levels of the γ-glutamyl hydrolase I gene predict Vitamin B9 content in potato tubers. Agronomy 9(11):734. https://doi.org/10.3390/agronomy9110734

Romero P, Rose JKC (2019) A relationship between tomato fruit softening, cuticle properties and water availability. Food Chem 295:300–310. https://doi.org/10.1016/j.foodchem.2019.05.118

Roth M, Florez-Rueda AM, Städler T (2019) Differences in effective ploidy drive genome-wide endosperm expression polarization and seed failure in wild tomato hybrids. Genetics 212(1): 141–152. https://doi.org/10.1534/genetics.119.302056

Rothan C, Diouf I, Causse M (2019) Trait discovery and editing in tomato. Plant J 97(1):73–90. https://doi.org/10.1111/tpj.14152

Schiessl K, Muiño JM, Sablowski R (2014) Arabidopsis JAGGED links floral organ patterning to tissue growth by repressing Kip-related cell cycle inhibitors. Proc Natl Acad Sci U S A 111(7): 2830–2835. https://doi.org/10.1073/pnas.1320457111

Serrani JC, Ruiz-Rivero O, Fos M et al (2008) Auxin-induced fruit-set in tomato is mediated in part by gibberellins. Plant J 56(6):922–934. https://doi.org/10.1111/j.1365-313X.2008.03654.x

Sevestre F, Facon M, Wattebled F et al (2020) Facilitating gene editing in potato: a Single-Nucleotide Polymorphism (SNP) map of the Solanum tuberosum L. cv. Desiree genome. Sci Rep 10(1). https://doi.org/10.1038/s41598-020-58985-6

Shao G, Xie L, Jiao G et al (2017) CRISPR/CAS9-mediated editing of the fragrant gene Badh2 in rice. Chin J Rice Sci 31(2):216–222. https://doi.org/10.16819/j.1001-7216.2017.6098

Shi J, Habben JE, Archibald RL et al (2015) Overexpression of ARGOS genes modifies plant sensitivity to ethylene, leading to improved drought tolerance in both arabidopsis and maize. Plant Physiol 169(1):266–282. https://doi.org/10.1104/pp.15.00780

Shi J, Drummond BJ, Wang H et al (2016) Maize and Arabidopsis ARGOS proteins interact with ethylene receptor signaling complex, supporting a regulatory role for ARGOS in ethylene signal transduction. Plant Physiol 171(4):2783–2797. https://doi.org/10.1104/pp.16.00347

Shi J, Gao H, Wang H et al (2017) ARGOS8 variants generated by CRISPR-Cas9 improve maize grain yield under field drought stress conditions. Plant Biotechnol J 15(2):207–216. https://doi.org/10.1111/pbi.12603

Sun H, Jia Z, Cao D et al (2011) GmFT2a, a soybean homolog of flowering locus T, is involved in flowering transition and maintenance. PLoS One 6(12):e29238. https://doi.org/10.1371/journal.pone.0029238

Sun Z, Li N, Huang G et al (2013) Site-specific gene targeting using transcription activator-like effector (TALE)-based nuclease in Brassica oleracea. J Integr Plant Biol 55(11):1092–1103. https://doi.org/10.1111/jipb.12091

Sun Q, Lin L, Liu D et al (2018) CRISPR/Cas9-mediated multiplex genome editing of the BnWRKY11 and BnWRKY70 genes in brassica napus L. Int J Mol Sci 19(9). https://doi.org/10.3390/ijms19092716

Sun BR, Fu CY, Fan ZL et al (2019) Genomic and transcriptomic analysis reveal molecular basis of salinity tolerance in a novel strong salt-tolerant rice landrace Changmaogu. Rice 12(1):1–15. https://doi.org/10.1186/s12284-019-0360-4

Takagi H, Tamiru M, Abe A et al (2015) MutMap accelerates breeding of a salt-tolerant rice cultivar. Nat Biotechnol 33(5):445–449. https://doi.org/10.1038/nbt.3188

Teixeira PJP, Colaianni NR, Fitzpatrick CR et al (2019) Beyond pathogens: microbiota interactions with the plant immune system. Curr Opin Microbiol 49:7–17. https://doi.org/10.1016/j.mib.2019.08.003

Tieman D, Taylor M, Schauer N et al (2006) Tomato aromatic amino acid decarboxylases participate in synthesis of the flavor volatiles 2-phenylethanol and 2-phenylacetaldehyde. Proc Natl Acad Sci U S A 103(21):8287–8292. https://doi.org/10.1073/pnas.0602469103

Tiwari V, Chaturvedi AK, Mishra A et al (2015) An efficient method of agrobacterium-mediated genetic transformation and regeneration in local indian cultivar of groundnut (Arachis hypogaea) using grafting. Appl Biochem Biotechnol 175(1):436–453. https://doi.org/10.1007/s12010-014-1286-3

Tiwari JK, Devi S, Ali N et al (2018) Progress in somatic hybridization research in potato during the past 40 years. Plant Cell Tissue Organ Cult 132(2):225–238. https://doi.org/10.1007/s11240-017-1327-z

Tiwari JK, Devi S, Buckseth T et al (2020) Precision phenotyping of contrasting potato (Solanum tuberosum L.) varieties in a novel aeroponics system for improving nitrogen use efficiency: in search of key traits and genes. J Integr Agric 19(1):51–61. https://doi.org/10.1016/S2095-3119(19)62625-0

Upadhyaya DC, Bagri DS, Upadhyaya CP et al (2020) Genetic engineering of potato (Solanum tuberosum L.) for enhanced α-tocopherols and abiotic stress tolerance. Physiol Plant. https://doi.org/10.1111/ppl.13252

Vardi M, Levy NS, Levy AP (2013) Vitamin e in the prevention of cardiovascular disease: the importance of proper patient selection. J Lipid Res 54(9):2307–2314. https://doi.org/10.1194/jlr.R026641

Wani SH, Sah SK, Hossain MA et al (2016) Transgenic approaches for abiotic stress tolerance in crop plants. In: Advances in plant breeding strategies: agronomic, abiotic and biotic stress traits, vol 2. Springer, pp 345–396

Wen S, Liu H, Li X et al (2018) TALEN-mediated targeted mutagenesis of fatty acid desaturase 2 (FAD2) in peanut (Arachis hypogaea L.) promotes the accumulation of oleic acid. Plant Mol Biol 97(1–2):177–185. https://doi.org/10.1007/s11103-018-0731-z

Weng JK, Li X, Bonawitz ND et al (2008) Emerging strategies of lignin engineering and degradation for cellulosic biofuel production. Curr Opin Biotechnol 19(2):166–172. https://doi.org/10.1016/j.copbio.2008.02.014

Whaley CM, Wilson HP, Westwood JH (2007) A new mutation in plant ALS confers resistance to five classes of ALS-inhibiting herbicides. Weed Sci 55(2):83–90. https://doi.org/10.1614/ws-06-082.1

Xie K, Zhang J, Yang Y (2014) Genome-wide prediction of highly specific guide RNA spacers for CRISPR-Cas9-mediated genome editing in model plants and major crops. Mol Plant 7(5):923–926. https://doi.org/10.1093/mp/ssu009

Yang Y, Zhu K, Li H et al (2018) Precise editing of CLAVATA genes in Brassica napus L. regulates multilocular silique development. Plant Biotechnol J 16(7):1322–1335. https://doi.org/10.1111/pbi.12872

Yarra R (2020) Plastome engineering in vegetable crops: current status and future prospects. Mol Biol Rep 47(10):8061–8074. https://doi.org/10.1007/s11033-020-05770-3

Yi L, Chen C, Yin S et al (2018) Sequence variation and functional analysis of a FRIGIDA orthologue (BnaA3.FRI) in Brassica napus. BMC Plant Biol 18(1):1–13. https://doi.org/10.1186/s12870-018-1253-1

Zaman QU, Chu W, Hao M et al (2019) CRISPR/Cas9-mediated multiplex genome editing of jagged gene in brassica napus L. Biomol Ther 9(11). https://doi.org/10.3390/biom9110725

Zeng Z, Han N, Liu C et al (2020) Functional dissection of HGGT and HPT in barley vitamin e biosynthesis via CRISPR/Cas9-enabled genome editing. Ann Bot 126(5):929–942. https://doi.org/10.1093/AOB/MCAA115

Zhang A, Liu Y, Wang F et al (2019a) Enhanced rice salinity tolerance via CRISPR/Cas9-targeted mutagenesis of the OsRR22 gene. Mol Breed 39(3):1–10. https://doi.org/10.1007/s11032-019-0954-y

Zhang R, Liu J, Chai Z et al (2019b) Generation of herbicide tolerance traits and a new selectable marker in wheat using base editing. Nat Plants 5(5):480–485. https://doi.org/10.1038/s41477-019-0405-0

Zhang P, Du H, Wang J et al (2020) Multiplex CRISPR/Cas9-mediated metabolic engineering increases soya bean isoflavone content and resistance to soya bean mosaic virus. Plant Biotechnol J 18(6):1384–1395. https://doi.org/10.1111/pbi.13302

Zhao J, Chi Y, Zhang XJ et al (2019) Implication of whitefly vesicle associated membrane protein-associated protein B in the transmission of Tomato yellow leaf curl virus. Virology 535:210–217. https://doi.org/10.1016/j.virol.2019.07.007

Germplasm Conservation for Biotechnology and Plant Breeding

4

Mounawer Badri and Ndiko Ludidi

Abstract

The world population is expected to reach 8.5 billion people by 2030, and plant production will have to double the current level to meet corresponding food demand. To meet this need, it is necessary to exploit a wider range of plant genetic diversity throughout the world. As a result, plant genetic resources (PGR) conservation programs for food and agriculture have been launched, and gene banks have been set up in many countries. PGR are the fuel of biotechnology uses and for breeding that, in search of more efficient and adapted genotypes, manipulates genes to meet the needs of farmers. Conservation activities for PGR include collection, conservation, and assessment for future use. Advances in biotechnology, particularly in the field of in vitro culture techniques and molecular biology, are important tools for improving the conservation and management of PGR. A better understanding of genetic diversity is a prerequisite for its conservation and use. Molecular markers are increasingly used in germplasm characterization to study genetic diversity, identify redundancies in collections, and resolve taxonomic relationships. This review summarizes the current status of plant genetic resources and their use in biotechnology and breeding programs.

M. Badri (✉)
Laboratory of Extremophile Plants, Centre of Biotechnology of Borj Cedria, Hammam-Lif, Tunisia
e-mail: mounawer.badri@cbbc.rnrt.tn

N. Ludidi
Plant Biotechnology Research Group, Department of Biotechnology, University of the Western Cape, Bellville, South Africa

DSI-NRF Centre of Excellence in Food Security, University of the Western Cape, Bellville, South Africa
e-mail: nludidi@uwc.ac.za

Kamaluddin et al. (eds.), *Technologies in Plant Biotechnology and Breeding of Field Crops*, https://doi.org/10.1007/978-981-16-5767-2_4

This chapter deals with conservation methods and approaches for characterizing germplasm. This document focuses on the exploitation of genetic resources in biotechnology and the selection of genotypes with good agronomic performance under environmental constraints.

Keywords

Plant genetic resources · Genetic diversity · Germplasm · Conservation · Cryopreservation · Plant breeding

4.1 Plant Genetic Resources

According to the convention on biological diversity, genetic resources are genetic materials of actual or potential value. Genetic material is understood to be material of plant, animal, microbial, or other origin, containing functional units of heredity. Because of their hereditary nature, these resources are qualified as genetic resources, and they constitute a fundamental element of biodiversity.

Plant genetic resources play an increasingly important role in global food security and economic development. They enable agriculture to cope with changes, be they environmental or socioeconomic. They are an integral part of agricultural biodiversity as they are crucial for scaling up sustainable agricultural production and ensuring the livelihoods of many of the women and men who depend on agriculture.

Plant genetic diversity may also offer traits that can help meet future challenges, such as the need to adapt crops to changing climatic conditions or to the emergence of disease outbreaks. The loss of genetic diversity decreases options for sustainable and resilient agriculture, in the face of adverse environments, and very rapidly fluctuating weather conditions. Genetic erosion, a term coined by scientists to refer to the loss of individual genes and combinations of genes (such as those found in traditional varieties adapted to local conditions), is a threat to genetic resource diversity. The main cause of genetic erosion is the replacement of local varieties by modern varieties (van de Wouw et al. 2009). Only 30 crops feed the world, of which 5 cereal crops provide 60% of the energy intake of the world population (FAO 2015). Indeed, crop improvement provides 50% of human nutrition (FAO 2015). Moreover, the introduction of commercial varieties into traditional farming systems generally results in a reduction in the number of cultivated varieties. Other causes of genetic erosion include the emergence of new pests, new weeds and diseases, environmental degradation, urbanization, and land clearing through deforestation and bush fires.

Conservation and efficient use of plant genetic resources are essential to ensure food security and nutrition, now and in the future. Meeting this challenge will require a continuous flow of improved crops and varieties adapted to the particular conditions of agroecosystems. In addition, sustainable use of genetic resources also requires establishment of partnerships and networks that encompass all relevant stakeholders, from farmers to researchers and those in charge of gene banks. These

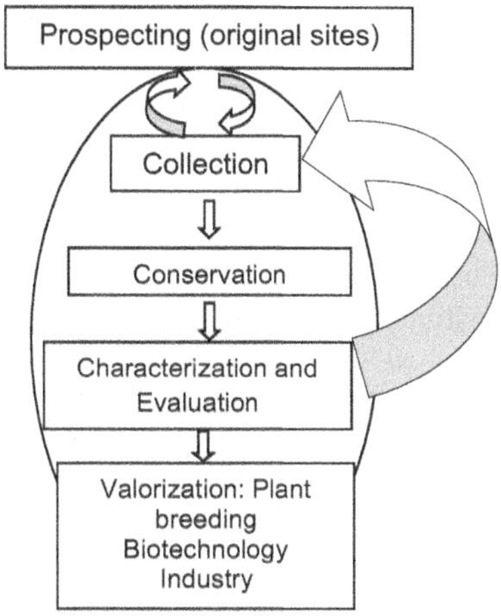

Fig. 4.1 Scheme illustrating the stages of genetic resources management from prospecting to their valorization

banks, which are well managed, preserve genetic diversity and make it available to breeders. Some 7.4 million crop plant samples are conserved in 1750 gene banks around the world.

The global approach for the management of the resources of a given species integrates five complementary stages, namely (1) prospecting, (2) collection, (3) conservation, (4) characterization and evaluation, and (5) utilization (Fig. 4.1). We will focus in this chapter on the different types of conservation, characterization, and use of biodiversity.

4.2 Conservation of Genetic Resources

Genetic resources are the building blocks of life on earth. By keeping them, we can protect them and the communities that depend on them. Depending on the biology of the species, one or other method of conservation is used. There are generally two types of preservation: In situ conservation, which consists of conserving the genetic resources of wild species in their natural environment, which will allow them to maintain their adaptive potential in the face of fluctuations in the environment, and ex-situ conservation, which refers to the conservation of plants gathered in conservatory collections in the field.

4.2.1 In Situ Conservation

In situ conservation refers to the maintenance and reconstitution of populations of viable species in their natural environment and, in the case of cultivated plant species, in the environment where their distinctive characteristics have developed. In situ conservation can be done in the farmer's field, in rangelands, and in national parks or other types of nature reserve and mainly concerns wild plants or wild relatives.

There are programs to help these farmers manage, conserve, and improve their plant genetic resources. In light of the report on the State of the World's Plant Genetic Resources, a few examples can be cited. In Ethiopia for example, the country's breeds of the main food crops, namely, barley, chickpea, sorghum, and common bean, are kept on the farm as part of a program undertaken by the National Institute of Biological Diversity, in cooperation with the African Program "Seeds of survival." In Sierra Leone, an on-farm rice and other crop conservation project was initiated by the Rokpur Rice Research Institute, as part of a community biodiversity conservation and development program.

4.2.2 Ex Situ Conservation

Resources can also be conserved outside of their natural environment. This method is particularly necessary when the environment is threatened.

4.2.2.1 Ex Situ in Vivo Conservation

Loss of much of the biological diversity in the field as a result of agricultural practices and the introduction of improved varieties led decision-makers in the 1970s to consider creating gene banks (Fig. 4.2) to conserve genetic material ex situ. Experts have estimated that there is very little time left to collect and conserve these endangered resources in the field.

Fig. 4.2 The National Gene Bank of Tunisia (http://www.bng.nat.tn/)

In the early 1970s, there were fewer than 10 gene banks that held perhaps little more than half a million samples. Today, 1750 gene banks (FAO 2010) are registered in the Global Information and Early Warning System database, and it is estimated that, globally, 7.4 million samples are currently stored in ex situ collections. However, it is estimated that many samples are already stored in duplicate.

It is estimated that 40% of the samples stored in gene banks are cereals, while 15% are food legumes. Vegetables, roots and tubers, fruits, and fodder each represent less than 10% of global collections. Aromatic, medicinal, and ornamental species are rarely present in public collections for long conservation. Aquatic plants which are of interest for food and agriculture are also absent.

However, ex situ conservation currently poses several problems, more particularly (1) the wear and tear of infrastructure, often built in developing countries by donor countries that have not made a long-term commitment to ensure maintenance of this infrastructure, and (2) the absence of censuses, inventories, and taxonomic studies, and no evaluation of the material present in gene banks. This type of knowledge is necessary for the valuation of the genetic resources conserved. Seed viability decreases even under optimal ex situ conservation conditions, requiring regeneration to replenish seed stocks.

It is also important to point out that dynamic management is a form of ex situ conservation. Indeed, the populations of a species have been cultivated under minimal anthropogenic pressure and in different environments to recreate the conditions for the evolution of genetic diversity by natural selection. More than 2500 botanical gardens, which also hold significant ex situ collections, are also listed (FAO 2010).

It is estimated that the world's food supply will one day depend on the collections of the Consultative Group on International Agricultural Research (CGIAR), a global network of international agricultural research centers. The member countries of the FAO Commission on Genetic Resources for Food and Agriculture reached an agreement in 1994 guaranteeing farmer's access to genetic resources in the CGIAR collections.

Ex Situ in Vitro Conservation

In vitro conservation refers to conservation in the form of organs, tissues, and cells. In vitro culture is used in the collection, propagation, and conservation of plant biodiversity, and it enables the large-scale production and multiplication of healthy material. The in vitro culture technique shows great potential for conserving (1) the genetic resources of species with recalcitrant seeds which are propagated vegetatively, (2) the products of biotechnology (elite genotypes, cell lines producing metabolites, GMOs), and (3) rare and threatened species. In addition, cryopreservation (liquid nitrogen, −196 °C) is used for long-term preservation, which allows the plant material to be stored without modifications or alterations for extended periods, protected from contamination, and with no maintenance required. Cryopreservation is well-developed suited for vegetatively propagating plants, while it is not the case with recalcitrant seed species. The latter have sensitivity to desiccation, structural complexity, and heterogeneity for stage of development and moisture content at maturity. The routine use of cryopreservation is still limited, but the number of examples for which it is used on a large scale is constantly growing.

4.3 Characterization and Evaluation of Genetic Resources

Diversity represents the primary material and the product of evolution. The characterization of plant genetic resources is a necessary prerequisite for any program aimed at their use, more particularly in biotechnology or breeding. Analysis of diversity within and between natural populations of plants can be performed using morphophenotypic traits and molecular markers. In light of the results of this characterization, core collections of accessions by species can be established, for which agronomic performances can be studied in the laboratory and/or in the field.

4.3.1 Phenotypic Characterization

Phenotypic characterization is carried out using phenotypic characters related to the aerial and root growth of plants. Although the phenotypic characters are affected by the environment as well as the stage of development of the plant, they are the first markers that were used and still continue to be used for the characterization of species.

Several works have used morphophenological characters for the characterization of natural populations of different plant species (Table 4.1). Badri et al. (2007, 2008a, b, 2016a, b, c, d) and Arraouadi et al. (2009) analyzed the variability within and between natural populations of annual *Medicago* species using quantitative traits. In addition, Neji et al. (2015) and Saoudi et al. (2019a, b) also explored the diversity level within and between populations of *Brachypodium hybridum* and *Hordeum marinum*, respectively, using morphophenological parameters.

4.3.2 Molecular Characterization

The molecular characterization of genetic resources allows their better management by constituting core collections. If we consider that there is a continuum between DNA, RNA, proteins/enzymes, and morphophysiological characters, the use of molecular markers amounts to zooming in on genotypes by increasing the power of discrimination between individuals and groups of individuals. The first molecular markers were biochemical markers that were used for the characterization of genetic resources. In addition, several studies have used dominant or codominant genetic markers for the characterization of germplasm (Table 4.2).

Table 4.1 List of plant species characterized using phenotypic characters

Plant species	Accessions number	Measured traits	Reference
Rice	217	Plant growth, grain yield, harvest index, seed weight	Li et al. (2010)
Stauntonia obovatifoliola	68	Fruit, seed and leaflet weight, length and width	Zou et al. (2020)
Allium sativum L.	34	Morphological descriptors and quantitative characters	Polyzos et al. (2019)
Common bean	183	Plant height, days to maturity, pods weight, seed length, 100-seed weight	Nadeem et al. (2020)
Sorghum (Sorghum spp.)	82	Photosynthetic parameters	Salas Fernandez et al. (2015)
Punica granatum L.	87	Pomological and leaf characters	Khadivi-Khub et al. (2015)
Olea europaea L.	561	Agromorphological traits	El Bakkali et al. (2013)
Elaeagnus angustifolia	100	Leaf length, leaf width, fruit weight, fruit width, fruit stem length	Khadivi (2018)
Apple (*Malus x domestica* Borkh)	19	Plant morphology and phenotype, leaf and fruit quality	Ganopoulos et al. (2017)
Rice (*Oryza sativa* L.)	130	Plant architecture, grain morphological and quality traits	Roy et al. (2014)
Sugarcane (*Saccharum* spp.)	300	Stalk height, stalk diameter, internode length, leaf sheath pubescence, inner stalk aerenchyma and pith, Brix, and presence of SCYLV	Todd et al. (2014)
Maize (*Zea mays* L.)	196	Agromorphological traits	Nelimor et al. (2019)
Bambara groundnut (*Vigna subterranea* [L.] Verdc.)	353	Seed coat color and pattern, seed eye color and pattern, and hilum color and pattern	Mohammed et al. (2016)
Fig (*Ficus carica* L.)	38	IPGRI and CIHEAM *Ficus carica* L. descriptors and introducing 14 new characters	Podgornika et al. (2010)
Medicago sp. (*Medicago laciniata, M. truncatula, M. ciliaris, M. polymorpha*)	308	Plant growth parameters, number, and weight of pods	Badri et al. (2007, 2008a, b, 2016a, b, c, d)
Medicago truncatula	48	Plant growth parameters, number, and weight of pods	Arraouadi et al. (2009)
Brachypodium hybridum	145	Vegetative and reproductive traits	Neji et al. (2014)
Hordeum marinum	150	Plant growth, number of seeds per spike, spike length and spike number per plant, spike weight	Saoudi et al. (2017)

Table 4.2 List of plant species characterized using molecular characters

Plant species	Accessions number	Markers used	Reference
Almond (*Prunus* sp.)	39	RAPD and SSR	Shiran et al. (2007)
Brachypodium hybridum	145	SSR	Neji et al. (2015)
Brassica rapa	93	AFLP	Warwick et al. (2008)
Durum wheat (*Triticum durum*)	184	Genotyping by sequencing (GBS)	Sieber et al. (2015)
Ginger (Zingiber barbatum Wall.)	19	P450-based analog (PBA) markers	Wicaksana et al. (2011)
Hordeum marinum	150	RAPD	Saoudi et al. (2019b)
Jatropha curcas	3	Intersimple sequence repeat (ISSR) markers	Senthil Kumar et al. (2009)
Jatropha curcas L.	28	RAPD and AFLP	Sudheer Pamidimarri et al. (2010)
Jatropha curcas L.	15	RAPD, AFLP, and SSR	Mastan et al. (2012)
Latvian apple (*Malus*)	109	SSR	Lacis et al. (2011)
Maize	450	Single-nucleotide polymorphism (SNP) markers	Semagn et al. (2012)
Medicago polymorpha	16	Isoenzymatic markers	Salhi-Hannachi et al. (1998)
Medicago polymorpha	120	SSR	Haddoudi et al. (2021)
Medicago sp.	33	RAPD	Brummer et al. (1995)
Medicago sp.	308	SSR	Badri et al. (2007, 2008a, b, 2016a, b, c, d)
Medicago truncatula	40	Single-nucleotide polymorphism (SNP) markers	Friesen et al. (2014)
Medicago truncatula	200	RAPD	Bonnin et al. (1996)
Medicago truncatula	48	SSR	Arraouadi et al. (2009)
Medicago truncatula	136	SSR	Lazrek et al. (2009)
Minor olive (*Olea europaea*)	31	SSR	Saddoud Debbabi et al. (2020)
Oil palm (*Elaeis guineensis* Jacq.)	6	SSR	Arias et al. (2012)
Olive (*Olea europaea* L.)	108	SSR	Koehmstedt et al. (2011)
Olive (*Olea europaea*)	26	AFLP	Taamalli et al. (2006)

(continued)

Table 4.2 (continued)

Plant species	Accessions number	Markers used	Reference
Pear landraces (*Pyrus communis* L.)	48	SSR	Queiroz et al. (2015)
Pepper (*Capsicum* sp.)	69	ISSR	Brilhante et al. (2021)
Red clover (*Trifolium pratense* L.)	46	SSR	Radinovic et al. (2017)
Ryegrass (*Lolium perenne* L.)	171	RAPD	Bolaric et al. (2005)
Safflower	131	ISSR	Ali et al. (2020)
Wheat	63	Simple sequence repeats (SSR)	Ijaz and Khan (2009)

4.3.3 Agronomic Performances Under Abiotic and Biotic Stresses

Several studies have analyzed the agronomic performance of natural populations and core collections of different species under environmental constraints (Table 4.3).

4.4 Use of Plant Genetic Resources

Genetic resources are a strategic component for several sectors, in particular the pharmaceutical, cosmetics, biotechnology, and agrifood industries. The stakes have become greater since the development of biotechnology. Genetic resources can be used for both commercial and noncommercial use. However, the distinctions between commercial and noncommercial use, as well as the nature of the actors involved, are not always clear.

Some examples of uses of genetic resources and associated traditional knowledge by sector are:

- Academic research: genetic resources can enable a better knowledge and understanding of the natural world, with activities ranging from taxonomic research to ecosystem analysis. These missions are generally entrusted to universities and public research institutes.
- Pharmaceutical industry: development of drugs for the treatment of diseases (including cancer, tropical diseases, and obesity);
- Biotechnologies: enzymes used by the textile, detergents, and human and animal food sectors to improve the quality of products and production processes, industrial use of bacteria (industrial water/soil depollution processes, methods of DNA analysis, etc.);

Table 4.3 List of plant species evaluated under environmental constraints

Plant species	Accessions number	Environmental constraint	Reference
Barely (*Hordeum vulgare* L.)	31	Salt stress	Allel et al. (2016, 2019)
Durum wheat	10	Salt stress	Momeni et al. (2021)
Hordeum marinum	150	Salt stress	Saoudi et al. (2019b)
Medicago sp.	193	Drought stress	Badri et al. (2011, 2016a, b, c)
Medicago truncatula	140	Salt stress	Arraouadi et al. (2011)
Medicago truncatula	40	Salt stress	Friesen et al. (2014)
Medicago truncatula	140	*Aphanomyces euteiches*	Djébali et al. (2013)
Medicago truncatula	20	Iron (Fe) deficiency	Kallala et al. (2019)
Mungbean	153	Phosphorus (P) deficiency	Reddy et al. (2020)
Sulla carnosa	150	Salt stress	Gandour et al. (2014)
Wheat (*Triticum aestivum* L.)	6	Salt stress	Ghonaim et al. (2021)

- Cosmetics industry: development of cosmetic products from plants or other organisms, including on the basis of traditional knowledge;
- Agriculture: development of phytopharmaceutical products (fungicides, insecticides, etc.) from plants, plant selection (new plant varieties), and biological control from control agents (e.g. insects) collected in the country of which the pest is native. The selection and genetic improvement of plants constitute the main research axis because they represent a low-cost solution for improving crop productivity;
- Ornamental horticulture: development of new ornamental varieties from exotic species. About 100–200 species of plants are used as genetic resources in commercial horticulture and 500 in home horticulture. This sector initially used spontaneous plants, but now the majority of resources come from nurseries, botanical gardens, or private collections. In South Africa, the National Botanical Institute (SANBI) and the Ball Horticultural Company formed a partnership in 1998 to market South African horticultural and floriculture products.

Overall, genetic resources are lost if they are not used.

4.5 Summary and Conclusion

Plant genetic resources represent a valuable tool in meeting the nutritional needs of a growing global human population in the context of climate change. Although there have been several conventions ratified between different member countries concerning access and the sharing of benefits arising from the use of plant genetic resources, their conservation and distribution continue to represent a challenge. The characterization of genetic resources is a necessary prerequisite for any variety selection and biotechnological development program. As in all fields, plant genetic resources take advantage of recent technological advances for their characterization and have several applications.

References

Ali F, Nadeem MA, Habyarimana E et al (2020) Molecular characterization of genetic diversity and similarity centers of safflower accessions with ISSR markers. Rev Bras Bot 43:109–121. https://doi.org/10.1007/s40415-019-00574-7

Allel D, Ben-Amar A, Badri M et al (2016) Salt tolerance in barley originating from harsh environment of North Africa. Aust J Crop Sci 10(4):438

Allel D, Ben-Amar A, Badri M et al (2019) Evaluation of salinity tolerance indices in North African barley accessions at reproductive stage. Czech J Genet Plant Breed 55(2):61–69

Arias D, Montoya C, Romero H (2012) Molecular characterization of oil palm *Elaeis guineensis* Jacq. Materials from Cameroon. In: Plant genetic resources: characterization and utilization, pp 1–9

Arraouadi S, Badri M, Abdul Jaleel C et al (2009) Analysis of genetic variation in natural populations of *Medicago truncatula* of Southern Tunisian ecological areas, using morphological traits and SSR markers. Trop Plant Biol 2(3):122–132

Arraouadi S, Badri M, Taamalli W et al (2011) Variability salt stress response analysis of Tunisian natural populations of *Medicago truncatula* (Fabaceae) using salt response index (SRI) ratio. Afr J Biotechnol 10(52):10636–10647

Badri M, Ilahi H, Huguet T et al (2007) Quantitative and molecular genetic variation in sympatric populations of *Medicago laciniata* and *M. truncatula* (Fabaceae): relationships with eco-geographical factors. Genet Res 89(2):107–122

Badri M, Zitoun A, Soula S et al (2008a) Low levels of quantitative and molecular genetic differentiation among natural populations of *Medicago ciliaris* Kroch.(Fabaceae) of different Tunisian eco-geographical origin. Conserv Genet 9(6):1509–1520

Badri M, Zitoun A, Ilahi H et al (2008b) Morphological and microsatellite diversity associated with ecological factors in natural populations of *Medicago laciniata* Mill. (Fabaceae). J Genet 87(3): 241–255

Badri M, Chardon F, Huguet T et al (2011) Quantitative trait loci associated with drought tolerance in the model legume *Medicago truncatula*. Euphytica 181(3):415–428

Badri M, Bouhaouel I, Arraouadi S et al (2016a) Variation in tolerance to drought among Tunisian populations of *Medicago truncatula*. Plant Genet Resour 14(1):41

Badri M, Ben Cheikh N, Mahjoub A et al (2016b) Morpho-phenological diversity among natural populations of *Medicago polymorpha* of different Tunisian ecological areas. Afr J Biotechnol 15(25):1330–1338

Badri M, Mahfoudh S, Abdelly C (2016c) Morph-physiological responses to water deficit in parental genotypes of *Medicago truncatula* recombinant inbred lines. Afr J Biotechnol 15(25):1339–1349

Badri M, Toumi G, Mahfoudh S et al (2016d) Diversity of response to drought in a collection of lines of *Medicago truncatula*, *M. ciliaris*, and *M. polymorpha*. Crop Sci 56(6):3125–3132

Bolaric S, Barth S, Melchinger AE et al (2005) Molecular characterization of genetic diversity in European germplasm of perennial ryegrass. Euphytica 146:39–44

Bonnin I, Huguet T, Gherardi M et al (1996) High level of polymorphism and spatial structure in a selfing plant species, *Medicago truncatula* (Leguminosae), shown using RAPD markers. Am J Bot 83(7):843–855

Brilhante BDG, Santos TO, Santos PHAD et al (2021) Phenotypic and molecular characterization of Brazilian *Capsicum* germplasm. Agronomy 11(5):854

Brummer EC, Bouton JH, Kochert G (1995) Analysis of annual *Medicago* species using RAPD markers. Genome. https://doi.org/10.1139/g95-047

Djébali N, Aribi S, Taamalli W et al (2013) Natural variation of *Medicago truncatula* resistance to *Aphanomyces euteiches*. Eur J Plant Pathol 135(4):831–843

El Bakkali A, Haouane H, Moukhli A et al (2013) Construction of core collections suitable for association mapping to optimize use of mediterranean olive (Olea europaea L.) genetic resources. PLoS One 8(5):e61265. https://doi.org/10.1371/journal.pone.0061265

FAO (2010) The second report on the state of the world's plant genetic resources for food and agriculture. http://www.fao.org/3/i1500e/i1500e.pdf

FAO (2015) Genetic resources and biodiversity for food and agriculture: A treasure for the future. http://www.fao.org/nr/cgrfa

Friesen ML, von Wettberg EJB, Badri M et al (2014) The ecological genomic basis of salinity adaptation in Tunisian *Medicago truncatula*. BMC Genomics 15(1):1–18

Gandour M, Neji M, Hessini K et al (2014) Assessing the salt tolerance of *Sulla carnosa* genotypes by agronomic indicators. Agron J 106:185–190

Ganopoulos I, Tourvas N, Xanthopoulou A et al (2017) Phenotypic and molecular characterization of apple (Malus × domestica Borkh) genetic resources in Greece. Sci Agric 75(6):509–518

Ghonaim MM, Mohamed HI, Omran AAA (2021) Evaluation of wheat (*Triticum aestivum* L.) salt stress tolerance using physiological parameters and retrotransposon-based markers. Genet Resour Crop Evol 68:227–242

Haddoudi L, Hdira S, Ben Cheikh N et al (2021) Assessment of genetic diversity in Tunisian populations of *Medicago polymorpha* based on SSR markers. Chil J Agric Res 81(1):53–61

Ijaz S, Khan IA (2009) Molecular characterization of wheat germplasm using microsatellite markers. Genet Mol Res 8(3):809–815

Kallala N, M'sehli W, Jelali K et al (2019) Biodiversity within *Medicago truncatula* genotypes toward response to iron deficiency: investigation of main tolerance mechanisms. Plant Spec Biol 34:95–109

Khadivi A (2018) Phenotypic characterization of Elaeagnus angustifolia using multivariate analysis. Ind Crop Prod 120(15):155–161

Khadivi-Khub A, Kameli M, Moshfeghi N et al (2015) Phenotypic characterization and relatedness among some Iranian pomegranate (*Punica granatum* L.) accessions. Trees 29:893–901. https://doi.org/10.1007/s00468-015-1172-9

Koehmstedt AM, Aradhya MK, Soleri D et al (2011) Molecular characterization of genetic diversity, structure, and differentiation in the olive (*Olea europaea* L.) germplasm collection of the United States Department of Agriculture. Genet Resour Crop Evol 58:519–531

Lacis G, Kota I, Ikase L et al (2011) Molecular characterization of the Latvian apple (Malus) genetic resource collection based on SSR markers and scab resistance gene Vf analysis. Plant Genet Res 9(2):189–192

Lazrek F, Roussel V, Ronfort J (2009) The use of neutral and non-neutral SSRs to analyze the genetic structure of a Tunisian collection of *Medicago truncatula* lines and to reveal associations with eco-environmental variables. Genetica 135:391–402

Li X, Yan W, Agrama H et al (2010) Genotypic and phenotypic characterization of genetic differentiation and diversity in the USDA rice mini-core collection. Genetica 138:1221–1230. https://doi.org/10.1007/s10709-010-9521-5

Mastan SG, Sudheer PDVN, Rahman H et al (2012) Molecular characterization of intra-population variability of *Jatropha curcas* L. using DNA based molecular markers. Mol Biol Report 39: 4383–4390

Mohammed MS, Shimelis HA, Laing MD (2016) Phenotypic characterization of diverse Bambara groundnut (*Vigna subterranea* [L.] Verdc.) germplasm collections through seed morphology. Genet Resour Crop Evol 63:889–899

Momeni M, Kalantar M, Dehghani-Zahedani M (2021) Physiological, biochemical and molecular responses of durum wheat under salt stress. Plant Genet Resour 19(2):93–103. https://doi.org/10.1017/S1479262120000416

Nadeem MA, Karaköy T, Yeken MZ et al (2020) Phenotypic characterization of 183 Turkish common bean accessions for agronomic, trading, and consumer-preferred plant characteristics for breeding purposes. Agronomy 10(2):272. https://doi.org/10.3390/agronomy10020272

Neji M, Geuna F, Taamalli W et al (2014) Morpho-phenological diversity among Tunisian natural populations of *Brachypodium hybridum*. J Agric Sci 153(6):1006

Neji M, Geuna F, Taamalli W et al (2015) Assessment of genetic diversity and population structure of Tunisian populations of *Brachypodium hybridum* by SSR markers. Flora 216:42–49

Nelimor C, Badu-Apraku B, Nguetta SPA et al (2019) Phenotypic characterization of maize landraces from Sahel and Coastal West Africa reveals marked diversity and potential for genetic improvement. J Crop Improv. https://doi.org/10.1080/15427528.2019.1674760

Podgornika M, Vuka I, Vrhovnik I et al (2010) A survey and morphological evaluation of fig (*Ficus carica* L.) genetic resources from Slovenia. Sci Hortic 125(3):380–389

Polyzos N, Papasotiropoulos V, Lamari FN et al (2019) Phenotypic characterization and quality traits of Greek garlic (*Allium sativum* L.) germplasm cultivated at two different locations. Genet Resour Crop Evol 66:1671–1689. https://doi.org/10.1007/s10722-019-00831-4

Queiroz A, Assuncão A, Ramadas I et al (2015) Molecular characterization of Portuguese pear landraces (*Pyrus communis* L.) using SSR markers. Sci Hortic 183:72–76

Radinovic I, Vasiljevic S, Brankovic G et al (2017) Molecular characterization of red clover genotypes utilizing microsatellite markers. Chil J Agric Res 76(1):41–47

Reddy VRP, Aski MS, Mishra GP et al (2020) Genetic variation for root architectural traits in response to phosphorus deficiency in mungbean at the seedling stage. PLoS One 15(6): e0221008

Roy S, Rathi RS, Misra AK et al (2014) Phenotypic characterization of indigenous rice (*Oryza sativa* L.) germplasm collected from the state of Nagaland, India. Plant Genet Resour 12(1): 58–66

Saddoud Debbabi O, Miazzi MM, Elloumi O, Fendri M, Ben Amar F, Savoia M, Sion S, Souabni H, Mnasri SR, Ben Abdelaali S, Jendoubi F, Mangini G, Famiani F, Taranto F, Montemurro C, Msallem M (2020) Recovery, assessment, and molecular characterization of minor olive genotypes in Tunisia. Plan Theory 9(3):382

Salas Fernandez MG, Strand K, Hamblin MT et al (2015) Genetic analysis and phenotypic characterization of leaf photosynthetic capacity in a sorghum (Sorghum spp.) diversity panel. Genet Resour Crop Evol 62:939–950. https://doi.org/10.1007/s10722-014-0202-6

Salhi-Hannachi A, Boussaid M, Marrakchi M (1998) Dynamic and gene flow of natural populations of *Medicago polymorpha*. Genet Sel Evol 30:121–135

Saoudi W, Badri M, Gandour M et al (2017) Assessment of genetic variability among Tunisian populations of Hordeum marinum using morpho-agronomic traits. Crop Sci 57:1–8

Saoudi W, Badri M, Taamalli W et al (2019a) Variability in response to salinity stress in natural Tunisian populations of *Hordeum marinum* ssp. marinum. Plant Biol 21(1):89–100

Saoudi W, Badri M, Gandour M, Smaoui A et al (2019b) Analysis of genetic diversity and spatial structure in Tunisian populations of *Hordeum marinum* ssp marinum based on molecular markers. J Agric Sci 157(5):399–412

Semagn K, Magorokosho C, Vivek BS et al (2012) Molecular characterization of diverse CIMMYT maize inbred lines from eastern and southern Africa using single nucleotide polymorphic markers. BMC Genomics 13:113

Senthil Kumar R, Parthiban KT, Govinda Rao M (2009) Molecular characterization of jatropha genetic resources through inter-simple sequence repeat (ISSR) markers. Mol Biol Rep 36:1951–1956

Shiran B, Amirbakhtiar N, Kiani S et al (2007) Molecular characterization and genetic relationship among almond cultivars assessed by RAPD and SSR markers. Sci Hortic 111:280–292

Sieber A-N, Longin CFH, Würschum T (2015) Molecular characterization of winter durum wheat (*Triticum durum*) based on a genotyping-by-sequencing approach. Plant Genet Resour 15(1): 36–44

Sudheer Pamidimarri DVN, Mastan SG, Rahman H et al (2010) Molecular characterization and genetic diversity analysis of *Jatropha curcas* L. in India using RAPD and AFLP analysis. Mol Biol Rep 37:2249–2257

Taamalli W, Geuna F, Banfi R et al (2006) Agronomic and molecular analyses for the characterization of accessions in Tunisian olive germplasm collections. Electron J Biotechnol 9(5). https://doi.org/10.2225/vol9-issue5-fulltext-12

Todd J, Wang J, Glaz B et al (2014) Phenotypic characterization of the Miami World Collection of sugarcane (Saccharum spp.) and related grasses for selecting a representative core. Genet Resour Crop Evol 61:1581–1596

van de Wouw M, Kik C, van Hintum T et al (2009) Genetic erosion in crops: concept, research results and challenges. Plant Genet Resour 8(1):1–15. https://doi.org/10.1017/S1479262109990062

Warwick SI, James T, Falk KC (2008) AFLP-based molecular characterization of *Brassica rapa* and diversity in Canadian spring turnip rape cultivars. Plant Genet Resour 6(1):11–21

Wicaksana N, Gilani SA, Ahmad D et al (2011) Morphological and molecular characterization of underutilized medicinal wild ginger (*Zingiber barbatum* Wall.) from Myanmar. Plant Genet Resour 9(4):531–542

Zou S, Yao X, Zhong C et al (2020) Phenotypic characterization of Stauntonia obovatifoliola Hayata subsp. urophylla germplasm: a potential new fruit crop. Genet Resour Crop Evol 67: 1037–1050

Recent Advances in Rice Breeding Using Biotechnological and Genomics Tools

5

Rajesh Kumar, Sarita Kumari, Sumeet Kumar Singh, Chandra Mohan Singh, and Sandeep Kumar Suman

Abstract

The area and production of rice have increased drastically since 1965, making it second major crop of the world and first for the prior countries. The major breakthrough in the productivity of rice resulted in green revolution which may be regarded as the threshold line for the advancement in rice cultivation. The development of technologies in the area of breeding, biotechnology and sequencing is the major driving force for boosting the rice production. The content of this chapter presents the glimpses of development of improved rice varieties through the intervention of biotechnological approaches such as genome sequencing, molecular marker technologies, etc. The varietal development through rice breeding programme involves pure line selections and hybrid rice technologies. The biotechnological approaches, DNA marker technologies, genome sequencing and genetic engineering, improve the rice breeding at parent selection, trait mapping and their introgression at early stages, thereby reducing the time, labour and cost of development of improved varieties with high precision. The involvement of marker technologies leads to the development of an area of marker-assisted breeding, marker-assisted backcross programme, marker-assisted gene pyramiding and marker-assisted recurrent selection. The genome sequencing increases the density of DNA markers at base level such as SNP improving the genotyping of mass population through chip and array type platform, genomic selection, association mapping and allele mining. The major genomic sites have been mapped for qualitative traits, viz., biotic and abiotic stress tolerances, nutritional improvement, and the quantitative traits such as yield-related traits like blast resistance, bacterial leaf blight resistance, gall midge resistance, salt

R. Kumar (✉) · S. Kumari · S. K. Singh · C. M. Singh · S. K. Suman
Dr. Rajendra Prasad Central Agricultural University, Pusa Samastipur, Bihar, India

Kamaluddin et al. (eds.), *Technologies in Plant Biotechnology and Breeding of Field Crops*, https://doi.org/10.1007/978-981-16-5767-2_5

tolerance, submergence tolerance, drought tolerance, high Fe and Zn content, dwarf stature, etc.

Keywords

Marker-assisted backcross · Marker-assisted gene pyramiding · *Oryza sativa* · High-yielding varieties · Hybrid technology · SNP

5.1 Introduction

The rice plant, *Oryza sativa*, belongs to family *Poaceae* and sub-family *Oryzoideae*. It has two cultivated species *Oryza sativa* L. and *Oryza glaberrima* Steud, both diploid and more than 20 wild species. The world production of rice is nearly 759.6 million tons over 167 million hectares (mha) area with the productivity of 4.44 tons/ha (http://www.fao.org/home/en/). Approximately 88.95% of the total rice-growing area is concentrated in Asia, which contributes 90.4% in the world production. In India, rice is grown in estimated area of about 143.8 mha with the production of 166.5 million tons, and its productivity is 2.39 tons/ha (https://www.statista.com/).

5.2 The Era of Green Revolution

Area and production of rice have been increased dramatically since 1960s to 2020s. Most of the earlier rice varieties released and cultivated were tall, low in fertilizer response, low yielding and susceptible to lodging with a productivity of less than 0.75 tons/ha. One event that changed the scenario was the introduction of dwarfing gene Dee-geo-woo-gen into the Indian varieties, which resulted in the release of popular IR series of rice varieties like IR8, IR20, etc. The existence of this one gene makes sure varieties with semi-dwarf character having good response to the fertilizer applied. The resultant non-lodging property enhanced yield, thus giving a productivity of more than 2 tons/ha when compared to the previously cultivated Indian rice cultivars. The physiological and genetic basis for action of this gene has been well documented, and it has been now proved beyond doubt that the change in plant architecture due to the presence of dwarfing gene *sd-1* has been due to non-responsiveness to endogenous Gibberellins.

In 1961, world rice production averaged 240 million metric tons, which increased to 492 million metric tons in 1986–1990. South Asia's rice production went from 72.2 to 135.9 million metric tons, China's went from 72.2 to 176.9 million metric tons over the same period, and South East Asia's rice production went from 4.91 to 106.9 million metric tons. But during the 1990s, production began to slow down. It is necessary to increase rice production from nearly 600 million tons annually to almost 800 million by the year 2025 to keep up with the population growth. It is essential to do it using less land, less labour, less water and less pesticide to make it sustainable.

5.3 Varietal Improvement

Before the onset of Green Revolution era, varietal improvement had largely been confined to pureline selection, resulting in identification of 445 varieties (ICAR, New Delhi 1960). Some of these varieties like 'Manoharsali', 'SR 26B' 'FR 13A' and 'Rongasali' in the fragile environments and 'Basmati 370' and 'NP 130' in north-western India continued to be grown. A large number of donors being used today for improving yield, stability and quality were pureline selections from these land races.

Exotic introductions further supplemented the early efforts in identifying successful varieties for cold and hill ecology of Himachal Pradesh, Kashmir and Uttar Pradesh. All early efforts had hardly helped raise the genetic yield potential appreciably to meet the growing demand. The first serious research effort to break the yield barrier of tropical rice was made through inter-varietal (*indica* × *japonica*) hybridization programme to combine fertilizer responsiveness of *japonica* varieties with adaptability and quality of *indica* ones. Although it was not that fruitful totally, this programme eventuated in the identification of Mahsuri in Malaysia and 'ADT 27' in Tamil Nadu, resulting in significant production advancement in India. This breeding programme convinced breeders that the key to higher yield lines is breeding for non-lodging plants.

By 1965, it was realized that plant type was the major yield barrier in rice. A major breakthrough was witnessed in the history of rice breeding in the form of 'IR 8' which bestowed practical expression to the concept of ideotype. As information on different components of plant type became available. India recognized new horizons for increasing productivity through identification of semi-dwarf varieties from introduced materials as well as selections from hybridization programmes. The coordinated research programme was geared up for this task to achieve the desired result at the earliest possible time, resulting in the release of 'IR 8' from the introduced material of the IRRI in 1966 and 'Jaya' from hybridization programme of the All India Coordinated Rice Improvement Programme (AICRIP) in 1968. This idea was followed further by the identification of several varieties which combine improved plant type with high yield for different maturity groups. The unique mechanism of multidiscipline-based multi-location testing of the AICRIP has thus facilitated rapid development of varietal and production technologies appropriate for varied agro ecologies (AICRIP, Hyderabad). The model, now adopted by several countries and international institutions, helped to evolve more than 1088 high-yielding varieties ensuring sustained self-sufficiency.

Thrust in breeding research varied with the changing needs and socioeconomic compulsions. If it was for high yield and general adaptability in the first decade after introduction of dwarf varieties, stability of yield by breeding for biotic and abiotic stress resistance or tolerance was the priority in the following decade. Some promising donors possessing biotic and abiotic stress resistance/tolerance are listed in Table 5.1.

Ecologies received increased research attention from the third decade onwards. The post-'Jaya' period of varietal development in India may broadly be grouped into four phases, viz., (1) the first decade of high-yielding varieties of different maturity period and varied grain quality, (2) the second decade of aggressive breeding for resistance to pests and diseases, (3) the third decade of breeding for high-yielding varieties adapted to diverse rain-fed ecologies and lastly (4) the development and use of hybrid technology.

5.4 Varietal Development for Yield, Adaptability and Quality

During the first decade (till 1979), the AICRIP, through its 3-tier system of evaluation, identified 17 centrally released and 73 state-released high-yielding varieties of short and medium durations for irrigated ecosystem. Relatively early-maturing, high-yielding dwarf varieties of rice replaced huge areas under millets and oilseed crops, making rice–wheat as the most productive and profitable cropping system. One of the notable achievements of Indian rice breeding programme is the successful development of early-maturing varieties like Pusa 2-21, Pusa 33 and Saket-4, capable of yielding as high as 90% yield of medium-duration varieties which enabled farmers to fit them into varied cropping systems substantially, thereby increasing cropping intensity as well as productivity. Some of the popular varieties among them were Jayanthi and 'IR20' in medium; 'Padma' and 'Ratna' in mid-early and 'Pusa 2-21', 'Saket-4' and 'Annapurna' in early-duration groups. Varieties like Jagannath and 'IR 5' (Pankaj) with long duration were released for rain-fed lowland ecosystem. Efforts continued during the second and third phases as well leading to the identification of many more promising varieties. Keeping in view the strong and varied consumer preference, efforts were directed towards quality improvement for incorporating fine-grained type, red grain type and semi-dwarf high-yielding aromatic types. 'Pusa 4-1-11' among the fine-grained types and 'Annapuran' and 'TKM 9' among the red grained types were widely accepted by the consumers. As for aromatic rice, concerted efforts continued for over long period helped to develop high-yielding, dwarf aromatic varieties ('Pusa Basmati 1', 'Kasturi') comparable with highly prized traditional basmati varieties like 'Basmati 370' and 'Taroari Basmati' in quality, aroma, elongation and texture on cooking besides 50–60% higher yield (Siddiq 1991). While early maturing varieties have helped crop intensification by ideally fitting as potential crop component in varied cropping systems, the fine-gained aromatic varieties have helped step up export of high-quality rice.

Table 5.1 Promising donors possessing biotic and abiotic stress resistance/tolerance

Traits		Biotic stress tolerance varieties
Blast	:	Tetep, Tadukan, Zenith, Carreon, IR 8, Raminad Str.3, Gampai, Dawn, Peta, Sigadis, Dular, PTB 10, Ceysron, Milyang 83, ARC 7098, Kalamdani, Tarabali, Mugisali, Madrisali, Sollpona, Rongaahu, Gajepsali, Beganbisi 2, Rongagutia, Kolimekuri, Rikhojoi 2.
Bacterial Leaf Blight	:	Lacrose X Zenith-Nira, Karjat, Pallavi, Zenith, M. Sung Song, Syntha, KM 6, Sigadis, DZZ 192, DV 85, ARC 5827, Dular, Pelita 1, ARC 18562, Dholamula, Moinasail, Japorisali, ajepsali, Jatiosali, AS 330, Kola ahu,
Rice Tungro Virus	:	Kataribhog, Latisail, ARC 14766, BJ 1, PTB 18, Gampai 15, Pankhari 203, Sigadis, Ambemohar, Habiganj.
Brown Spot	:	Ch 13, Ch 45, BAM 10, AC 2550, Bhatta Dhan
Other diseases	::	
• False smut		MNP 85, ARC 5378, BR 16, IR 24, IR 29
• Stem rot		Basmati 370, Bara 62
• Ragged stunt		PTB 21, PTB 33
Brown Plant Hopper	:	ARC 6650, Ptb 33, Leb Muey Nahng, Rathu Heenati, Sinna Sivappu, Suali, ARC 7080, ARC 14766, NCS 91, NCS 131, NCS 707
Gall Midge	:	Eswarakora, Siam 29, ARC 5984, Ptb 18, Ptb 21, OB 677, ARC 10660, ARC 6605, Leuang 152, ARC 5959, ARC 13516, ARC 14787, CR 94-1512-6, Shakti
White Backed Plant Hopper	:	ARC 5803, ARC 6064, ARC 7138, ARC 7318, ARC 10340
Stem Borer	:	TKM 6, CB I, CB II, ARC 6158, ARC 10386, ARC 10443, NCS 266, NCS 336, NCS 464, ARC 5500, W 1263, Manoharsali
Green Leaf Hopper	:	Ptb 2, Ptb 21, ARC 6606
Leaf Folder	:	ARC 1129, Ptb 12, Gorsa, Darukasali
TRAITS		ABIOTIC STRESS TOLERANCE VARIETIES
Drought Tolerance	:	N 22, MTU 17, Lalnakanda 41, Kalakeri, JBS 508, CR 143-2-2, Saria, Sathchali, Nepalikalam, Dular, Kodibudama, NC 487, Mettamolagolukulu, NC 488, ARC 10372, Janki, NC 492, AS 180, Hasakumra, Maibi, Koijapori, AS 313/11, AS 47, Bairing, Ahu joha, Dagaranga, ARC 10372, Noga ahu, Soraituni, Lakhi.
Cold Tolerance	:	Dunghansali, AC 540, Sita, Rajai, CB 1, Boro 33, Dholiboro, IRGC 100081, IRGC 100114, IRGC 100028.
Submergence Tolerance	:	FR 13A, FA 43B, Chakia 59, CN 540, S 22, Madhukar, S 24, S 25, S 28, Solpona, Sail badal, Dhola badal, Kolasali, Boga bordhan, Rongasali, Khajara, Dhusara, Nali Baunsagaja
Deep water tolerance	:	HBJ 1, Nageri Bao, Kekoa bao, Jalmagna, Jaladhi 1, Jaladhi 2
Coastal Saline/ Alkaline Tolerance	:	SR 26 B, Getu, Dasal, Patnai 23, Pokkali, Kala farm, Rahas Panjar, Nona Bokra, Katla, Hamilton, Damodar
Water logging Tolerance	:	Jhingasail, Tilakkachari, NC 496, Patnai 23, Kalakhersail

5.5 Hybrid Rice Technology

Unlike other cross-pollinated crops like maize, sunflower, etc., in rice, being a self-pollinated crop, it was thought earlier that heterosis levels might be too low for commercial exploitation. The developments in China predominantly due to the efforts of a single individual LP Yuan during the 1970s have significantly changed this opinion and also the course of Chinese rice production. The dogged determination of the Chinese has showed that hybrid rice technology can be commercially made successful and can address the concerns related to rice yield stability and plateauing. The development and utilization of cytoplasmic male sterile (CMS) lines from the wild abortive (WA) cytoplasm background have resulted in the popular CMS lines currently used all over the world like Zhenshan A, IR58025A, IR62829A, etc. in the popular three-line hybrid rice technology. It is anticipated that the popularization of hybrid rice technology will go a long way towards food security in spite of the deficiencies witnessed in implementation of the technology in the past decade. Every new technology needs to be fine-tuned before popularization, and hybrid rice technology is no exception to this. The recent developments in China and also at the International Rice Research Institute, Philippines, with respect to two-line rice hybrids hold great promise, and it is expected that the two-line hybrid breeding system eliminating necessity of a restorer line will slowly replace the three-line system and will increase heterosis levels to the tune of 30%. India, with its wide array of climatic conditions, stands to gain if proper impetus is given to the development of two-line rice hybrids suited for our conditions.

5.6 Biotechnology in Rice Improvement

The efforts of traditional and hybrid rice breeders have no doubt brought the rice yield to such a stage where at least for the present, food production rates would outrace the population growth. But we should not be complacent as the vagaries of monsoon and disturbing trend with respect to soil health are bound to destabilize rice production. Since the mid-1980s, rice biotechnology has gained momentum with focused research world over. Tools of molecular biology have enabled us to unravel the mysteries of genetic information contained in the rice genome, and genetic engineering protocols for rice have been perfected to move desirable genes across the genomic and sexual barrier. From a breeder's perspective, biotechnological tools provide breeding process more accurately with great precision as compared to traditional breeding approaches/methods. Biotechnological approaches improve rice breeding process by following through:

- Transfer of desirable traits across the rice germplasms (genus/species), i.e., broadening of genetic diversity.
- Manipulation of desired trait without disturbing the non-target rice genome sequence (i.e. enhancing the efficiency in selection process)
- Shortening the breeding cycle

These three biotechnological approaches can directly or indirectly contribute in rice genomic improvement.

1. DNA/Molecular marker technology for precision in rice breeding by non-transgenic methods
2. Genetic engineering for the traits that cannot be transferred by conventional means
3. Application of genomic tools for identifying new and useful genes in germplasm.

5.6.1 DNA/Molecular Marker Technology

Selection of desired individuals for the desired trait(s) has been the hallmark of all plant breeding activities since the beginning. Earlier, plant breeders were using morphological markers and statistical methods to select superior segregants. This method has many impediments like influence of environment on the expression of the trait, less abundance, non-uniform distribution of the morphological markers across the genome, time taking, labour-intensive, pleiotropy and polygenic nature that create difficulty to map and introgress the traits in desire genotypes. Recent advancements in molecular biology have led to the development of molecular markers that, i.e., basically landmark over the genome, can be identified with ease and confidence and form the basis for selection and transfer of desirable traits to crop plants with precision. The development of molecular markers and their applications has led to the development of new transition area between biotechnology and breeding science, i.e., named as Molecular Breeding. Molecular breeding involves the utilization of molecular tools and techniques such as PCR, molecular markers, gel documentation, molecular hybridization, sequencing, etc. for analysing the polymorphism across genotypes. The sequence polymorphisms detected by these techniques are collectively called molecular markers or DNA markers technology and themselves referred as DNA markers. DNA markers are broadly classified into two groups: RFLP- and PCR-based markers. Among the different type of markers used in rice genome analysis, PCR-based DNA markers like RAPDs, microsatellites, Inter-Simple Sequence Repeats (ISSRs) and sequence-tagged sites (STSs) have been widely used for genetic analysis and tagging different traits due to their simplicity, robustness and speed of assay. Nowadays, advancement of sequencing technology and reduction in time and cost of sequencing have developed SNP (single-nucleotide polymorphism) markers. The development of SNP markers across the genome is used to genotype the population in single run using chip/array like platform. Several SNP chip-like platforms such as 44 K Affymetrix array, 50 K Infinium array and 50 K SNP chip have been developed in rice. The development of molecular markers accelerates the breeding programme for improvement of crop. Breeders are using molecular markers as a tool to select the desired genotype for desire trait at early stage with ease. The opportunity to select desirable lines based on genotype rather than phenotype is really a benediction to plant breeder.

Genetic maps have been constructed in many crop plants using these markers, and rice geneticists too have used these markers to map genes controlling quantitative and qualitatively inherited traits. More than 25 agronomically important rice genes have already been tagged with markers and can readily be deployed by breeders in breeding programme.

5.6.1.1 Application of Molecular Marker Technology in Rice Improvement

Yield of rice is compromised by climatic factor, responses of genotype to the agronomic practices that commonly lead to different biotic and abiotic stresses. Introduction of the molecular markers in rice breeding programme opens the avenue for marker-assisted selection (MAS), marker-assisted backcrossing (MABC), marker-assisted gene pyramiding (MAGP) and marker-assisted recurrent selection. The development of genetic linkage map and physical map of genome is easily performed with the molecular markers. Nowadays, it is being widely used for high-density linkage map development and mapping of gene for quantitative and qualitative traits of rice.

Marker-Assisted Breeding for Biotic Stress Resistance in Rice

Rice crop is attacked by many insect, pests and diseases. The major insects/pests are stem borer, leaf folder, brown plant hopper, green leaf hopper, gall midge, etc., and devastating diseases are blast, bacterial blight, sheath blight, etc. Even though traditional breeding efforts have resulted in development and release of many pest and disease-resistant varieties, the evolution of new biotypes/pathotypes of insect pests/pathogens has often resulted in the breakdown of resistance. Further, breeding varieties with multiple pest resistance is cumbersome and difficult. Through the biotechnological tools, development of durable multiple pest-resistant varieties is far simpler now. Moreover, for pests like stem borer and diseases like sheath blight for which there is no known source of complete resistance in the rice germplasm, it is hoped that deployment of foreign genes through transgenic rice may help to develop resistance. Once the genetics of a resistance is established, molecular markers can be deployed to tag and map the gene so that the gene-linked marker can be used for tracking the introgression of the gene across segregating populations. Many pest and disease-resistant genes have been tagged (Table 5.2).

Pyramiding different genes conferring resistance to diseases and pests is a practical approach for ensuring broad-spectrum and durable resistance. Different resistance genes confer resistance to different races, isolates, biotypes, etc. as the case may be. When several genes confer the same phenotype, markers can be used effectively to select for the gene pyramids.

Pyramiding of Bacterial Leaf Blight Resistance Genes into Elite Indica Rice

Bacterial leaf blight (BLB) is a devastating disease of rice caused by a bacterial pathogen. A successful use of marker-assisted selection (MAS) has been shown in pyramiding four Xa-genes for bacterial blight resistance. A great achievement in the development of bacterial leaf blight-resistant rice cultivars through marker-assisted

Table 5.2 Important pest and disease resistance genes tagged and mapped with molecular markers in rice

S. No.	Gene	Trait	Chromosome	Marker	Reference
1	Pi-1	Blast resistance	11	Npb181	Yu et al. (1991)
2	Pi-2 (t)	Blast resistance	6	RG64	Hittalmani et al. (1995)
3	Pi-4	Blast resistance	12	RG869, XNpb294	Yu et al. (1991), Mew et al. (1994)
4	Pi-ta	Blast resistance	12	RZ397	Yu et al. (1991)
5	Xa-1	Bacterial blight	4	Npb235	Yoshimura et al. (1992)
6	Xa-4	Bacterial blight	11	Npb181	Yoshimura et al. (1995)
7	xa-5	Bacterial blight	5	RG556	McCouch et al. (1991)
8	Xa-7	Bacterial blight	6	G1091	Kaji and Ogawa (1995)
9	xa-13	Bacterial blight	8	RZ390 RG136	Yoshimura et al. (1995), Zhang et al. (1996)
10	Xa-21	Bacterial blight	11	PTA248	Ronald et al. (1992)
11	RTSV	Rice Tungro spherical virus resistance	4	RZ262	Sebastian et al. (1996)
12	Bph-1	Brown plant hopper resistance	12	XNpb248	Hirabayashi and Ogawa (1995)
13	Bph-10(t)	Brown plant hopper resistance	12	RG457	Ishii et al. (1994)
14	Gm-1	Gall midge resistance	9	RM219, RM316, RM444	Biradar et al. (2004)
15	Gm-2	Gall midge resistance	4	RG329	Mohan et al. (1994)
16	Gm-4	Gall midge resistance	8	F43	Nair et al. (1996)
17	Gm-6 (t)	Gall midge resistance	4	OPM06, RG214, RG476	Katiyar et al. (2001)
18	Gm7	Gall midge resistance	4	SA598, F8	Sardesai et al. (2002)

selection was reported by scientific group of Punjab Agricultural University, Ludhiana.

IIRR, Hyderabad, in collaboration with CCMB, Hyderabad, initiated a programme for development of durable BLB resistance in the background of elite rice cultivar BPT5204. One dominant (*Xa21*) and two recessive (*xa5* and *xa13*) BLB-resistant genes were introgressed into the genetic background of BPT5204 with the help of tightly linked PCR-based DNA markers. Locus-specific SSR markers were used for accelerated recovery of the recurrent parent genome at each backcross generation, ultimately reducing the backcrossing to only BC4. Analysis of the selfed progeny at BC_4F_2 clearly shows that the selected plants are not only resistant to multiple races of BLB but also much similar to the recurrent parent (BPT5204) in terms of yield and grain quality traits. Some of these lines have been evaluated at different locations under the All India Coordinated Rice Improvement Programme. In 42nd Annual Rice Group Meeting, the Varietal Identification Committee (VIC) has recommended the culture RPBio-226 (IET 19046) for release in selected states. Similarly, another culture IET 18990 (improved Pusa Basmati-1) developed at IARI, New Delhi, using MAS has been already approved by the CVRC for release. These efforts clearly show the utility of MAS in backcross breeding programmes.

Tagging, Mapping and Utilization of Genes for Gall Midge Resistance in Rice
Rice gall midge is one of the most serious insect/pests of rice in India. There are only few reliable methods of chemical control known for the pest. The deployment of resistant varieties is considered as the only feasible option for its management. More than nine gall midge-resistant genes (*Gm1*, *Gm2*, *gm3t*, *Gm4*, *Gm5*, *Gm6*, etc.) have been identified in rice. Out of which, four genes, *Gm2*, *Gm 4(t)*, *Gm 6(t)* and *Gm 7*, have been mapped using molecular markers. The effectiveness of the DNA markers linked to the gall midge-resistant genes in marker-aided selection, has been demonstrated. Even though many resistant varieties with single resistance gene are being grown widely, the existence and emergence of new biotypes of the insect pest have complicated resistance deployment strategies. Thus, pyramiding of more than one resistance gene in the background of an elite cultivar is considered as the best strategy for durable resistance. Towards this objective, a major, dominant gall midge-resistant gene *Gm1* has been tagged with the help of microsatellite markers and identified the tentative chromosomal location of the gene on Chr 9 of rice. The SSR markers RM219, RM316 and RM444 which are flanking *Gm1* are being used for introgression of the gene from various donors like W1263, Kavya, Mahamaya, etc. into the genetic background of popular rice cultivar BPT5204 and Swarna. Similarly, gall midge-resistant gene *Gm2* has been mapped using flanking molecular markers, RM317 and F8 by Nair et al. (1995), and hence has been useful in marker-assisted breeding programme.

MAS for Quantitative Traits

Most traits of agronomic importance such as yield, abiotic stress tolerance, etc. are complex in nature, controlled by several genes and deeply influenced by environmental conditions. Compared to the traits controlled by major genes, improvement of quantitative traits thorough MAS raises more questions. A lot more concerted effort is required to reliably map the QTL controlling complex traits. In mapping QTL, the percent phenotype variation under the control of each locus is generally used to assess the effect of specific locus on the trait. For any specific QTL, the percentage of phenotype explained value varies depending on the environmental conditions to which the population has been subjected.

The difficulty in manipulating the quantitative traits is due to their complexity in number and interaction between the loci (i.e. epistasis). Since several genes located in different regions of the genome with small individual effects and their interaction are involved in the expression of the trait, several regions (QTL) are to be manipulated simultaneously. QTLs for yield have been identified in wild relatives of rice (*O. rufipogon*). This demonstrates that DNA marker technology might be useful for novel goals in rice breeding besides performing the tasks of conventional breeding with added precision. However, the success of utilization of QTLs is very limited. Future research in this direction therefore should focus on (1) identification of major QTLs (with more than >40% influence on trait) which could be exploited in different environments, (2) exploitation of complementary QTL to isolate transgressive segregants particularly from interspecific crosses, (3) identification of orthologous QTL among different species, as the conservation of such QTL among species may provide new opportunities for manipulation of economic traits, (4) high resolution of QTL to determine whether QTL are single gene or clusters of tightly linked genes and whether over-dominance plays a significant role in heterosis, and (5) cloning of QTL based on high-resolution mapping.

Application of Markers for Hybrid Rice Breeding

Purity Analysis of Hybrids Seed and Parental Lines

Maintenance of purity is vital for sustenance of hybrid rice technology. Conventionally, purity of hybrid seed is assessed through a morphology-based assay called 'Grow-out test' which has several limitations. IIRR, Hyderabad, in collaboration with the Centre for Cellular and Molecular Biology (CCMB), Hyderabad, developed a test for the assessment of purity of seeds of hybrid rice and also the parental lines using SSR markers. SSR markers are co-dominant and hence can distinguish a pure hybrid from its parental lines and other contaminants. The methodology involves isolation of DNA from a representative seed sample (~400 seeds) representing a seed-lot (of about 1–2 tons), use of extracted DNA in a PCR assay involving SSR and/or STS markers, gel electrophoresis of the amplified fragments, gel documentation and analysis of the marker alleles to identify the impurities among the seedlings assayed. The entire assay right from DNA isolation to gel analysis takes just about 4–5 h and can be carried out even in a moderately equipped laboratory. Thus, this DNA marker-based assay can serve as a replacement to the conventional assay of Grow-Out Test (GOT), which suffers from many limitations, but still commonly used by hybrid seed companies for seed purity estimation. Another assay for distinguishing the cytoplasmic male sterile (CMS) lines from maintainer and restorer lines has also been developed with the help of CMS-mitochondria-specific markers. Together, these assays are helping hybrid rice breeders and seed companies to rapidly and reliably monitor parental and hybrid seed purity.

Mapping Fertility Restoration Gene(s)

One of the important requirements for stable performance of rice hybrids is stability in fertility restoration behaviour across the locations, and this could be achieved by the identification and use of new and diverse restorer lines and further transfer of restorer gene(s) from these restorers to desirable genetic background, resulting in the development of genetically diverse restorers, which could result into heterotic hybrids. The identification of new restorers from the germplasm and transfer of restorer gene(s) could be greatly facilitated by understanding the genetics of fertility restoration and identifying marker(s) linked to fertility restorer gene(s). Identifying a DNA marker closely linked to fertility restorer gene(s) can greatly increase the efficiency of restorer breeding as it would help in:

- Marker-aided selection (MAS) for restorer gene(s) without having sterile cytoplasmic background and without resorting to extensive test crossing.
- Identification of plants with restorer genes at early seedling stage in a backcross breeding programme for transfer of restorer gene(s) to desirable agronomic background and thus saving of resources.
- Developing restorers with normal cytoplasm. This would solve the problems faced when isocytoplasmic restorers are used.
- Marker-based pyramiding of restorer genes for enhancing restoration ability, particularly when partial restoration is a constraint.
- Selection for restorer genes independent of environmental influence on restoration.
- Marker-assisted identification of new restorer lines from the germplasm.

Mapping Genes Governing Genic Male Sterility in Rice

One of the major anticipated constraints to the sustenance of heterosis breeding in rice is the availability or quality seed within inexpensive range. The three lines are very difficult to maintain that are utilized in rice breeding programme using cytoplasmic male sterility technique. The widely employed cytoplasmic genic male sterility system based on three-line breeding is often difficult to maintain. Problems with the maintenance of A lines, the lack of diversity in A and R lines and the presence of minor fertility gene in B lines have to be addressed constantly in breeding programmes, ultimately leading to high seed production costs. Following the landmark finding by Prof. Shi Mang Shang of the Hubei University that sensitive rice genotypes reversibly turn male fertile to male sterile and vice versa with change in environmental factors, such as temperature and /day length during critical phases of plant growth, the concept of two-line breeding emerged as an alternative to the three-line approaches. The major benefit of two-line hybrid breeding incorporates the ability to extend genotypes as male parents and absence of negative effects associated with male sterility-inducing cytoplasm. At the same time, separate maintainer lines are not required.

In photoperiod-sensitive genic male sterility, fertility is controlled by oligoene. Thus, it should be relatively easy to develop new PGMS lines by transferring the PGMS lines from one genetic background to another, particularly if marker-aided system of transfer can be developed. The trait temperature-sensitive genic male sterility is known to be influenced by fertility differential genes, and epistatic interactions make the selection of appropriate TGMS lines under field/controlled conditions very difficult. The identification and use of suitable molecular markers closely linked to TGMS could be an easy and reliable approach to solving this problem. As many as seven TGMS genes *tms1*, *tms2*, *tms3*, *tms4*, *tms4(t)*, *tms5* and *ms-h* have been mapped. The identification of these markers linked to various TGMS and PGMS genes will greatly facilitate the transfer of these genes to desirable genetic backgrounds through marker-assisted selection. Molecular mapping of these genes will also help in gene pyramiding of non-allelic TGMS sources, which are sensitive under a wide range of temperature conditions, thereby increasing the stability of the sterile phase during hybrid seed production, each under conditions of unpredicted low temperature.

Grain Quality Traits

The grain quality of rice is a complex trait means composed of many components, viz., nutritional value, physical appearance, cooking and eating quality. Quality rice production is a major problem in the world, especially in the field of hybrid rice production. The major problems are eating, cooking, appearance and milling quality. Eating and cooking quality in large extent is determined by physical and chemical texture of the photosynthates, starch content of amylose, gel consistency and gelatinization temperature present in the endosperm. The opacity in the endosperm is also important feature that affects the grain appearance as well as breakage, ultimately regulating the grain quality too. In a major breakthrough, Zhou et al. (2003) improved all the four quality traits of the hybrid parent line Zhenshan 97A simultaneously by molecular marker-assisted selection.

Aroma is another important grain quality character. Conventionally, the test for aroma is done using test or by chemical method. These methods are subjective and are sometimes harmful. A co-dominant PCR-based marker SCU-Rice-SSR 1 linked to fragrance gene has been recently identified. Similarly, another group identified two SSR markers namely SCU015RM (4 cM) and SCU017RM linked to fragrance gene. These markers can be effectively used for MAS for fragrance gene. A perfect marker for fragrance has recently been identified in Australia and can be used for selecting for aroma.

Use of Molecular Markers in Development of Rice with Abiotic Stress Tolerance

Various abiotic stresses (salinity, drought, alkalinity and cold) limit the production of rice. Traditional breeding methods contributed significantly in the development of varieties against salinity tolerance or resistance in India such as CSR10, CSr11, CSR27 and CSR30. But unlike biotic stress (pest/disease) resistant single gene provides resistance against the biotic stress, abiotic stress controlled by many genes is complicated in nature. Considering the complexities in the mechanisms involved in abiotic stress tolerance, a thorough understanding of these pathways is necessary for enhancing abiotic stress tolerance either through traditional or through molecular breeding. Molecular marker technology, transgenics and genomics can offer viable solutions to breeders to develop rice cultivars with durable abiotic stress tolerance.

Studies using molecular markers basically aim at tagging and mapping of genes/QTLs associated with abiotic stress tolerance. Tanksley (1993) identified QTL associated with salt tolerance and has tagged them with markers C 1232 and RZ698. Similarly, the marker RG 1 was identified to be linked to the QTL for osmotic tolerance, an important trait associated with abiotic stress tolerance. So many such QTLs have been identified and for other abiotic stress-tolerant mechanisms also. Drought tolerance is usually associated with modifications in the root morphology and root penetration ability and distribution. Many QTLs linked with these traits have been identified. Once tightly linked molecular markers are identified for such QTLs associated with salt tolerance, these traits can be pyramided in the background of a popular high-yielding cultivar.

5.6.2 Genetic Engineering for Rice Improvement

Biotechnological tools mediated genetic transformation that can provide revolution in rice production of India in present scenario. The most important feature of genetic transformation technology is to transfer desirable gene between kingdom with least disruption of rice genome. First successful transgenic rice development reported in 1988 was Japonica. After that a large number of transgenic varieties were developed bearing agronomically and economically important traits. Biolistic/Gene gun and Agrobacterium-mediated gene transformation methods are the most common methodology for directly and indirectly gene transfer in rice, respectively. So many transgenic types of rice were developed and tested by scientific group worldwide for biotic and abiotic stress tolerance. Transgenic rice incorporated with herbicide-resistant gene, and Bt and Xa21 genes have also been tested under field conditions.

5.6.2.1 Engineering Rice Against Biotic Stresses

Transformation studies are involved in the standardization of various gene transfer techniques, and subsequently, genes were targeted that confer resistance against pest/disease. The first transgenics were for herbicides resistance, i.e., Kanamycin resistance followed by virus resistance for TMV in tobacco. Few years later, Nayak et al. (1997) reported the development of first transgenic rice with Bt gene. Since then, several groups started working on transfer of different genes into important genotypes of rice, most notably the introduction of *Bt* genes such as *cry1A(b)*, *cry1A(c)* to obtain resistance against yellow stem borer. Research groups in India have recently succeeded in transferring *Bt* genes into *indica* rice cultivars such as IR64, Karnal Local and Pusa Basmati using *Agrobacterium mediated transformation* strategy. Similarly, transgenic rice, resistant to bacteria leaf blight and sheath blight using constructs with *Xa21* and Thaumatin like proteins are being developed. The Indian Institute of Rice Research (IIRR), Hyderabad, has been working on production of transgenic elite *indica* rice varieties resistant to diseases and insects. Transgenics have also been developed in the background Chaithanya possessing *gna* lectin gene which confers resistance against sucking insect/pest of rice. A list of

Table 5.3 Genes for tolerance to biotic stress in rice

rice variety	Transformation methodology used	Gene transferred	Trait	Reference
Indica/ Japonica	Biolistic	*Bar/gus*	Resistance to herbicide	Christou et al. (1991)
IR72	Protoplast (PEG mediated)	*Bar*	Resistance to herbicide	Datta et al. (1992)
Japonica	Protoplast	*CP-stripe virus*	Resistance to stripe virus	Hayakawa et al. (1992)
Indica	Protoplast (PEG)	*Chitinase (Chi11)*	Resistance to sheath blight	Lin et al. (1995)
Japonica	Biolistic/Protoplast	*pinII*	Resistance to Insect	Duan et al. (1996)
Indica	Biolistic/Protoplast	*Bt*	Stem borer resistance	Datta and Baltazar (1996)
Indica	Biolistic	*Bt*	Stem borer Multiple resistance	Tu et al. (1998a)
Indica/ Japonica	Biolistic/Protoplast	*Bt*	Stem borer resistance, Tissue specific	Datta et al. (1998)
Indica	Biolistic	*Xa21*	Bacterial leaf blight resistance	Tu et al. (1998b)
Japonica	Biolistic	*HVA1*	Osmoprotectant	Xu et al. (1996)
Indica/ Japonica	Biolistic/Protoplast	*adh/pdc*	Submergence tolerance	Quimio et al. (2000)
Japonica (T309)	*Agrobacterium*	Ferritin	Iron improvement	Goto et al. (1999)
Japonica (T309)	*Agrobacterium*	*Psy, crt1, lyc*	β-Carotene	Potrykus (1999)
Indica (IR64, IR68144, BR29)	*Agrobacterium* (Antibiotic free, PMI selection system)	*Psy, crt1, lyc*	β-Carotene	Datta et al. (2003)

genes that is importance for biotics stress introduced in rice along with the transformation methodology adopted is given in Table 5.3.

Transgenic rice lines, for agronomically important genes have been already developed both in India as well as other countries. Now it requires critical evaluation and assessment of their suitability under different climatic zones of India. IARI, New Delhi is the first Indian Institute to conduct field evaluation of transgenic varieties IR64, possessing *Bt* gene. IIRR, Hyderabad is also undertaking the field evaluation of transgenic indica rice lines developed at IIRR and IRRI for stem borer and bacterial leaf blight resistance for the past 2 years and is in the process of documentation of the data generated.

5.6.2.2 Engineering Rice for Abiotic Stress Resistance

The branch of Genetic engineering provides another biotechnological approach to develop rice varieties with desirable stress tolerance property. The main objective is to unravel the cross-talk puzzle between different types of stress-signalling biochemical pathways and develop the strategies for bioengineering pathway to combat the stress. Multidisciplinary actions such as biotechnologist, molecular biologist, plant breeder, Plant pathologist, physiologist and agronomist require for precise genetic improvement as well as for their critical evaluations in the field of rice production. Osmotic adjustment is very effective technique to develop drought or salinity stress tolerance in rice. Plants under osmotic stress maintain their turgidity, by accumulating the osmo-protectants in the cytoplasm. Thus, introduction of genes involved in accumulation of osmo-protectants may provide tolerance to both drought and salinity stresses. Several stress-related genes were successfully cloned and integrated into rice genome that enhanced the production of osmolytes to combat against the salinity/drought conditions (Table 5.4).

In India, research work towards development of transgenic rice with abiotic stress tolerance has begun in the right earnest. Research groups at University of Delhi, South Campus (UDSC) and IARI, New Delhi, are developing abiotic stress tolerant lines in rice. Attempts are being made to develop transgenic rice plants resistant/tolerant to drought by introducing genes that express over-production of Proline like Pyroline-5-carboxylate synthase (p5CS). Similarly, *Arabidopsis* dehydration responsive gene, *rd22* and *AtMyb2* are also being introduced and tested for drought

Table 5.4 Important transgenes conferring abiotic stress tolerance in rice

Gene	Gene action	Phenotypic expression	Reference
Adc	Arginine decarboxylase	Reduced chlorophyll loss under drought stress	Capell et al. (1998)
CodA	Choline oxidase (glycine betaine synthesis)	Increased salinity and cold tolerance	Sakamoto and Alia (1998)
GS2	Chloroplastic glutamine synthetase	Increased salinity resistance and chilling tolerance.	Hoshida et al. (2000)
P5cs	Pyrroline carboxylate synthetase (praline synthesis)	Increased biomass production under drought and salinity stress	Zhu et al. (1998)
pdc1	Pyruvate decarboxylase	Increased submergence tolerance	Quimio et al. (2000)
HVA1	Group 3 Late Embryogenesis Abundant (LEA) protein	Increased tolerance of drought and salinity stress	Xu et al. (1996)
OsCDPK7	Transcription factor	Increased tolerance of cold, salinity, and drought	Saijo et al. (2000)
DREB1A	Transcription factor	Increased tolerance of cold, salinity, and drought	Shinozaki and Yamaguchi-Shinozaki 2007
TPSP	Trehalose biosynthesis	Increased tolerance of cold, salinity, and drought	Garg et al. (2002)

tolerance in rice. Engineering rice for improved flooding tolerance through alterations in the levels of PDC and ADH proteins has been reported and a large number of transgenic rice plants, which over-express PDC sub-units have been developed at UDSC by Grover and coworkers. Beyond doubt transgenic technology compared to traditional breeding approaches offer more powerful solution for incorporation of abiotic stress tolerance. Proper application of functional genomics, transgenics and DNA marker technology can certainly result in the development of abiotic stress tolerance in rice.

5.6.2.3 Biotechnological Approaches for Nutritional Enhancement of Rice

Genetic engineering can also play critical role in nutritional quality improvement of rice. Considering the deficiency of rice with respect to human nutrition and the non-availability of enough genetic variation in rice gene pool with respect to nutritional traits, researchers worldwide have targeted deployment of transgenes from other taxa for nutritional improvement of rice.

Nearly 400 million people worldwide are vulnerable to infections and blindness due to deficiency of vitamin A. Iron deficiency affects 3.7 billion people worldwide, particularly women, leading to high maternal deaths and infant mortality. Thus, developing micronutrient-rich rice, with higher amounts of iron, zinc and Vitamin A, can have tremendous impact on the health of the low-income people. Conventional breeding, together with biotechnology, provides the most powerful tools to achieve this goal. The recent breakthrough in genetic engineering of biosynthesis pathways by Ingo Potrykus and Peter Beyer and their teams gives confidence that biotechnology can be used effectively to address the critical nutritional problem for the poor (Potrykus 1999). Three genes—two from daffodil and one from a bacterium *Erwinia uredovora*—have been used to provide the biosynthesis pathway for the production of beta-carotene, a precursor of Vitamin A. Transgenic seeds have already been produced through transformation on a japonica rice variety, T309, and recently in indica rice IR64. These can now be used in the breeding programme to transfer the genes of interest to popular indica rice consumed by the masses. Since the inventors of the technology have donated it free of cost to developing countries like India, Department of Biotechnology and Indian council of Agricultural Research have formalized a programme to transfer the beta-carotene biosynthetic traits to locally popular Indian rice varieties through marker-assisted backcross breeding and genetic transformation. Directorate of Rice Research, Hyderabad, Indian Agricultural Research Institute, New Delhi, University of Delhi, South Campus, New Delhi, and Tamil Nadu Agricultural University, Coimbatore, have been assigned the responsibility of developing Golden rice in India.

Iron deficiency anaemia is another problem widespread among rice consumers of Asia. To address this problem, a ferric chelate reductase gene that allows plants to absorb more iron from soil has been isolated and tested, thus widening the scope of rice varieties with high iron uptake. Similarly, Ferretin gene has been cloned from soybean and introduced into rice which is reported to have twofold increase in iron content in rice grains. A thermotolerant phytase gene from *Aspergillus fumigatus* has been transferred to rice resulting in tremendous increase in iron content in rice grains due to degradation of iron chelating phytic acid by the phytase enzyme. Similarly, overexpression of a cysteine-rich protein, which increases the cysteine content, that may substantially degrade phytate during food preparation and digestion is another exciting development for nutritional improvement using biotechnology.

5.6.3 Application of Genomics for Rice Improvement

Similar to DNA marker technology and rice transgenics, rice genomics is another area full of prospects. The advancement in the sequencing technologies in the last 5 years has been explosive, and very soon we are expected to have a complete sequence of all the 11 types of rice genome (Jacquemin et al. 2013; Lu et al. 2009). As the rice genome is being completely sequenced, biotechnologists have started a reverse genetic approach for the systematic assessment of phenotypes resulting from the disruption of putative gene sequences with genetic resources such as mutants, near-isogenic lines, permanent mapping populations, and elite germplasm. Functional genomics, to a large extent, is analogous to the extensive germplasm screening that has allowed the extraction of useful traits in conventional breeding programmes, with DNA sequence level precision on a genome scale. The judicious utilization of the sequence information through functional genomic analyses will certainly offer solutions to many breeding problems through means that were not possible till date. Functional dissection of dwarfing gene has shown that a single mutation in the Gibberellin response pathway is responsible for the alteration in plant morphology and the subsequent gains made in productivity through the development of semi-dwarf varieties. The information of rice genome has helped in identification of orthologous genes in cereals and also facilitates the sequencing of other cereal genomes. An international collaboration was established for completion of rice genome sequencing and to coordinate the concerted utilization of sequence information for the benefit of humankind. This initiative called the International Rice Genome sequencing Project (IRGSP) is publicly funded and has eight countries as its members. IRGSP has recently completed the rice genome sequencing to tenfold redundancy (http://rgp.dna.affrc.go.jp/rgp/pres_ releas20011225.htm).

Broadly, the science of genomics has two components—Structural and Functional genomics. Structural genomics deals with large scale sequencing of genomes and understanding how genes are arranged in the genome. Whereas functional genomics deals with utilization of genome sequence to understand; how genes function, their interactions, how genes are expressed and how their expression is regulated spatially and temporally. Shortly speaking the goal of functional geneticists is to deduce the function of complete set of 40,000 rice genes assigning the roles to each and every rice gene for various cellular, structural and regulatory functions like transcription factors, mRNA synthesis, splicing, protein structure and their sites of action. Studies on secondary metabolites and their action in rice are also important aspect to be addressed through functional genomics.

The rice genome information is vital starting point for mining new genes and various pathways. The benefits anticipated from rice genome sequence information are:

1. Geneticists and breeders can acquire an almost unlimited number of DNA markers and genes for crop improvement
2. Molecular biologists can use the full set of genes to permit comprehensive characterization of gene expression by several high-throughput approaches
3. Physiologists, developmental biologists and biochemists can inspect the gene complements in rice and related species to see which pathways are shared and which are unique, and how these pathways may have been modified.
4. Quantitative geneticists can gain instantaneous access to the genes in their segregating population that may be responsible for the traits like yield, abiotic stress tolerance etc. that are controlled by multiple loci.
5. Functional geneticists can survey unlimited rice germplasm through single nucleotide polymorphisms (SNPs) for the study of allelic variability and distribution in rice
6. Structural biologists can use rice genome database for complete set of predicted and known peptides to identify those that are of most interest for three-dimensional characterization.

India, being an active partner in the International Rice Genome Sequencing Project (IRGSP) can exploit the rice genome sequence information made available to it by development of more reliable gene specific DNA markers, identification of critical pathways and metabolic networks related to yield, biotic and abiotic stress resistance and identification of novel rice genes for deployment through transgenic technology. The field of nutritional genomics needs proper impetus. Understanding and manipulation of the phenomenon of heterosis is another area where genomics can help.

5.7 Conclusion

Through the judicious and pragmatic application of marker technology, transgenics (GM) and genomic tools, it should be possible to develop different rice lines which are high yielding, resistant to biotic and abiotic stresses, that provide balanced human nutrition and, more importantly, will not deplete the soil and water resources. But all these depend on the successful development of inter-disciplinary collaboration between plant breeders, molecular biologists, plant protection specialists, physiologists and biochemists and focused problem-solving approaches in specific areas.

References

Biradar SK, Sundaram RM, Thirumurugan T et al (2004) Identification of flanking SSR markers for a major rice gall midge resistance gene *Gm1* and their validation. Theor Appl Genet 109:1468–1473

Capell T, Escobar C, Liu H et al (1998) Over-expression of oat arginine decarboxylase cDNA in transgenic rice (*Oryza sativa* L.) affects normal development pattern *in vitro* and results in putrescine accumulation in transgenic plants. Theor Appl Genet 97:246–254

Christou P, Ford TL, Kofron M (1991) Production of transgenic rice (Oryza sativa L.) plants from agronomically important indica and japonica varieties via electric discharge particle acceleration of exogenous DNA into immature zygotic embryos. Biotechnology 9:957–960

Datta SK, Baltazar AM (1996) Integrated weed management in rice in Asia. In: Naylor R (ed) Herbicides in Asian Rice: Transitions in Weed Management. Institute for International Studies, Stanford University and Manila, Palo Alto (California). International Rice Research Institute, Philippines, pp 145–166

Datta SK, Datta DK, Soltanifar N et al (1992) Herbicide resistant Indica rice plants from IRRI breeding line IR72 after PEG-mediated transformation of protoplasts. Plant Mol Biol 20:619–629

Datta K, Vasquez A, Tu J et al (1998) Constitutive and tissue-specific differential expression of cryIA(b) gene in transgenic rice plants conferring enhanced resistance to rice insect pests. Theor Appl Genet 97:20–30

Datta K, Baisakh N, Olivia N et al (2003) Bioengineered '*golden*' indica rice cultivars with β-carotene metabolism in the endosperm with hygromycin and mannose selection systems. Plant Biotechnol J 1:81–90

Duan X, Li X, Xue Q et al (1996) Transgenic rice plants harbouring an introduced potato proteinase inhibitor II gene are insect resistant. Nat Biotechnol 14:494–498

Garg AK, Kim JK, Owens TG, Ranwala AP, Choi YD, Kochian LV, Wu RJ (2002) Trehalose accumulation in rice plants confers high tolerance levels to different abiotic stresses. Proc Natl Acad Sci U S A 99:15898–15903

Goto F, Yoshihara T, Shigemoto N et al (1999) Iron fortification of rice seed by the soybean ferritin gene. Nat Biotechnol 17:282–286

Hayakawa T, Zhu Y, Itoh K et al (1992) Genetically engineered rice resistant to rice stripe virus, an insect-transmitted virus. Proc Natl Acad Sci U S A 89:9865–9869

Hirabayashi H, Ogawa T (1995) RFLP mapping of Bph-1(brown planthopper resistance gene) in rice. Breed Sci 45:369–371

Hittalmani S, Foolad MR, Mew T et al (1995) Development of a PCR-based marker to identify rice blast resistance gene, *Pi-2(t)*, in a segregating population. Theor Appl Genet 91:9–14

Hoshida H, Tanaka Y, Hibino T et al (2000) Enhanced tolerance to salt stress in transgenic rice that overexpresses chloroplast glutamine synthetase. Plant Mol Biol 43:103–111

Ishii T, Brar DS, Multani DS et al (1994) Molecular tagging of genes for brown plant hopper resistance and earliness introgressed from *Oryza australiensis* in to cultivated rice *O. sativa*. Genome 37:217–221

Jacquemin J, Bhatia D, Singh K et al (2013) The international *Oryza* Map alignment project: development of a genus-wide comparative genomics platform to help solve the 9 billion-people question. Curr Opin Plant Biol 16:1–10

Kaji R, Ogawa T (1995) Identification of the located chromosome of the resistance gene Xa-7, to bacterial leaf blight in rice. Jpn J Breed 45(Suppl. 1):79

Katiyar SK, Tan Y, Huang B et al (2001) Molecular mapping of gene *Gm6(t)* which confers resistance against four biotypes of Asian rice gall midge in China. Theor Appl Genet 103:953–961

Lin W, Aunratha CS, Datta K et al (1995) Genetic engineering of rice for resistance to sheath blight. Bio/Technology 13:686–691

Lu F, Ammiraju JS, Sanyal A et al (2009) Comparative sequence analysis of *MONOCULM1* orthologous regions in 14 *Oryza* genomes. PNAS 106:2071-2076

McCouch SR, Abenes ML, Angeles R et al (1991) Molecular tagging of a recessive gene, *xa-5*, for resistance to bacterial blight of rice. Rice Genet Newslett 8:143–145

Mew TW, Parco AR, Hittalmani S et al (1994) Fine mapping genes for blast resistance in rice. Rice Genet Newslett 11:126–128

Mohan M, Nair S, Bentur JS et al (1994) RFLP and RAPD mapping of rice Gm2 gene that confers resistance to biotype 1 of gall midge (*Orseolia oryzae*). Theor Appl Genet 87:782–788

Nair S, Bentur JS, Rao UP et al (1995) DNA marker tightly linked to gall midge resistance gene (Gm2) are potentially useful for Marker aided Selection in rice breeding. Theor Appl Genet 91: 68–73

Nair S, Kumar A, Srivastava MN et al (1996) PCR-based DNA markers linked to a gall midge resistance gene, Gm4t has potential for marker-aided selection in rice. Theor Appl Genet 92: 660–665

Nayak P, Basu D, Das S et al (1997) Transgenic elite indica rice plants expressing CryIAc delta-endotoxin of Bacillus thuringiensis are resistant against yellow stem borer (*Scirpophaga incertulas*). Proc Natl Acad Sci U S A 94:2111–2116

Potrykus I (1999) Vitamin-A and iron-enriched rices may hold key to combating blindness and malnutrition: a biotechnology advance. Nat Biotechnol 17:37

Quimio CA, Torrizo LB et al (2000) Enhancement of submergence tolerance in transgenic rice overproducing pyruvate decarboxylase. J Plant Physiol 156:516–521

Ronald PC, Albano B, Tabien R et al (1992) Genetics and physical analysis of the rice bacterial blight disease resistance locus, Xa21. Mol Gen Genet 236:113–120

Saijo Y, Hata S, Kyozuka J et al (2000) Overexpression of a single Ca2+ dependent protein kinase confers both cold and salt/drought tolerance on rice plants. Plant J 23:319–327

Sakamoto A, Alia MN (1998) Metabolic engineering of rice leading to biosynthesis of glycine betaine and tolerance to salt and cold. Plant Mol Biol 38:1011–1019

Sardesai N, Kumar A, Rajyashri KR et al (2002) Identification and mapping of an AFLP marker linked to Gm7, a gall midge resistance gene and its conversion to a SCAR marker for its utility in marker-aided selection in rice. Theor Appl Genet 105(5):691–698

Sebastian LS, Ikeda R, Huang N et al (1996) Molecular mapping of resistance to rice tungro spherical virus and green leafhopper. Phytopathology 86:25–30

Shinozaki K, Yamaguchi-Shinozaki K (2007) Gene networks involved in drought stress response and tolerance. J Exp Bot 58:221–227

Siddiq EA (1991) Genes and rice improvement. Oryza 28:1–7

Tanksley SD (1993) Mapping polygenes. Annu Rev Genet 27:205–223

Tu J, Datta K, Alam MF et al (1998a) Expression and function of a hybrid Bt toxin gene in transgenic rice conferring resistance to insect pests. Plant Biotechnol 15(4):195–203

Tu J, Ona I, Zhang Q et al (1998b) Transgenic rice variety IR72 with Xa21 is resistance to bacterial blight. Theor Appl Genet 97:31–36

Xu D, Duan X, Wang B et al (1996) Expression of a late embryogenesis abundant protein gene, HVA1, from barley confers tolerance to water deficit and salt stress in transgenic rice. Plant Physiol 110:249–257

Yoshimura S, Yoshimura A, Saito A et al (1992) RFLP analysis of introgressed chromosomal segments in three near-isogenic lines of rice for bacterial blight resistance genes, *Xa-1, Xa-3*, and *Xa-4*. Jpn J Genet 67:29–37

Yoshimura S, Yoshimura A, Iwata N et al (1995) Tagging and combining bacterial blight resistance genes using RAPD and RFLP markers. Mol Breed 1:375–387

Yu ZH, Mackill DJ, Bonman JM et al (1991) Tagging genes for blast resistance in rice via linkage to RFLP markers. Theor Appl Genet 81:471–476
Zhang G, Angeles ER, Abenes MLP et al (1996) RAPD and RFLP mapping of the bacterial blight resistance gene xa-13 in rice. Theor Appl Genet 93:65–70
Zhou PH, Tan YF, He YQ et al (2003) Simultaneous improvement for four quality traits of Zhenshan 97, an elite parent of hybrid rice, by molecular marker-assisted selection. Theor Appl Genet 106:326–331
Zhu B, Su J, Chang M et al (1998) Overexpression of a D1-pyrroline-5- carboxylate synthetase gene and analysis of tolerance to water- and salt-stress in transgenic rice. Plant Sci 139:41–48

Application of Genomics and Breeding Technologies to Increase Yield and Nutritional Qualities of Rapeseed-Mustard and Sunflower

6

Vivek K. Singh, Pratik Istari Bhoyar, Anu, and Vijay Sharma

Abstract

The oilseed crop plays a pivotal role in the agricultural economy. The main breeding objectives of the oilseed crop are to increase the yield of seeds and oil, improve the quality of the oil and meal according to its use, and developing stable biotic and abiotic-resistant/tolerant varieties. Adding to the yield potential of varieties is a natural way to enhance the quantity of both oil and meal. The fatty acid composition determines the oil's quality, whereas the desired fatty acid profile is determined by the oil's consumption. Reduced erucic and eicosenoic acid content has significantly improved the quality of edible oil in rapeseed and mustard species, whereas high oleic acid and tocopherol content has been achieved in sunflower. With the advancement in genetics, genomics and availability of sequenced information of genomes, the traditional breeding is replaced by genomics-assisted breeding which will lead to the effective development of new-generation high-yielding oil genotypes. In this review, we aim to provide an overview of the various advanced approaches for genomics-assisted breeding to enhance genetic gain in yield and nutritional quality.

Keywords

Genomics · GWAS · Conventional breeding · Oil qualities · Rapeseed-mustard · Sunflower

V. K. Singh · Anu
CCS Haryana Agricultural University, Hisar, Haryana, India

P. I. Bhoyar (✉)
Indian Agricultural Research Institute, Regional Station, Indore, Madhya Pradesh, India

V. Sharma
Banda University of Agriculture and Technology, Banda, Uttar Pradesh, India

Kamaluddin et al. (eds.), *Technologies in Plant Biotechnology and Breeding of Field Crops*, https://doi.org/10.1007/978-981-16-5767-2_6

6.1 Introduction

Oil-bearing crops like soybean, rapeseed-mustard, sunflower, groundnut, sesame, safflower, niger (for edible oil) and linseed, castor (non-edible oil) are mainly cultivated for the oil preserved in their seeds. In addition to this, a wide variety of minor oilseed crops and oil-bearing trees have also been cultivated in India and other countries. Next to the United States of America (USA), China and Brazil, India is the world's fourth-largest vegetable oilseed producers. Oilseeds are predominantly used as a source of dietary oil and also as a high-protein feed for livestock but are also used in the manufacturing of industrial products, for example lubricants, emulsifiers, plasticizers, detergents, cosmetics, pharmaceutical products etc. Subsequently, the oils of the crops like rapeseed-mustard, sunflower and soybean have been identified as potential oils for biodiesel (Qiu et al. 2011). Globally, rapeseed-mustard (12.5%) and sunflower (9.2%) ranked third and fourth, respectively, in total vegetable oil economy after palm (36.5%) and soybean oil (27.4%). Vegetable oils are in high demand across the world due to the increase in per capita oil intake in our diets and their use as biofuels. The global market for vegetable oils is estimated to more than double by 2050, compared to current production (Savadi et al. 2017). The nutritional and functional properties of the oil are determined by the fatty acid composition and distribution pattern within the triacylglycerol molecule (Pham and Pham 2012). From a nutritional point of view, oils of sunflower and rapeseed-mustard are believed to be high-quality oil owing to the existence of unsaturated fatty acids, essential fatty acids and tocopherols which reduce the risk of cardiovascular diseases and lower blood cholesterol (Iocca et al. 2016). However, the oil of rapeseed-mustard also contains a high amount of erucic acid which is undesirable for health; therefore, there is a need to develop more varieties like Pusa mustard 24, RLC 1 and RLC 2 etc. which have erucic acid $<2\%$. Furthermore, the possibility of using its oil as a raw material for the production of biodiesel has piqued the interest of farmers, agricultural experts and companies across the globe. Thus, to meet the increasing demands, there is a need to increase oil yields. One option is to enhance the oil content in oilseeds crop to improve oil yields without expanding the area under cultivation and to save the inputs used to increase the additional crops needed to meet the upcoming global oil crises.

Improvement in oil yields can be achieved by enhancing the amount of oil per seed, higher test weight or improvement in seed yield per plant. The development of varieties/hybrids with higher seed yield along with increased oil content has been led by classical plant breeding approaches such as pure-line selection, heterosis and mutation breeding. In terms of optimism to improve traditional breeding approach constantly, modern tools are required to considerably improve efficacy, accuracy and save time, resource and efforts. Still, the level of effectiveness relies on the objectives and key points in species-specific pathways and the techniques used to exploit those targets. The seed yield is complex and relies on many other morphological features, which are mostly quantitatively inherited and strongly influenced by the environment, so direct selection for seed yield alone decreases the efficacy of the selection and ultimately results in limited success in improving it. The genome regions containing certain genes linked with a specific quantitative trait are referred to as quantitative trait loci (QTL). DNA markers may be used for the identification of allelic disparity for QTL or major gene underlying the trait of interest due to genetic linkage. In many oilseeds crop, several QTLs regulating seeds oil have been reported (Savadi et al. 2017). Moreover, there is a need for stacking of new gene controlling traits like increased oil content, seed yield and oil yield with each improved variety. This chapter addresses developments in the genetic improvement of oilseed crops and various techniques that can be used in the cultivation of oilseeds to achieve sustainability.

6.2 Origin and Distribution

6.2.1 Sunflower

Botanically the term "sunflower" is originated from the Greek words "Helios" = "sun" and "anthos" = "flower". Archaeological and botanical findings suggested that it has been domesticated by American tribes in the temperate region of North America, probably about 1000 years BC; it was then transported to America's eastern and southern parts (Putnam et al. 1990). However, evidence procured from San Andres archaeological site in Mexico suggested that it may have been grown as far back as 2600 BC. It was brought from North America to Europe in 1510 and has been used exclusively as an ornamental plant by Spanish explorers for over two centuries (Putt 1997). In the middle of eighteenth century, sunflower cultivated as an oilseeds crop when it reached Russia (USSR) from Europe, where it was easily domesticated in local environments and was exploited as food, source of oil (28–50%) and for medicinal purposes (anti-inflammatory effects). Thereafter, sunflower reintroduced in the United States of America (USA) from Russian immigrants (Davey and Jan 2010). Russia, Argentina, Ukraine, United States, China, India and Turkey are the major sunflower-producing countries in the world. Sunflower has acquired the rank of an important commercial oilseeds crop in India after the introduction of Russian varieties such as Peredovick (EC 68414) and Armavirskii (EC 68415) during the 1960s and due to its day length neutrality, broader adaptability and responsiveness to added inputs, it is cultivated in a variety of climatic and geographical regions.

6.2.2 Rapeseed-Mustard

The family Brassicaceae comprises almost 3500 species and 350 genera. *Brassicas* are one of the most primitive crops domesticated by man since 5000 BC. The findings of Allchin (1969) suggested that mustard was in cultivation since Channhu-daro of Harrapan civilization ca. 2300–1750 BC. Mediterranean basin and south-western Asia are considered as the origin of the genus *Brassica* and its wild relatives. However, it is geographically distributed in the south-western Mediterranean region.

It is evident from earlier reports that *Brassica* originated from the Miocene's genus *Sinapidendron* through *Diplotaxis-Erucastrum*. Among the various species of oilseeds brassica, four species namely *B. juncea* (Indian mustard), *B. napus* (winter and spring rape), *B. rapa* (*B. compestris*) and *B. carinata* (Ethiopian mustard) are of economic importance. The wild form of *B. juncea* thought to be evolved in the Middle East, since its putative parental species, viz., *B. nigra* and *B. rapa* coincided geographically in wild. In the Indian subcontinent, it is grown as an oilseeds crop whereas in China, as a leafy vegetable. Vavilov (1949) suggested that Central Asia and Afghanistan are the primary centres of origin, whereas Asia minor, eastern India and western China included as a secondary centres of origin. The *B. nigra* (black mustard), also known as table mustard, consumed as spices and widespread in southern Europe. *B. rapa* also known as *B. compestris* (L.) is probably native of uplands near the Mediterranean Sea. From here, it was introduced into Germany and Europe. It spread into China as an agricultural crop from Mangolia. In India, it is grown as an oilseeds crop without any known wild relatives. Bengal region of the Indian subcontinent exhibited large diversity of yellow sarson (Chauhan et al. 2011). It is believed that the wild form of *B. napus* has been native to the coastal region of Sweden, Gothland, the Netherlands and Britain, whereas *B. carinata* (Braun) is native to the Ethiopian plateau. *Eruca sativa* (taramira) is widespread in southern Europe/northern Africa and recently introduced in India. Indian mustard, toria, yellow sarson, brown sarson, gobhi sarson, karan rai, black mustard and taramira are eight distinct varieties of rapeseed-mustard that are grown in 53 countries around the world.

6.3 Taxonomical Classification

6.3.1 Sunflower

Botanically, sunflower is known as *Helianthus annuus* (L.), belongs to tribe: *Heliantheae* (Table 6.1), diploid ($2n = 34$), interfertile, an annual herb, 60–300 cm in height with hairy stem, broad and rough leaves, circular head called as disk of diameter of 7.0–30 cm, widely adopted due to short growing season among other oilseeds crops. The cultivated sunflower is usually unbranched, solitary head with several ray florets and bigger achenes. *Helianthus annuus* L. consist of three main races: (1) *H. annuus* ssp. *Macrocarpus*, cultivated for seed or fodder purpose; (2) *H. annuus* ssp. *Annuus*, weed sunflower, grown as ornament plant and (3) *H. annuus* ssp. *Lenticularis*, wild and uncultivated sunflower.

6.3.2 Rapeseed-Mustard

Rapeseed-mustard belongs to *Brassicaceae* (Syn. *Cruciferae*) family of plants which holds above 338 genera and 3709 species. The genus *Brassica* of *Brassicaceae* family comprises over 150 species; however, there is conflict about the exact group

Table 6.1 Botanical classification of *H. annuus* L. and *B. juncea* L.

Taxa		
Domain	Eukaryota	Eukaryota
Kingdom	Plantae	Plantae
Subkingdom	Tracheobionta	Tracheobionta
Superdivision	Spermatophyta	Spermatophyta
Division	Magnoliophyta	Magnoliophyta
Class	Magnoliopsida	Magnoliopsida
Subclass	Asteridae	Dilleniidae
Order	Asterales	Capparales
Family	Asteraceae	Brassicaceae
Tribe	Heliantheae	Brassiceae
Genera	*Helianthus*	*Brassica*
Species	*annuus*	*juncea*

and nomenclature of various species. It includes several cultivated vegetables such as cabbages, cauliflower, broccoli, turnip and rutabaga, along with cultivated oilseeds species. Species such as *B. juncea*, *B. napus*, *B. rapa* and *B. carinata* are grown as oilseeds crops whereas *B. oleracea* and *B. nigra* for seed condiments. Among them, *B. juncea* contributes significantly in terms of the area and production of edible oilseeds. The taxonomic classification of *B. juncea* is given in Table 6.1. Massive diversity in the company of a wide range of wild relative unveils by the genera *Brassica*. However, most of the wild accessions show incompatibility barriers due to the availability of these species in secondary and tertiary gene pools. Table 6.2 contains some of the economically important species of the genus *Brassica*.

6.3.2.1 Genomic Evolution of Brassica

The naturally occurring polyploidy relationship found in the *Brassica* species is interesting and we will discuss it in detail here. This includes three collective *Brassica* diploid species, namely *B. rapa*, *B. nigra* and *B. oleracea*, having haploid chromosome numbers of 10, 8 and 9, respectively. Genome designations A, B and C have been assigned to these (Fig. 6.1). Pairwise crossing among these diploid species after the chromosome doubling led to the evolution of the three amphidiploid species *B. juncea*, *B. napus* and *B. carinata*. *B. juncea* (AABB) contains the genomes of the two species, *B. rapa* (AA) and *B. nigra* (BB). *B. napus* (AACC) share the genomes of *B. rapa* (AA) and *B. oleracea* (CC), whereas *B. carinata* (BBCC) combines the genomes of *B. nigra* (BB) and *B. oleracea* (CC).

The cytological studies of Morinaga (1934) and his associates fundamentally explained the evolutionary relationship among cultivated *Brassica* species. For example, when *B. juncea* is hybridized with *B. rapa*, ten chromosomes of *B. juncea* are seen to pair with the ten of *B. rapa* leaving the other eight as univalent; however, when *B. juncea* is hybridized with *B. nigra*, eight bivalents are usually detected at meiosis which showed that *B. juncea* evolved from these two ancestor species. Similar results were also observed for *B. carinata* and *B. napus*. The

Table 6.2 List of some economic important species of genus *Brassica*

Botanical name	Chromo-some No.	Genome	common name	Usages
Diploid species				
Brassica rapa	20	AA		
B. rapa spp. *Oleifera*	20	AA	Turnip rape	Oilseed
B. rapa var. *brown sarson*	20	AA	Brown sarson	Oilseed
B. rapa var. *yellow sarson*	20	AA	Yellow sarson	Oilseed
B. rapa var. *toria*	20	AA	Toria	Oilseed
B. rapa spp. *Rapifera*	20	AA	Turnip	Fodder, vegetable (root)
B. rapa spp. *japonica*	20	AA	Curled mustard	
B. rapa spp. *Chinensis*	20	AA	Chinese mustard	Vegetable (leaves), fodder
B. rapa spp. *pekinensis*	20	AA	Chinese cabbage	Vegetable (leaves)
B. rapa spp. *nipposinica*	20	AA	–	Vegetable (leaves)
B. rapa spp. *Parachinensis*	20	AA	–	Vegetable (leaves)
Brassica nigra	16	BB	Black mustard, Banarasi rai	Condiment (seed)
Brassica oleracea	18	CC		
B. oleracea var. *acephala*	18	CC	Kale	Vegetable, fodder (leaves)
B. oleracea var. *capitata*	18	CC	Cabbage	Vegetable (head)
B. oleracea var. *sabauda*	18	CC	Savoy cabbage	Vegetable (terminal buds)
B. oleracea var. *gemmifera*	18	CC	Brussels sprouts	Vegetable (head)
B. oleracea var. *gongilodes*	18	CC	Kohlrabi	Vegetable, fodder (leaves)
B. oleracea var. *botrytis*	18	CC	Cauliflower	Vegetable (inflorescence)
B. oleracea var. *italic*	18	CC	Broccoli	Vegetable (inflorescence)
B. oleracea var. *fruticosa*	18	CC	Branching bush kale	Fodder (leaves)
B. oleracea var. *alboglabra*	18	CC	Chinese kale	Vegetable (stem, leaves)
Eruca sativa	22	EE	Taramira, rocket salad	Vegetable, non-edible oilseed
Raphanus sativus	18	RR	Radish	Vegetable, fodder

(continued)

Table 6.2 (continued)

Botanical name	Chromo-some No.	Genome	common name	Usages
Sinapis alba	24	SS	White mustard	Oilseed
Brassica tournefortii	20	DD	Asian/African mustard, Jangali rai	–
Tetraploid species				
Brassica juncea	36	AABB	Indian Mustard, rai	Oilseeds, vegetable
B. juncea var. *Cuneifolia*	36	AABB	Vegetable mustard	Vegetable
Brassica napus	38	AACC	Rape	Oilseed
B. napus spp. *oleifera*	38	AACC	Rapeseed, gobhi sarson	Oilseed
B. napus spp. *Rapifera*	38	AACC	Rutabaga, swede	Fodder
B. napus spp. *pabularia*	38	AACC	Siberian kale, leaf rape	–
Brassica carinata	34	BBCC	Ethiopian mustard, Karan rai	Vegetable, oilseed

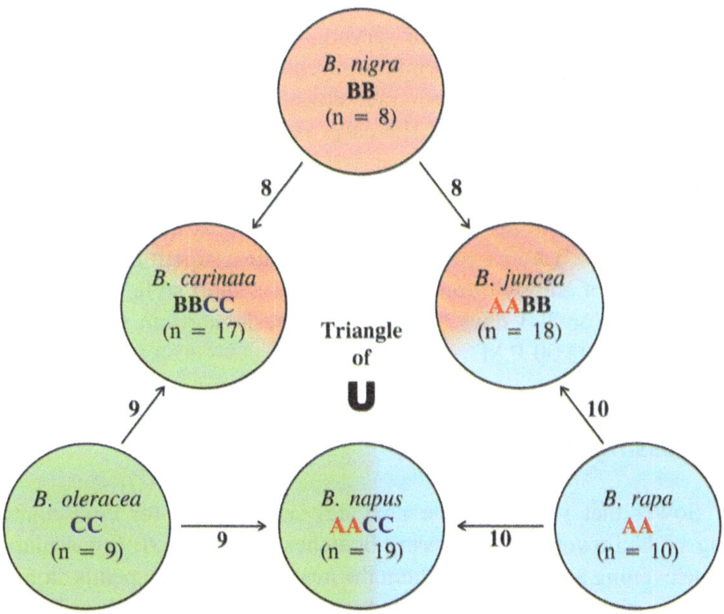

Fig. 6.1 The "Triangle of U", demonstrating the genomic connections between the six *Brassica* species. AA, BB and CC are the genome symbols of *B. rapa*, *B. nigra* and *B. oleracea*, respectively. AABB, AACC, BBCC are the three tetraploid species (*B. juncea*, *B. napus* and *B. carinata*). The numbers on the lines connecting two genomes indicate the maximum of bivalents possible among the respective tetraploid species

genomic evolutionary relationship among six species of Brassica may be represented in triangular form, known as “U” triangle (Fig. 6.1).

6.4 Floral Biology

The inflorescence of sunflower is a capitulum or head which may be convex, concave or flat attached to the main stem at different angles (Fig. 6.2a). The capitulum is enclosed by an involucral bract of variable shape and size (Fig. 6.2a). Each capitulum is composed of two distinct forms of flowers. The peripheral flowers are ray florets, usually yellow and sterile or pistillate type which attract the insect pollinators. The central/inner part of the capitulum is disc florets, hermaphrodite and fertile, which are organized as curves radiating from the centre of the capitulum. The two papus scales are modified form of calyx, and the corolla tube is made up of five petals that are joined together. The stamens are free and connected to the corolla’s base. Five anthers come together to form an anther tube, and the style is inside the tube, with a bilobed stigma. Stigma remains receptive for 2–4 days. The flowering process begins from periphery to centre of the capitulum (Fig. 6.2a). Generally, anthesis starts in the morning hours between 07:00 and 10:00 A.M.

In the case of rapeseed-mustard, the basic floral biology of all the *Brassica* species is almost the same, except for varying flowers size (Fig. 6.2b). The inflorescence of brassica is typically a corymbose raceme; indeterminate in flowering starts at the base of the main shoot and continue upward. Flowers of *Brassica* species are regular, bisexual and hypogynous. The calyx is pale green, contains four sepals in two whorls each. Corolla consists of four free cruciform petals which are clawed and regular. Tetradynamous androecium comprises six stamens arranged in two whorls, viz., two stamens with shorter filaments on the lateral side and four median stamens with longer filaments. Two carpels combine to create a superior ovary with a “false” septum, two rows of campylotropous ovules, and a short style and stigma. Stigma remains receptive for 2–3 days. Generally, anthesis starts in the mid of the day between 11:00 and 02:00 P.M.

6.5 Crossing Techniques

Buds of flower that will open the next day are selected for emasculation. The remaining buds, flowers and flowering branches are detached. Emasculation takes place in the evening and pollinates them the next morning. The petals along with the stamens of the picked buds are detached with the help of forceps and emasculated buds are bagged. In the next morning, ripe anthers are collected from desired male parent, and pollination is carried out by dusting pollen over the stigma. The buds are again bagged after pollination. Pollen grains remain viable up to 7 days, whereas stigma is receptive up to 3 days after emasculation. Moreover, sunflower is predominantly a cross-pollinated crop due to its protandrous nature, and the position of stigma above the anther makes it self-incompatible. Emasculation is carried out in

Fig. 6.2 (**a**) Floral biology of *H. annuus* (**b**) Floral biology of *B. juncea*

the early morning by using forceps to remove the anthers of the disc florets in 2–3 whorls. The pollen from the preferred male parent is collected the next morning and dusted on the emasculated head with a small piece of cotton. This procedure is repeated for another 2–3 days. Undeveloped central florets are removed. To avoid undesirable cross-pollination, hands and forceps have to be surface sterilized with ethanol. Emasculation can also be carried out by spraying 100 ppm GA_3 at the bud-initiation stage consecutively for 3 days in the morning.

6.6 Breeding History and Breeding Objectives

6.6.1 Breeding History

Breeding in sunflower began around 1912 in the former Soviet Union and the most successful early breeding programme was that of V.S. Pustovoit at VNIIMK. The concerted efforts for four decades resulted in increasing oil content from 30% to almost 52%. High oil sunflower varieties, such as Peredovik, Armavirski 3497, Mayak, VNIIMK 8931, VNIIMK 6540 and Smena developed by V.S. Pustovoit and his associates enabled the spread of sunflower crop not only in the Soviet Union but also in other continents. In the 1940s, Putt in Canada developed shorter, early maturing cultivars (Miller et al. 1992). Rust resistance was incorporated from wild species. Argentina and so many other countries began breeding programmes at the same time. In the USA, Kinman began a breeding programme around 1950. Intensive breeding programmes were pursued in several countries around the world, as a result of which sunflower is now grown over a large area in many countries. On another hand, at the beginning of the nineteenth century, research work started for the improvement of rapeseed-mustard through a collection of landraces and their purification at Pusa (Bihar). Scientific work began in Lyallpur (now Faisalabad, Pakistan), then in Punjab (India), for the varietal development of Indian *oleiferous* Brassiceae (NRCRM 2000). At the initial efforts, RL18, a variety of Indian mustards, was identified and released for cultivation in 1937 and L1, a variety of yellow sarson through selection, whereas RT11, another strain of Indian mustard from Uttar Pradesh demonstrated considerably higher yield than local check and released in 1936. After independence, high-yielding varieties of Indian mustard (Laha 101, Varuna, Durgamani, Patan Pustard); yellow sarson (T 151, Patan sarson, YSPb 24, T 42); brown sarson (BSA, BSG, BSH 1, BS 2, BS 65, BS 70); toria (Abohar, BR 23, M 27, T 9, T 36, DK 1) and taramira were developed from 1947 to 1967 (NRCRM 2007). In January 1981, a separate unit of All India Coordinated Research Project (AICRP) on rapeseed-mustard was accordingly established at the campus of the CCS Haryana Agricultural University, Hisar. However, Australia started the first public rapeseed breeding programme in 1970 in Victoria, followed by New South Wales and Western Australia in 1973. The two species *B. napus* and *B. rapa* were initially developed for Australia, with short duration *B. rapa* directed especially at lower rainfall environments. In the late 1970s and early 1980s, breeding programmes of *B. juncea* for canola quality were started in Canberra, Victoria and Western Australia. Kirk and Oram recognized *ZEM 1* and *ZEM 2*, low erucic acid line of *B. juncea*, which were widely distributed to mustard breeder everywhere in the world, including India.

The development of hybrids in maize and other crops in the 1930s stimulated sunflower and rapeseed-mustard breeders to work towards developing hybrids in sunflower and rapeseed-mustard. Although a high percent of hybrid seed could be obtained on the female line, the *percent* seed set varied with the lines and environmental conditions. The discovery of cytoplasmic male sterility (*PET-1*) in the progeny of a cross between *Helianthus petiolaris* Nutt and cultivated sunflower by

Leclercq (1969) was a turning point in the development of commercial sunflower hybrids. Evolving hybrids using genetic male sterility was carried out by Varanceanu and Stoenescu (1982) in Romania. Kinman in USA, Putt in Canada and several other sunflower workers started a hybrid production scheme taking advantage of the self-incompatibility system in the crop (Fick and Miller 1997). Male fertility (MS ms) was linked with red anthocyanin pigmentation enabling rouging of fertile plants in the female line in the seed production plots. Hybrid seed production cost was high because of the labour required to remove fertile plants. The system was stable as evidenced by sterility obtained in the progeny of male sterile plants crossed with fertile cultivated sunflower plants. In 1970, Kinman discovered genes for genetic restoration of fertility (*Rf1*) in wild species. The first commercial hybrid based on cytoplasmic male sterility was made available in 1972 in the USA. Subsequently, the cultivation of sunflower hybrids spread throughout the globe. In India, with the initiation of AICRP on Sunflower in 1972–1973, the importance of hybrids and heterosis breeding was recognized. In 1974 and 1975, experimental hybrids have been developed in Bangalore with the help of four CMS lines (CMS2, CMS124, CMS 204 and CMS234) and two restorer lines (RHA 266 and RHA 274) introduced from the USA. The achene and oil yield in all the hybrids were significantly higher than the check variety EC 68415. Thus, BSH-1 (CMS 234A X RHA 274), the first sunflower hybrid was released in 1980 for commercial cultivation.

Moreover, the heterosis breeding strategy may be a feasible alternative for crop variety improvement. Singh and Mehta (1954) were the first to report heterosis in brown sarson. In *B. juncea*, 13–91% heterosis has been reported by many workers for seed yield (Verma et al. 1998). A significant level of heterosis for yield contributing parameters was observed among the crosses obtained from the genetically diverse group as compared to the same group. Heterosis for increased branch number, siliqua number and 1000-seed weight are the major contributors to the heterosis for yield (Chauhan et al. 2011). Besides this, in *B. juncea*, there have been many reports of *genetic male sterility* (Banga and Labana 1985). The majority of them are originated spontaneously and have a monogenic inheritance. Due to the lack of linkage between a morphological marker and GMS controlling gene, we could not distinguish male fertile plants before flower initiation. Therefore, the GMS system could not be economically feasible for the development of a commercial hybrid.

Out of various male sterility systems, *cytoplasmic male sterility* (CMS) is maternally inherited and most effective due to its easier maintenance. This male sterility system remains intact in a variety of environmental conditions. Rawat and Anand (1979) gave the first report on CMS in *B. juncea*. The most significant drawback of this system is the depletion of sterility at higher temperatures. Later on, in 1985, Raphanus-based CMS was introduced into *B. juncea* using *Ogura* system (Banga and Labana 1985) resulting in highly thermostable male sterile lines. However, it has two major limitations, viz., (1) chlorophyll deficiency at low temperature (<13 °C) and (2) lack of fertility restoration. Kirti et al. (1993) fixed the chlorosis from *Ogura*, *Mori* and *Oxy* system through protoplast fusion between chlorotic sterile and normal green plant. Fertility restorers for most of the male sterile systems are not available. *Ogura*-based CMS line of *B. juncea* (RLM 198) has been obtained from male sterile *B. napus* through repetitive backcrossing and selection (Kirti et al. 1995). To overcome the problem of the male sterility/restoration system of *B. juncea*, transgenic lines having the Barnase/Barstar gene (both from *Bacillus amyloliquefacieus*) have been developed using spacer DNA and CaMV35S promoter (Jagannath et al. 2001).

6.6.2 Breeding Objectives

Traditional goals for crop improvement primarily focus on endeavours to produce, preserve and modify the biomass as per human and industrial requirements. In Asia, higher seed/achne yield is the major breeding goal, although breeding for oil quality trait and meal quality received greater attention in European countries. In India, improvement in seed yield of rapeseed-mustard is largely determined by the number of primary and secondary branches per plant, number of siliquae on the main shoot, number of seeds per siliqua and 1000-seed weight; however, head diameter, seed filling percent and 100-seed weight are the major yields contributing traits in sunflower. Short-duration and high-yielding cultivars/varieties are necessary for the relay, multiple and intercropping systems to work. Hybrid development and deployment are some of the utmost feasible alternatives for overcoming yield barriers. At present, there is almost 15–20% yield enhancement in the case of *Brassica*, while in sunflower two times higher achene yield is recorded than open-pollinated varieties (Yadava et al. 2012). The diversity of parental lines must also be improved to reach a higher degree of heterosis. Under the three-line system (i.e., A, B and R) of hybrid development programmes, diverse CMS sources may offer high-heterotic hybrids. The CMS/restorer system introgression should be carried out by marker-assisted backcross breeding to save time. The development of cultivars resistant to disease and insect pests is also an important goal of oilseeds breeder. To achieve the stable yield, source of resistance should be identified and their introgression in parental stocks against major diseases namely powdery and downy mildew, Alternaria blight and rust and insects like head borer, Bihar hairy caterpillar, green semilooper, cut worms, leafhoppers, mustard aphid and mustard sawfly for their exploitation in hybrid breeding programmes. It is important to explore techniques of plant biotechnology such as the gene pyramiding/QTLs for different biotic and abiotic stresses. Besides this, enhancing oil content and improving oil quality such as "double zero" in rapeseed-mustard and "tocopherols" in sunflower is also a major breeding objective of oilseeds crops.

6.7 Breeding Strategies

6.7.1 Exploitation of Genetic Resource

The exploitation of prevailing or newly created genetic variation is a matter of plant breeding. The plant breeder can use naturally existing genetic variation through germplasm collections, typically placed in germplasm banks. For the initiation of a breeding programme, assessment of germplasm based on a novel trait is necessary. Utilizing indigenous collections of *B. juncea*, several donor parents have been identified for various yield components traits, quality components in addition to different biotic and abiotic stresses (Kumar and Chauhan 2005). Though, the least variability existed for oil quality (Chauhan et al. 2011). Besides, exotic lines introduced from Canada (L4 and L6, white rust-resistant) and Australia (ZEM 1 and ZEM 2, low erucic acid) have been exploited in the development of rust-resistant (JM 1 and JM 2) and low erucic acid (LES 1-27, LET 17, LET 18 and Pusa Karishma) varieties of Indian mustard through hybridization. The crops wild relatives were being commonly used as a resistance source of diseases such as powdery mildew, downy mildew, *Sclerotinia* wilt, *Verticillium* wilt, charcoal rot and *Phoma* black stem in the sunflower breeding programme (Seiler 2010). Resistant to broomrape was identified in germplasm collections of the sunflower which broaden the genetic base for resistance to new races (Seiler and Jan 2014). Interspecific and intergeneric crosses may be attempted if such desirable traits are not present in the same gene pool but present in related species. *H. argophyllus* is commonly known as silver sunflower because of extreme hairiness and lathery leave which support them to adapt in drought conditions; however, *H. paradox* has been used as a genetic source for salinity tolerance. To upsurge genetic diversity, *H. annuus* (cultivated sunflower) were crossed with *H. argophyllus*, *H. annuus*, *H. petiolaris* and *H. debilis* (Sujatha et al. 2008). Based on the existence of the various genomes in the species, diploid annual and perennial species of *Helianthus* did not cross easily. However, interspecific hybridization is more effective for *Brassica* if a female parent is an amphidiploid that shares one similar genome with a parent of pollen (Zhang et al. 2006). The blackleg-resistant gene has been successfully transferred from *B. juncea* to *B. napus* due to apparent recombination among the A and C genome in *B. juncea* crosses (Sacristan and Gerdemann 1986). Similarly, the resistant gene for triazine was successfully introgressed from *B. napus* to *B. oleracea* (Ayotte et al. 1987). If those approaches fail, an alternative method is to create new variation in the absence of natural variation or where its transition to the cultivated species is impracticable. Generally, identification of tightly linked QTL, mapping of gene and gene transfer, despite regulatory hurdles, is the last possibility.

6.7.2 Breeding for Yield

The success of yields enhancement depends on the ability to substitute more desirable alleles in a genotype. Rapeseed-mustard includes several species that have breeding mechanism varied from self-pollination to cross-pollination. Moreover, the sunflower is a highly cross-pollinated species. So, from a breeding point of view, these are quite interesting material. In the case of cross-pollinated species selection procedure ranging from mass selection to recurrent selection, whereas in self-pollinated species, desirable plants are chosen from germplasm collections, gene pools and segregating populations. The recurrent selection programmes are designed to upsurge the frequency of desirable alleles in the population. The end product can be utilized in two ways: (1) the improved population can be used for commercial cultivation if it is superior to the existing cultivated varieties and (2) it forms the reservoir of genotypes for developing inbred lines for heterosis breeding programme. The efficacy of recurrent selection to improve various traits has been well documented in sunflower, but a very limited attempt has been made to exploit this technique in *B. compestris*. In the self-pollinated *Brassicas*, the yield improvement in classical breeding programmes has been achieved by various types of pedigree methods. Handling segregating populations from bi-parental or multiple crosses as bulk populations, before extraction of pure lines, is a common practice. Visual selection in F_2 on a distinct plant basis for seed yield and oil content has proved inefficient. However, the influence of environment on yield component relationship modulates the response to selection (Thurling 1993). In India, several high-yielding varieties (namely RH 725, RH 749 and RH 761) have been released through the pedigree method. Chauhan and Singh (2004) argued that most of the *B. juncea* varieties were pure line and had a narrow genetic base due to a common ancestor, and a limited number of donors were exploited in breeding programmes. Transfer of desirable gene or new traits from one strain to another line or well-adapted cultivars has been achieved by backcross breeding methods. Dihaploid techniques have helped to shorten the breeding cycle in *B. napus* but not successful in *B. compestris*, due to inbreeding depression. Since most agronomic traits have a quantitative inheritance pattern, their identification, mapping and characterization would be essential for heterosis' ability to exploit favourable genes for achieving future production goals.

6.7.3 Hybrid Breeding

Hybrids are superior in yield potential and uniform maturity over synthetic and open-pollinated varieties. It is mainly because of the manifestation of heterosis. Mostly, heterosis is higher in crosses involving parents with diverse genetic background (Hladni et al. 2018). Sunflowers have been shown to have a high degree of heterosis in addition to *Brassicas*. Efforts to generate desirable hybrids are stimulated by the availability of a broad range of CSM sources. Of the various CMS source *PET 1*, *PET 2* (in sunflower) and *Mori*, *Ogura* (in Indian mustard) have been extensively used in the development of hybrids in India. Several sunflower hybrids, viz., KBSH-53, PSFH-569, DCS-107, RSFH-130 and CO-2 have also been released for commercial cultivation in different parts of India. PGSH-51, the first CMS-based gobhi sarson hybrid has recommended for cultivation in Punjab by the

Punjab Agricultural University, Ludhiana (India). Thereafter, two hybrids, viz., NRCHB-560 and DMH-1, have been identified for release by CVRC. Also, to overcome the limitation associated with CMS systems of *Brassicas*, a transgenic system of pollination control has been advocated. Of these, *barnase-barstar* appears to be very promising.

6.7.4 Breeding for Quality Traits

The concept of quality oil in vegetable oil is assessed by their fatty acid content and level of tocopherols, sterols, glucosinolates and other nutrients which rely on the ultimate use of the oil, i.e., food and industrial purpose. Hence, the key goal of oilseeds breeders in the development of improved quality oil producing lines and hybrids. Oils of rapeseed-mustard comprised of two major anti-nutritional factors namely erucic acid (C22:1) and a sulphur-containing compound, i.e., glucosinolate (Agnihotri et al. 2007). Intake of mustard oil with high erucic acid can cause myocardial fibrosis, lipidosis as well as enhanced blood cholesterol (Ackman et al. 1977). Most of the varieties of Indian mustard have erucic acid content. Glucosinolate is the major anti-nutritional factor found in seed meal of rapeseed-mustard which reduces the feed palatability and iodine uptake in animals, particularly in non-ruminants (Fenewick et al. 1983). A polish *B. napus* strain, "Bronowski", is the only known source of low glucosinolate ($\leq$12 μmol/g of the defatted meal) till 2015 (Gupta 2016). In India, breeding efforts have been carried out since 1970 for the development of canola type (erucic acid <2% in oil and glucosinolate <30 μmol/g in the defatted meal) cultivars, i.e., "00" type. Pusa Karishma (LES-39), the first low erucic acid variety of Indian mustard and GSC 5, the first "00" variety of *B. napus* (gobhi sarson) were released in 2005. Successively, five more, low erucic varieties have been released from IARI, New Delhi. Recently, PDZ-1 commonly known as Pusa Double Zero Mustard (PDZ-31), the first "00" variety of *B. juncea* developed by crossing Pusa Mustard-21 (LES-1-27) × NUDHYJ-3 following the pedigree method of selection was released in 2017 for Zone-II. Nowadays, breeding efforts have been carried out in the development of "00" varsities with improved oil content and enhanced vitamin levels.

On the other hand, achenes of sunflower have high oleic acid (C18:1) and tocopherols (a component of vitamin E) which makes it desirable for cooking (Škoric 2012). The biosynthesis of oleic and linoleic acids is significantly affected by genetic factors along with environmental factors such as high temperature during the seed filling period (Škoric 2012). Soldatov (1976) developed a mutant having a high level of oleic acid (80%) through induced mutation using dimethyl sulphate solution. Later on, it has also been extensively used as a source of the gene by sunflower breeders throughout the world. The inheritance of high oleic acid is controlled by one to three major genes with multiple alleles (Velasco et al. 2000); however, two non-allelic unlinked genes namely *Tph-1* (increased β-tocopherol) and *Tph-2* (increased γ-tocopherol) controlled the composition of tocopherol. Of the four derivatives of tocopherols, α-tocopherol (90%) showed maximum vitamin E activity while γ-tocopherol is the best antioxidant (Fernández-Martínez et al. 2007). Velasco et al. (2004) identified few lines which have high γ-tocopherol, δ-tocopherol (>65%) and β-tocopherol (>75%). Moreover, combining genes for high oleic acid and derivatives tocopherols endorsed the development of hybrids with different oil quality types (Skoric et al. 2008).

6.8 Genomics-Assisted Breeding

For the past plant, breeding has always aided with the latest technologies in biology and genetics to accelerate crop improvement. Phenotypic selection has been employed for several centuries to harvest and increase the productivity of crop varieties. With the discovery of DNA and its structure and properties, the technology has advanced at a nucleic acid level in every aspect. In the same way, the identification of DNA markers such as RFLP, RAPD, SSR, DArt, SNP etc., has led to the foundation of marker-assisted breeding. With the advancement of DNA technologies, the generation of markers also evolved with their efficiency and efficacy in breeding and dissecting the genes. The linkage map gave the tool to identify and locate the genes through the markers which are linked and segregate together on the chromosome. The linkage mapping helps to identify the QTL for different traits of interest and to locate them on the chromosome and use them efficiently in the breeding programme (Perez-de-Castro et al. 2012). The bi-parental mapping has overshadowed by the association mapping in the present years due to its advantages of large-scale selection on the germplasm, which consider prehistoric hierarchy in the germplasm. The marker has aided several breeding studies in different crops, proving their efficiency for breeding for different traits of interest. With the advancement in sequencing technologies and reduction in sequencing, cost leads to several sequencing programmes in different crops and an increase in available information on different crops, which accelerates the precision in breeding strategies at the nucleotide level (Kole et al. 2015; Bevan et al. 2017). These new technologies are capable of comprehensive large-scale gene studies and at a high pace parallelly these can allow studying several genes at a time.

Genomic-assisted breeding has emerged as the advanced breeding strategy with the advancement in the DNA technologies such as makers, DNA sequencing and reduced cost of these technologies. Genomics-assisted breeding is an integrated approach in which genomic information is applied to the breeding to improve the genotypes for different traits (Bevan et al. 2017). The selection has been employed at the genomic level in addition to phenotype and the traits were studied throughout the germplasm, and variation for the specific traits were identified and are employed for improvement of the candidate genotypes (Kole et al. 2015). The sequencing has given insight into the genome of the several crop plants in the same way the sequencing of the genome of rapeseed-mustard and sunflower will provide the information about the genome which can help to identify the variation present in the germplasm for the specific nutritional quality or yield traits. The identified variation can be integrated into the genotypes which are lacking in the quality of oil but having the yield potential (Perez-de-Castro et al. 2012). The genomics-assisted breeding approaches include marker-assisted selection, genotyping by sequencing (GBS), association mapping and genomic selection which are discussed in the context of rapeseed and sunflower.

6.8.1 Marker-Assisted Selection

Marker-assisted selection (MAS) evolved with the identification of different molecular markers and played a very significant role in accelerating the traditional breeding programmes. MAS has made a selection at the genomic level with a specific gene combination of desirable traits possible. The linkage mapping and QTL mapping has proved very beneficial in several crops and can be employed to enhance the oil quality in rapeseed-mustard and sunflower. The linkage map gives an insight into DNA sequences concerning these important traits which can help to plant breeder improve the present genotypes (Rauf et al. 2020). Several QTL responsible for the oil quality have been identified and mapped which can be further used for improvement in other genotypes which need to be improved (Table 6.3). The genetic map has been constructed using different markers such as RFLP, AFLP, SSR and DArt in both crops and can be applied to identify other important QTL for oil quality and yield traits (Chen et al. 2010; Pushpa et al. 2015). The advancement in sequencing technology and increased available data on the genomics of species helps to carry out a rigorous examination to identify the candidate genes responsible for oil quality. The available sequence data help to identify the SNPs available for traits and correlate with the quality of oil and further employed in the improvement of different genotypes for the quality.

Sunflower was also studied for marker association with different traits such as yield, head diameter, and hull content. Also, to yield traits, oil composition traits like

Table 6.3 Summary of linkage map and QTL identified for different oil quality traits in sunflower and rapeseed

Sr. No.	Trait	Marker	Reference
Rapeseed-mustard			
1	Oil content and seed yield	SSR and SRAPs	Chen et al. (2010)
2	Erucic acid content	AFLP	Saini et al. (2016)
3	Oil composition	AFLP	Singh et al. (2013)
4	Oil content	SSR and SRAPs	Huang et al. (2016)
5	Glucosinolate content	SSR	Pushpa et al. (2015)
Sunflower			
1	SSR markers	SSR	Tang et al. (2002)
2	Fatty acid composition	SSR	Pérez-Vich et al. (2004)
3	Oil content	SSR and SRAPs	Wang et al. (2013)
4	Oleic acid content	SLAFseq	Zhou et al. (2018)
5	β-Tocopherol	SSR	Vera-Ruiz et al. (2005)

oleic acid content, omega acid content, tocopherol content, linoleic acid content and linolenic acid content linkage maps were constructed with different types of markers such as SSR, RFLP and AFLP (Table 6.3). The investigated linkage map has given the interrelationship between marker, and traits of interest as well as QTL were mapped for the same. Primer sets such as NI-3F/N2-IR serve to identify defective FAD2-ID gene which is responsible for high oleic acid in sunflower seed (Rauf et al. 2020). A fine-genetic map of 5019 SNP via RAD sequencing was constructed (Talukder et al. 2014). The oil properties of most of them are quantitative, whereas Oleic Acid Content (OAC) could be considered as a semi-qualitative trait due to its dependency on the genetic background of the receiver in addition to the environment (Ferfuia et al. 2015). Recently, Premnath et al. (2016) identified two additional QTL for OAC on LG8 and LG9. OAC is the choice of study for the identification of different linked markers due to the importance of oleic acid content in the seeds of sunflower (Bilgen 2016). F4-R-1 is most effective in MAS for OAC (Dimitrijević et al. 2017). Three SSR, viz., ORS-1093, ORS-222, and ORS-598, found tightly linked with Tph1 suggesting a role in β-tocopherol accumulation in the seed. Rauf et al. (2020) successfully parted the lines having low β-tocopherol from high β-tocopherol using ORS716. Recently developed fine linkage maps in sunflower triggered the sunflower breeding at the molecular level and allow the identification of the genes responsible for these traits. These linkage maps help to link the markers with different traits, which helps in achieving precision in sunflower breeding for these traits.

In rapeseed, seed yield was directly correlated with test weight, the number of pods/plant and the number of seeds/pod (Cai et al. 2014). Cai et al. (2014) also suggested the indirect role of biomass yield, plant height, first effective branch height, first effective branch number, length of the main inflorescence and the number of pods on the main inflorescence in seed yield. Several attempts were made for the identification of QTL associated with yield and yield-related traits in rapeseed (Chen et al. 2010; Saini et al. 2016; Huang et al. 2016). The nutritional quality depends on the fatty acid composition of the seed which includes oleic acid, erucic acid, glucosinolate, linoleic acid and linoleic acid. The oleic acid content in the seed decides the oil quality, and several markers linked to oleic acid content were identified (Singh et al. 2013), percent of glucosinolate in oil plays a key role in the seed oil composition, several markers including AFLP, RFLP, SSR and other were mapped on the rapeseed-mustard genome (Pushpa et al. 2015). Erucic acid is an unwanted component of the oil, its increased percentage in seed oil is harmful to human and cattle as well for consumption. Markers that are linked with erucic acid content in the rapeseed were mapped (Saini et al. 2016), use of the makers in breeding for selection can help to reduce the genotypes which have higher erucic acid. Linoleic acid is another important component of the seed oil, its quantity is the other parameter for the oil quality and the increased linoleic acid content is favoured over low linoleic acid content. The linkage maps were constructed for the linoleic acid content of rapeseed-mustard (Huang et al. 2016). The identified linkage maps for different traits have significant use in plant breeding for the nutritional quality of the seed oil in rapeseed-mustard. These can be employed for the selection of the traits in the population with the desirable composition of the oil quality. QTL mapping with the help of different DNA markers like RAPD, AFLP, SSR and SNP discloses that the oleic acid and linolenic acid content linked to loci present on A04 and C10 chromosome of *B. napus* (Yang et al. 2012). The identified allele-specific DNA markers play a key role in the selection of homozygous genotypes for the content of oleic acid in rapeseed breeding.

The traditional bi-parental mapping has a weakness for identifying QTL with small effect, and methods applied for the identification of QTL may also hinder crop improvement. The bi-parental mapping in addition to these also have few limitations: (1) bi-parental population do not possess the same level of allelic diversity throughout the breeding programme which makes them unsuitable as representative of the populations, (2) developing population and its maintenance become costly affair, (3) identified QTL are needed to be validated which require further efforts and (4) QTL with small effects entirely missed due to stringent significant threshold. With the availability of the NGS, the bi-parental mapping slowly replaced association mapping which is more cost-effective as well as precise for QTL mapping and trait investigations (Table 6.3).

6.8.2 Genotyping by Sequencing

The genome-wide association studies require many molecular markers distributed throughout the genome of the plant. The GBS become a tool to identify new molecular marker which accelerates the breeding through the use of such markers present throughout the genome (Celik et al. 2016). With a gradual reduction in the sequencing cost, GBS has triggered genomics breeding in the present era. GBS is an innovative strategy where SNPs discovered and genotyped using NGS in different crop genome or population (Mondon et al. 2018). GBS helps in constructing the SNP array for the crops using NGS which widely employed in genome-wide studies which allow the breeder to characterize the available germplasm of any crops in a very short time to identify the available allele diversity. In GBS, the genomic DNA is digested using restriction enzymes, the digested fragments ligated with barcode adaptor and multiplex libraries generated which subjected to NGS further. The GBS can produce more than thousands of molecular markers. The GBS is flexible to species, populations and research objective which makes it the ideal tool for plant genomics studies. GBS stands as an ideal MAS tool due to its cost-effectiveness, extensively employed for genome-wide association study (GWAS), genomic diversity study, genetic linkage analysis, molecular marker discovery and genomic selection (Fig. 6.3).

Pioneered use of GBS for large-scale SNP detection in sunflower was carried by Celik et al. (2016); they produced a linkage map of 817 SNP-markers covering all 17 LG by analysing an F_2 obtained from the cross RHA 436-H08 M1. Mondon et al. (2018) used GBS using SNP for genotyping of 182 samples of sunflower from 11 sites in Argentina to find the actual population source of the Argentinian samples and to find admixture. However, they surprisingly come across two distinguishing

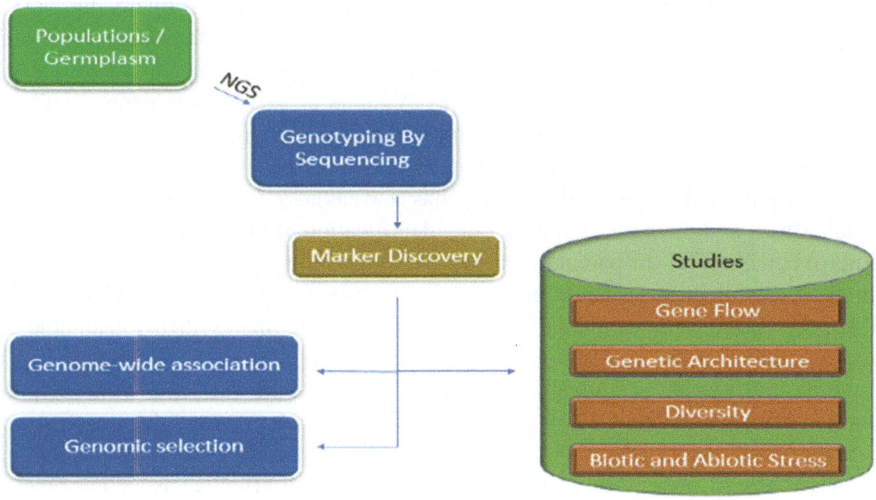

Fig. 6.3 Applications of genotyping by sequencing (GBS)

forms of *H. petiolaris* in Argentina, one from *H. petiolaris* subsp. *petiolaris* as expected, but the other from an unknown source. Ma et al. (2017) conducted a GBS study which suggested that Pl_{20} gene located on linkage group 8 of the sunflower genome belongs to wild *Helianthus argophyllus* which can be used for marker-assisted selection of resistance gene. Ma et al. (2018) used GBS approach to identify new rust resistance gene R_{15} in cultivated sunflower HA-R8. SNP markers closely linked to R_{15} were identified, facilitating marker-assisted selection of resistance genes. Baute et al. (2016) conducted genome-wide GBS to study wild *Helianthus* diversity, genetic structure and interspecies gene flow. The GBS has widely adapted for study gene flow, genetic architecture, diversity or disease resistance in sunflower which makes it a tool for future use in the study of genes responsible for yield and its components in addition to oil quality.

In rapeseed, several genome-wide studies have conducted to study the genetic architecture of different stresses, diversity and disease resistance (Lee et al. 2016; Fu et al. 2020; Ding et al. 2020). The scope of acceleration of the GBS in rapeseed-mustard due to *Brassica* 60K Illumina Infinium™ array, which can genotype about 52, 157 SNPs of *B. napus*, opens the door to study the gene architecture for yield and its components and oil quality responsible alleles (Mason et al. 2017). A genome-wide study conducted on Canadian canola/rapeseed to identify resistance loci for blackleg disease caused by fungus *Leptosphaeria maculans* (*Lm*) using GBS. They identified 32 and 13 SNPs from the Canadian and Chinese accessions, respectively, which are tightly associated with blackleg resistance with p values $<1 \times 10^{-4}$ (Fu et al. 2020). In another study, waterlogging tolerance QTL were identified using GBS in rapeseed (Ding et al. 2020). Lee et al. (2016) characterize the 79 genotypes of *B. napus* using GBS in association with the neighbour-joining clustering method. Zhai et al. (2020) identified 148 SNP loci significantly associated with fatty acid content traits and 20 orthologs of the candidate genes regulating the fatty acid biosynthesis out of 201,187 SNP markers developed from SLAFseq (specific locus amplified fragment sequencing). In the future, GBS has given the pace to genomic-aided breeding in crop species and will help to conduct GS on new genotypes with or without prior molecular tools available and to find the population structure without known diversity.

Association Mapping The association mapping approach has picked up the pace in crop improvement before it was widely adopted in human disease to identify responsible genes. Recently, association mapping (AM) approach is the choice of genetic study to identify QTL in several major crops like maize, wheat, barley, rice and beans (Nambeesan et al. 2015). AM has advantages over bi-parental mapping as it uses the variation present throughout the germplasm or natural population which helps to construct the fine genetic map with the precise marker-trait association. AM involves establishing marker-trait association using several thousand markers throughout the genome which used to find out the allelic diversity for different traits present in germplasm or natural population; henceforth, it makes more beneficial than traditional bi-parental mapping. The identified allelic diversity can be used to improve the genotypes further as per need. In general, AM has been two approaches, viz., GWAS and candidate gene (CG). The AM has greater power over the bi-parental mapping as loci form similar parents contributing to the traits that are unable to map. AM considers the LD event throughout the germplasm which helps to map the allele at a higher resolution to associated with traits (Nambeesan et al. 2015). Previously the bi-parental mapping was widely applied in many crop species, as the whole genome sequences have become available recently. The high-throughput marker system provided the whole genome information which made the recent advances possible such as GWAS, QTLSeq and genome selection.

In sunflower, LD decays rapidly which affects the resolution level of QTL for genes of interest; henceforth, the population structure plays a key role in establishing correct marker-trait association (Kolkman et al. 2007). Mandel et al. (2013) pioneered the GWAS in sunflower whereas most of the studies conducted were candidate gene-based (Cadic et al. 2013; Nambeesan et al. 2015). GWAS was performed on a population of 271 lines for flowering time, branching and heterotic groups using 5359 SNP marker from the Illumina Infinium Beadchip (Mandel et al. 2013). The variability for LD throughout the genome was observed but the marker-trait association was significant suggesting that initial domestication and variety selection for the disease might be the cause of variable LD profile throughout the genome. The candidate gene-based approaches have been well documented in sunflower for branching (Nambeesan et al. 2015) and flowering time (Cadic et al. 2013). The use of CG approach for flowering time and branching is well notified but its implication to trace quality traits is a matter of future study in sunflower. In recent years, AM has proved a very efficient breeding approach to study linkage mapping especially for quantitative traits in different crop species, in the same manner, it is needed to explore sunflower in future.

In rapeseed, several attempts were made to study linkage disequilibrium; however, the homologous nature of subgenomes and limitation of earlier technologies was not enough to develop the high-density molecular marker. With the development of SNP-array in rapeseed, the availability of higher density marker has increased at a genome-wide level. About 4300 SNP markers were developed from 313 inbreeds of *B. napus* (Delourme et al. 2013). A high-density 26841 SNP was developed from 472 accessions of *B. napus* (Li et al. 2014a). A GWAS reported a novel locus for seed oil content on chromosome A05 of *B. napus* (Liu et al. 2013). The AM has two prerequisites, the extent of LD and population or family structure in diverse genetic panel (Sorkheh et al. 2008). A comprehensive study revealed a high-density SNP-based genetic map in four different segregating DH populations using 5764 SNP markers in oilseed rape. GWAS can be performed with a low number of SNP but evenly spaced in the genome of oilseed rape (Delourme et al. 2013). Even in rapeseed, the use of GWAS and CG has been limited, the other quality traits investigation in the area of future research in this crop.

AM has disclosed significant marker-trait association in several important crops such as maize, wheat, rice and barley which is well documented. The GWAS has been used to identify the marker associated with morphological traits (Honsdorf et al. 2010) and seed quality traits (Li et al. 2014b). CG is used to trace the gene polymorphism for glucosinolate biosynthesis (Hasan et al. 2008), tocopherol content and composition gene (Fritsche et al. 2012). Two types of these AM, viz., GWAS and CG, have separate advantages but for quantitative characters, GWAS has the potential to find out the allelic diversity, and in CG, it can help to identify the allelic diversity for the single gene itself (Honsdorf et al. 2010). With the advancement in NGS platforms, the cost and time of sequencing have reduced substantially which increases the number of available markers for GWAS. Both these AM approaches have been applied in both crops which are considered of importance here.

Genomics Selection (GS) Genomic selection (GS) is widely used in crops nowadays which explore available molecular markers to predict genomic-estimated breeding values based on new marker-based models (Bhat et al. 2016; Xu et al. 2020). GS approach comprises two populations, viz., training population (reference population) and breeding population (testing population). Training population is used to predict the genomic-estimated breeding values for testing population based on a marker-based statistical model developed using phenotypic and genotypic information of the training population (Xu et al. 2020). GS for self- and cross-pollinated crops follows the different skim (Fig. 6.4) as the training population and breeding population varies. The GS has two advantages over traditional MAS as there is no need to unearth the QTL related to target traits, and phenotyping for a breeding population can be exempted which reduce the time for GS. GS provides the opportunities to enhance the genetic gain of multigenic traits per unit time and cost. The high-throughput techniques of the genome-wide association have become cheaper and several new markers have been developed in a large population with or without the reference genome sequence is the most important consideration for GS implementation in crop species (Bhat et al. 2016). The next-generation

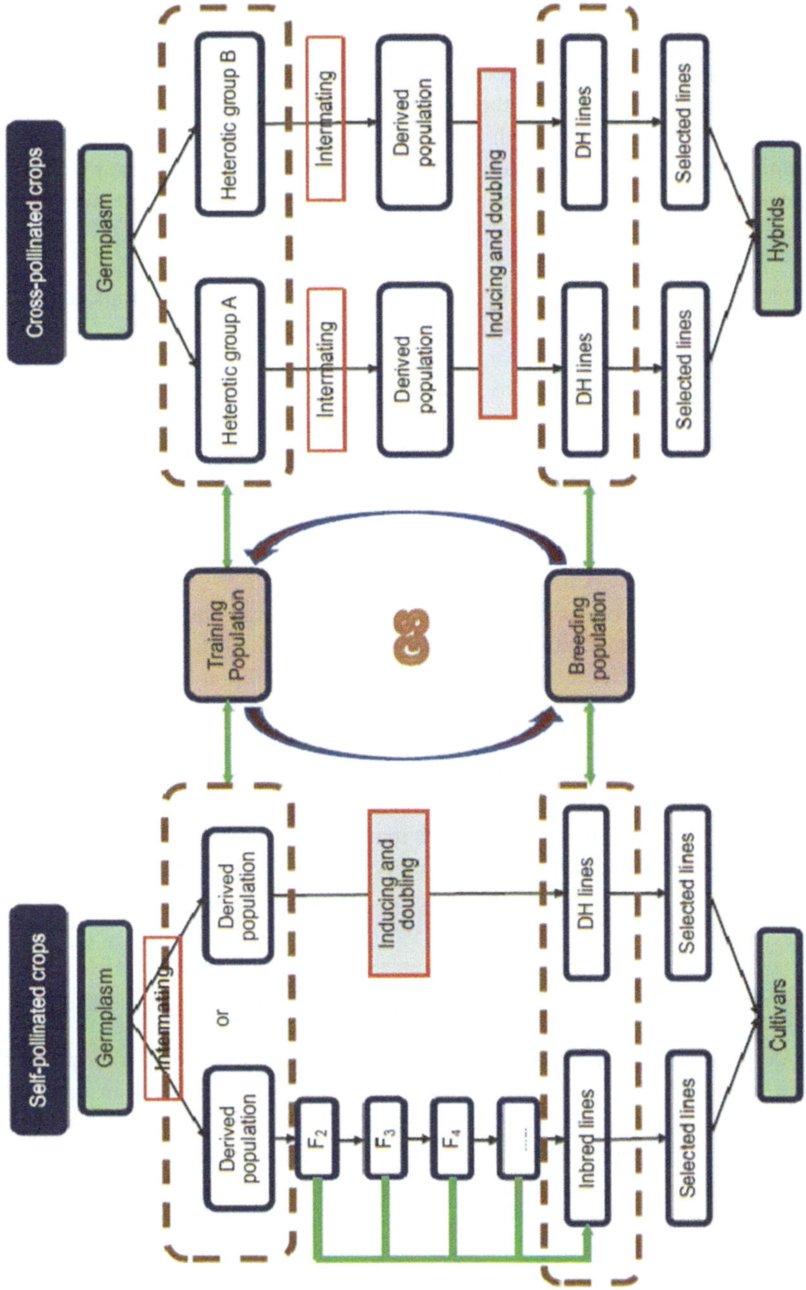

Fig. 6.4 A flow chart representing the GS for self-pollinated (left) and cross-pollinated (right) crops (Xu et al. 2020)

sequencing has provided an SNP genotyping platform through GBS; hence, the availability of the SNP markers for genome-wide studies has increased, so the precision in the marker-trait relation has increased. Availability of such high precision molecular marker and its platform made the GS routine work for crop improvement in both model and non-model crop species. The GBS using NGS has increased the precision in the prediction of the genomic-estimated breeding values (Xu et al. 2020). The GS must combine with high-throughput phenotyping to acquire maximum genetic gain from complex traits. The gradual decrease in sequencing cost has made sequencing of complete genome possible for all important crops which will accelerate the genomic selection in present and future also.

The genomic selection has employed widely in a variety of major crops due to the generation of vigorous SNP markers through sequencing data for the SNP array construction and development of a suitable statistical model for prediction (Würschum et al. 2014). Owing to its advantages, the GS is getting the attention of breeders in crops like rapeseed-mustard and sunflower. GS has been used to study the performance of hybrid in sunflower (Mangin et al. 2017). GS can predict the performance of the hybrid for oil content in sunflower using GCA-based model for prediction from genotypic and phenotypic data which strongly suggest that GS increases the breeding efficiency, although the parents are not phenotyped when compared to classical GCA modelling (Mangin et al. 2017). Würschum et al. (2014) reported the potential of the GS for complex agronomic traits in rapeseed breeding. The traits which are quantitative and regulated by several QTL, the GS allows the breeder to capture the effects of the small-effect QTL in addition to large-effect QTL. A high-density SNP array is the basis of GS in many crops but crops with a narrow gene pool with a low-density SNP array can be employed for better prediction. Even a few hundreds of low-density marker set can enable prediction with precision in breeding population with strong LD in Asian rapeseed. GS becomes a choice of breeder due to the efficiency of NGS which made high-throughput genotyping more easy and economical; henceforth, it is widely adopted by breeders in many crop species.

6.9 Conclusion

The modern-day genomic tool has given a plant breeder the power to exploit the available plant genetic resources which could lead to improvements in rapeseed-mustard and sunflower concerning yield-related traits and oil quality and whatnot. The available allelic diversity present in the natural germplasm can be used to improve the traits. The GBS helps to develop new markers which can be further used for genome-wide studies and genomic selection. The genome-wide studies add the available allelic diversity for the traits with their position in the genome. Genomic selection can help to predict the genetic values of the population which enhances the breeding approach and its advantages. These will give the inside of genetic architecture of the trait and available positive and negative diverse allele.

Such variation can be incorporate in the genotypes further to make the next-generation crops with better climate-resilient, better yield and better oil quality.

References

Ackman RG, Eaton CA, Sipos JC et al (1977) Comparison of fatty acids from high levels of erucic acid of RSO and partially hydrogenated fish oil in non-human primate species in a short-term exploratory study. Nutr Diet 25:170–185

Agnihotri A, Prem D, Gupta K (2007) The chronicles of oil and meal quality improvement in rapeseed. In: Gupta SK (ed) Advances in botanical research–rapeseed breeding. Academic Press/Elsevier Ltd, San Diego, CA, pp 50–99

Allchin FR (1969) Early cultivated plants in India and Pakistan. In: Ueko PJ, Dimbleby GW (eds) The domestication and exploitation of plants and animals. Duckworth, London, pp 323–328

Ayotte R, Harney PM, Machado VS (1987) Transfer of triazine resistance from *Brassica napus* to *B. oleracea*. I. Production of F1 hybrids through embryo rescue. Euphytica 36:615–624

Banga SS, Labana KS (1985) Male sterility in Indian mustard (*Brassica juncea* (L.) Coss). IV. Genetics of MS-4. Can J Genet Cytol 27(5):487–490

Baute GJ, Owens GL, Bock DG et al (2016) Genome wide genotyping by sequencing data provide a high resolution view of wild *Helianthus* diversity, genetic structure, and interspecies gene flow. Am J Bot 103(12):2170–2177

Bevan MW, Uauy C, Wulff BB et al (2017) Genomic innovation for crop improvement. Nature 543(7645):346–354

Bhat JA, Ali S, Salgotra RK et al (2016) Genomic selection in the era of next generation sequencing for complex traits in plant breeding. Front Genet 7:221

Bilgen BB (2016) Characterization of sunflower inbred lines with high oleic acid content by DNA markers. In: Proceedings of the 19th International Sunflower Conference ISA, Edirne, pp 662–668

Cadic E, Coque M, Vear F et al (2013) Combined linkage and association mapping of flowering time in Sunflower (*Helianthus annuus* L.). Theor Appl Genet 126(5):1337–1356

Cai D, Xiao Y, Yang W et al (2014) Association mapping of six yield-related traits in rapeseed (*Brassica napus* L.). Theor Appl Genet 127(1):85–96

Celik I, Bodur S, Frary A et al (2016) Genome-wide SNP discovery and genetic linkage map construction in sunflower (*Helianthus annuus* L.) using a genotyping by sequencing (GBS) approach. Mol Breed 36(9):1–9

Chauhan JS, Singh NB (2004) Breeding approaches in rapeseed mustard varietal improvement. In: Singh NB, Kumar A (eds) Rapeseed-mustard research in India. National Research Center on Rapeseed-Mustard, Sewar, Bharatpur, pp 51–64

Chauhan JS, Singh KH, Singh VV et al (2011) Hundred years of rapeseed-mustard breeding in India: accomplishments and future strategies. Indian J Agric Sci 81(12):1093–1109

Chen G, Geng J, Rahman M et al (2010) Identification of QTL for oil content, seed yield, and flowering time in oilseed rape (*Brassica napus*). Euphytica 175(2):161–174

Davey MR, Jan M (2010) Sunflower (*Helianthus annuus* L.): genetic improvement using conventional and in vitro technologies. J Crop Improv 24:349–391

Delourme R, Falentin C, Fomeju BF et al (2013) High-density SNP-based genetic map development and linkage disequilibrium assessment in *Brassica napus* L. BMC Genomics 14(1):1–18

Dimitrijević A, Imerovski I, Miladinović D et al (2017) Oleic acid variation and marker-assisted detection of Pervenets mutation in high-and low-oleic sunflower cross. Crop Breed Appl Biotechnol 17(3):235–241

Ding XY, Xu JS, Huang H et al (2020) Unraveling waterlogging tolerance-related traits with QTL analysis in reciprocal intervarietal introgression lines using genotyping by sequencing in rapeseed (*Brassica napus* L.). Journal of Integrative. Agriculture 19(8):1974–1983

Fenewick GR, Heaney RK, Mullin J (1983) Glucosinolates and their breakdown products in food and food plants. CRC Crit Rev Food Sci Nutr 18:123–201

Ferfuia C, Turi M, Vannozzi GP (2015) Variability of seed fatty acid composition to growing degree-days in high oleic acid sunflower genotypes. Helia 38(62):61–78

Fernández-Martínez JM, Pérez-Vich B, Velasco L, Domínguez J (2007) Breeding for speciality oil types in sunflower. Helia 30(46):75–84

Fick GN, Miller JF (1997) Sunflower breeding. In: Scheiter AA (ed) Sunflower production and technology. American Society of Agronomy, Madison, WI, pp 395–440

Fritsche S, Wang X, Li J et al (2012) A candidate gene-based association study of tocopherol content and composition in rapeseed (*Brassica napus*). Front Plant Sci 3:129

Fu F, Zhang X, Liu F et al (2020) Identification of resistance loci in Chinese and Canadian canola/rapeseed varieties against *Leptosphaeria maculans* based on genome-wide association studies. BMC Genomics 21(1):1–11

Gupta SK (2016) Brassicas in breeding oilseed crops for sustainable production. Elsevier Inc, Amsterdam

Hasan M, Friedt W, Pons-Kühnemann J et al (2008) Association of gene-linked SSR markers to seed glucosinolate content in oilseed rape (*Brassica napus* ssp. napus). Theor Appl Genet 116(8):1035–1049

Hladni N, Zorić M, Terzić S et al (2018) Comparison of methods for the estimation of best parent heterosis among lines developed from interspecific sunflower germplasm. Euphytica 214(7): 1–19

Honsdorf N, Becker HC, Ecke W (2010) Association mapping for phenological, morphological, and quality traits in canola quality winter rapeseed (*Brassica napus* L.). Genome 53(11): 899–907

Huang XQ, Huang T, Hou GZ et al (2016) Identification of QTLs for seed quality traits in rapeseed (*Brassica napus* L.) using recombinant inbred lines (RILs). Euphytica 210(1):1–16

Iocca AFS, Dalchiavon FC, Malacarne BJ et al (2016) Content and oil productivity in sunflower genotypes produced in campo novo do parecis – Mt, Brazil. In: 19th International Sunflower Conference, Edirne, Turkey, pp 1136–1141

Jagannath A, Bandhopadhyay P, Arumugam N et al (2001) The use of a spacer DNA fragment insulates the tissue specific expression of a cytotoxic gene (*barnase*) and allows high frequency generation of transgenic male sterile lines in *Brassica juncea* L. Mol Breed 8:11–23

Kirti PB, Narasimhulu SB, Mohapatra T et al (1993) Correction of chlorophyll deficiency in aloplasmic male sterile *Brassica juncea* through recombination between chloroplast genomes. Genet Res 62:11–14

Kirti PB, Banga SS, Prakash S et al (1995) Transfer of *Ogu* cytoplasmic male sterility to *Brassica juncea* and improvement of the male sterile line through somatic cell fusion. Theor Appl Genet 91(3):517–521

Kole C, Muthamilarasan M, Henry R et al (2015) Application of genomics-assisted breeding for generation of climate resilient crops: progress and prospects. Front Plant Sci 6:563

Kolkman JM, Berry ST, Leon AJ et al (2007) Single nucleotide polymorphisms and linkage disequilibrium in sunflower. Genetics 177(1):457–468

Kumar A, Chauhan JS (2005) Status and future thrust areas of rapeseed-mustard Research in India. Indian J Agric Sci 75:621–635

Leclercq P (1969) Une sterilité male chez le tournesol. Ann Amélior des Plantes 19:99–106

Lee CJ, Li G, Duncan RW (2016) Characterization of *Brassica napus* L. genotypes utilizing sequence-related amplified polymorphism and genotyping by sequencing in association with cluster analysis. Mol Breed 36(11):1–13

Li F, Chen B, Xu K et al (2014a) Genome-wide association study dissects the genetic architecture of seed weight and seed quality in rapeseed (*Brassica napus* L.). DNA Res 21(4):355–367

Li N, Shi J, Wang X et al (2014b) A combined linkage and regional association mapping validation and fine mapping of two major pleiotropic QTLs for seed weight and silique length in rapeseed (*Brassica napus* L.). BMC Plant Biol 14(1):1–14

Liu L, Qu C, Wittkop B et al (2013) A high-density SNP map for accurate mapping of seed fibre QTL in *Brassica napus* L. PLoS One 8(12):e83052

Ma GJ, Markell SG, Song QJ et al (2017) Genotyping-by-sequencing targeting of a novel downy mildew resistance gene Pl 20 from wild *Helianthus argophyllus* for sunflower (*Helianthus annuus* L.). Theor Appl Genet 130(7):1519–1529

Ma GJ, Song QJ, Markell SG et al (2018) High-throughput genotyping-by-sequencing facilitates molecular tagging of a novel rust resistance gene, R 15, in sunflower (*Helianthus annuus* L.). Theor Appl Genet 131(7):1423–1432

Mandel JR, Nambeesan S, Bowers JE et al (2013) Association mapping and the genomic consequences of selection in sunflower. PLoS Genet 9(3):e1003378

Mangin B, Bonnafous F, Blanchet N et al (2017) Genomic prediction of sunflower hybrids oil content. Front Plant Sci 8:1633

Mason AS, Higgins EE, Snowdon RJ et al (2017) A user guide to the *Brassica* 60K Illumina Infinium™ SNP genotyping array. Theor Appl Genet 130:621–633

Miller JF, Seiler GJ, Jan CC (1992) Introduced germplasm use in sunflower inbred and hybrid development. In: Use of plant introductions in cultivar development, part 2. Crop Science Society of America, Madison, WI, pp 151–156

Mondon A, Owens GL, Poverene M et al (2018) Gene flow in Argentinian sunflowers as revealed by genotyping-by-sequencing data. Evol Appl 11(2):193–204

Morinaga T (1934) On the chromosome number of *Brassica juncea* and *B. napus*, on the hybrid between the two, and on offspring line of the hybrid. Jpn J Genet 9(3):161–163

Nambeesan SU, Mandel JR, Bowers JE et al (2015) Association mapping in sunflower (*Helianthus annuus* L.) reveals independent control of apical vs. basal branching. BMC Plant Biol 15(1): 1–12

NRCRM (2000) Rapeseed-mustard varieties of India. NRCRM, Bharatpur. 115 p, V

NRCRM (2007) Perspective plan. Vision 2025. National Research Centre on Rapeseed-Mustard, Bharatpur, pp 1–46

Perez-de-Castro MA, Vilanova S, Cañizares J et al (2012) Application of genomic tools in plant breeding. Curr Genom 13(3):179–195

Pérez-Vich B, Knapp SJ, Leon AJ et al (2004) Mapping minor QTL for increased stearic acid content in sunflower seed oil. Mol Breed 13(4):313–322

Pham LJ, Pham PJ (2012) Biocatalyzed production of structured olive oil triacylglycerols. In: Olive oil-constituents, quality, health properties and bioconversion. InTech, Rijeka, pp 447–456

Premnath A, Narayana M, Ramakrishnan C et al (2016) Mapping quantitative trait loci controlling oil content, oleic acid and linoleic acid content in sunflower (*Helianthus annuus* L.). Mol Breed 36(7):106

Pushpa HD, Yadava DK, Singh N et al (2015) Validation of molecular markers linked to low glucosinolate QTLs for marker assisted selection in Indian mustard (*Brassica juncea* L. Czern & Coss). Ind J Genet Plant Breed 76:64

Putnam DH, Oplinger ES, Hicks DR et al (1990) Sunflower. Field crops manual. University of Wisconsin Extension Program

Putt ED (1997) Early history of sunflower: sunflower technology production. In: Agronomy monograph. American Society of Agronomy : Crop Science Society of America, Madison, WI, p 35

Qiu F, Li Y, Yang D et al (2011) Heterogeneous solid base nanocatalyst: preparation, characterization and application in biodiesel production. Bioresour Technol 102:4150–4156

Rauf S, Warburton M, Naeem A et al (2020) Validated markers for sunflower (*Helianthus annuus* L.) breeding. OCL 27:47

Rawat DS, Anand IJ (1979) Male sterility in Indian mustard. Ind J Genet Plant Breed 39:412–415

Sacristan ND, Gerdemann M (1986) Different behavior of *Brassica juncea* and *B. carinata* as sources of interspecific transfer to *B. napus*. Plant Breed 97:304–314

Saini N, Singh N, Kumar A et al (2016) Development and validation of functional CAPS markers for the FAE genes in *Brassica juncea* and their use in marker-assisted selection. Breed Sci 66: 831

Savadi S, Lambani N, Kashyap PL et al (2017) Genetic engineering approaches to enhance oil content in oilseed crops. Plant Growth Regul 83(2):207–222

Seiler G (2010) Utilization of wild Helianthus species in breeding for disease resistance. In: International Symposium "Sunflower Breeding on Resistance to Diseases"; Krasnodar, Russia, pp 37–51

Seiler GJ, Jan CC (2014) Wild sunflower species as a genetic resource for resistance to sunflower broomrape (*Orobanche cumana* Wallr.). Helia 37(61):129–139

Singh D, Mehta TR (1954) Studies in breeding of brown sarson I. comparison of F1's and theirs parents. Ind J Genet Plant Breed 14:74–77

Singh S, Mohapatra T, Singh R et al (2013) Mapping of QTLs for oil content and fatty acid composition in Indian mustard [*Brassica juncea* (L.) Czern. and Coss.]. J Plant Biochem Biotechnol 22(1):80–89

Škoric D (2012) The genetics of sunflower. In: Kovacevic Z, Škorić D, Sakac Z (eds) Sunflower genetics and breeding-international monogram. Serbian Academy of Sciences, Serbia, pp 1–125

Skoric D, Jocic S, Sakac Z, Lecic N (2008) Genetic possibilities for altering sunflower oil quality to obtain novel oils. Can J Physiol Pharmacol 86(4):215–221

Soldatov KI (1976) Chemical mutagenesis in sunflower breeding. In: Proceedings of the 7th International Sunflower Conference. International Sunflower Association, Krasnodar, pp 352–357

Sorkheh K, Malysheva-Otto LV, Wirthensohn MG et al (2008) Linkage disequilibrium, genetic association mapping and gene localization in crop plants. Genet Mol Biol 31(4):805–814

Sujatha M, Prabakaran AJ, Dwivedi SL et al (2008) Cytomorphological and molecular diversity in backcross-derived inbred lines of sunflower (*Helianthus annuus* L.). Genome 51(4):282–293

Talukder ZI, Gong L, Hulke BS et al (2014) A high-density SNP map of sunflower derived from RAD-sequencing facilitating fine-mapping of the rust resistance gene R 12. PLoS One 9(7): e98628

Tang SJYK, Yu JK, Slabaugh MB et al (2002) Simple sequence repeat map of the sunflower genome. Theor Appl Genet 105(8):1124–1136

Thurling N (1993) Physiological constraints and their genetic manipulation. In: Labana KS, Banga SS, Banga SK (eds) Breeding oilseed brassicas. Springer, Berlin, pp 44–68

Varanceanu AV, Stoenescu FM (1982) Achievements and prospects of sunflower genetics, breeding and induced mutation utilization. In: Improvement of Oilseed and Industrial Crops by Induced Mutations. IAEA, Vienna, pp. 81–87

Vavilov NI (1949) The origin, variation, immunity and breeding of cultivated plants. Chron Bot 13: 1–364

Velasco L, Pérez-Vich B, Fernández-Martínez JM (2000) Inheritance of oleic acid content under controlled environment. In: Proceedings of the 15th International Sunflower Conference, vol 1. International Sunflower Association, Toulouse, pp A31–A36

Velasco L, Perez-Vich B, Fernandez-Martinez JM (2004) Development of sunflower germplasm with high delta-tocopherol content. Helia 27(40):99–106

Vera-Ruiz EM, Pérez-Vich B, Fernández-Martínez JM et al (2005) Comparative genetic study of two sources of beta-tocopherol in sunflower. Helia 28(42):1–8

Verma OP, Singh HP, Singh PV (1998) Heterosis and inbreeding depression in Indian mustard. Cruciferae Newsl 20:75–76

Wang X, Wang H, Long Y et al (2013) Identification of QTLs associated with oil content in a high-oil Brassica napus cultivar and construction of a high-density consensus map for QTLs comparison in *B. napus*. PLoS One 8(12):e80569

Würschum T, Abel S, Zhao Y (2014) Potential of genomic selection in rapeseed (*Brassica napus* L.) breeding. Plant Breed 133(1):45–51

Xu Y, Liu X, Fu J et al (2020) Enhancing genetic gain through genomic selection: from livestock to plants. Plant Commun 1(1):100005

Yadava DK, Vasudev S, Singh N et al (2012) Breeding major oil crops: present status and future research needs. In: Gupta SK (ed) Technological innovations in major world oil crops, vol 1: breeding. Springer Science and Business Media, New York, NY, pp 17–51

Yang Q, Fan C, Guo Z et al (2012) Identification of FAD2 and FAD3 genes in *Brassica napus* genome and development of allele-specific markers for high oleic and low linolenic acid contents. Theor Appl Genet 125(4):715–729

Zhai Y, Yu K, Cai S et al (2020) Targeted mutagenesis of BnTT8 homologs controls yellow seed coat development for effective oil production in *Brassica napus* L. Plant Biotechnol J 18(5): 1153–1168

Zhang GQ, Zhang DQ, Tang GX et al (2006) Plant development from microspore-derived embryos in oilseed rape as affected by chilling, desiccation and cotyledon excision. Biol Plantarum 50: 180–186

Zhou F, Liu Y, Liang C et al (2018) Construction of a high-density genetic linkage map and QTL mapping of oleic acid content and three agronomic traits in sunflower (*Helianthus annuus* L.) using specific-locus amplified fragment sequencing (SLAF-seq). Breed Sci 68:596

Accelerated In Vitro Propagation of Sweetpotato Clones (*Ipomoea batatas* L.)

7

R. Vollmer, J. Espirilla, J. C. Sánchez, L. Arroyo, G. Flores, A. Rojas, N. L. Anglin, J. Kreuze, and D. Ellis

Abstract

Fast and reliable propagation of plant material is an important need in different stages of breeding programs and production systems. In vitro propagation ensures that pathogen- and virus-free plants stay phytosanitary clean over time while providing high multiplication rates. Using liquid instead of solid culture medium can reduce the interval of individual propagation cycles and contributes to speeding up of the process (1.5–2.5 times), especially during the initial growth phase of the plants. Sophisticated immersion systems have been developed for many plant species, but they are difficult to apply when hundreds or thousands of different genotypes are propagated simultaneously. Additionally, these systems require a high input of technical equipment, know-how and experience to avoid bacterial or fungal contamination during the propagation process. The following protocol describes a low-input suspension technique that combines the use of liquid and solid medium, and permits the successful propagation of genetically diverse sweetpotato genotypes [*Ipomoea batatas* (L.) Lam.] with a high multiplication rate. As sweetpotato is an important staple crop in low-income/technology countries of Africa, Asia, and South America, the described method may find valuable application for the breeding programs in these regions.

R. Vollmer (✉) · J. Espirilla · J. C. Sánchez · L. Arroyo · G. Flores · A. Rojas · J. Kreuze · D. Ellis
International Potato Center (CIP), Lima, Peru
e-mail: r.vollmer@cgiar.org; j.espirilla@cgiar.org; juan.sanchez@cgiar.org; l.arroyo@cgiar.org; g.flores@cgiar.org; a.rojas@cgiar.org; j.kreuze@cgiar.org; d.ellis@cgiar.org

N. L. Anglin
International Potato Center (CIP), Lima, Peru
e-mail: Noelle.Anglin@usda.gov

Kamaluddin et al. (eds.), *Technologies in Plant Biotechnology and Breeding of Field Crops*, https://doi.org/10.1007/978-981-16-5767-2_7

Keywords

In vitro propagation · Sweetpotato · *Ipomoea batatas* · Liquid medium

7.1 Introduction

The efficient large-scale propagation of genetic resources of important staple crops such as sweetpotato is a crucial component for industrial production, breeding programs, and research projects. Plant material can be propagated clonally in the greenhouse (cuttings), field, or laboratory, but due to practical and economic reasons, propagation in the greenhouse is frequently preferred over in vitro propagation in the laboratory. However, in vitro propagation has the distinct advantage that it is performed under controlled environmental and sterile conditions within glass test tubes or other vessels, which allows a space-efficient and continuous propagation that guarantees that pathogen-free material stays clean over time. The use of virus-free material not only contributes to higher yields and thus increased quality of storage roots produced in the field, but also satisfies phytosanitary criteria required for international exchange of plant genetic resources (PGR). Further, the use of phytopathogen-free in vitro material eliminates the effect of virus infection as a source of variation in field, greenhouse, and laboratory experiments as virus-infected plants may show a different pattern of growth, development, and crossing ability. Although plant tissue culture methods were developed in the first half of the twentieth century, based on pioneering work of Haberlandt, Kotte, Robbins, Went, White, Skoog, Miller, and others (Haberlandt 1902; Kotte 1922; Robbins 1922; Went 1926; White 1939; Skoog and Miller 1957), a major general breakthrough was achieved in the early 1960s, through the development and publication of the probably most famous plant culture medium, the Murashige & Skoog medium (MS) (Murashige and Skoog 1962). Over the following decades, it was shown that thousands of species grow well on solid MS medium, generally with full salt strength concentration, but sometimes also with a reduced (½ or ¾) salt strength. Haberlandt already used suspension culture in liquid medium for his pioneer experiments (Haberlandt 1902), but only after White published the use of liquid culture media for the culture of tomato roots tips in 1933/1934 (White 1933, 1934), did liquid medium become established as an alternative to solid culture medium. Later, in the 1980s and 1990s, more sophisticated systems were developed, both for complete and temporary immersion of plant explants (Harris and Mason 1983; Tisserat and Vandercook 1985; Aitken-Christie and Davies 1988; Simonton et al. 1991; Alvard et al. 1993), principally with the objective to increase oxygenation in the culture medium, remove phytotoxic compounds, maintain the nutrient balance (renewal of medium), and avoid hyperhydration. All these are excellent propagation methods, but they involve a high level of technical knowledge and experience, as well as specialized equipment, i.e. inputs developing economies and economies in transition do not always have easy access to obtain. Additionally, the risk of contamination increases when liquids are continuously pumped, moved from one vessel to another,

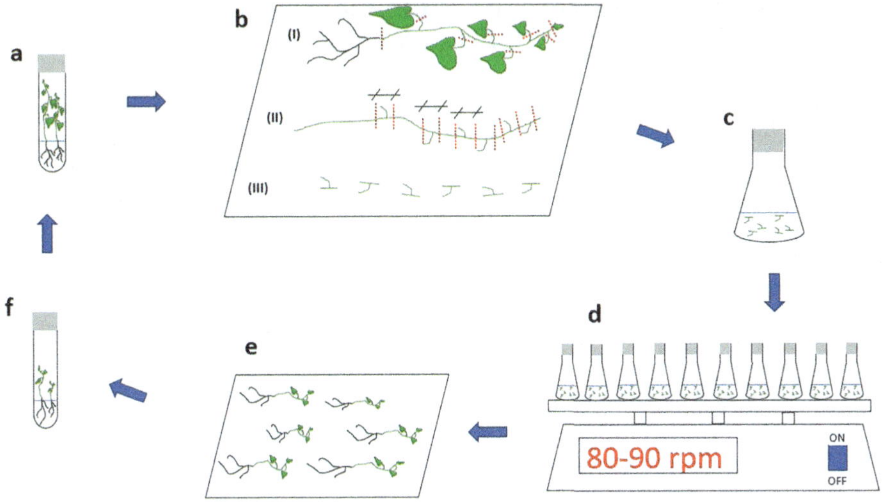

Fig. 7.1 (**a**) Five to 8-week-old in vitro sweetpotato plants are used as starting material. Plants should be completely developed, have minimum 5–6 nodes (leaves), and show a normal aspect (no signs of yellowing, deformation, hyperhydration, etc.) (**b**) (**I**) In vitro plants placed on sterile paper sheet. Roots, leaves, and tip are cut off using forceps for plant tissue culture and scalpel No. 10. (**II**) Cutting of uni-nodal stem segments of ~0.5–1.5 cm. Do not cut too close to the axillary bud to avoid damaging it. (**III**) Stem segments ready to be transferred to liquid culture medium. (**c**) Uni-nodal stem segments placed in liquid culture medium (50 mL Erlenmeyer flask, ~22 mL of culture medium). Place 5–15 stem segments per flask, close flask with sterile aluminum paper and seal with saran wrap or parafilm®. (**d**) Incubate Erlenmeyer flasks under rotation (80–90 rpm) at 25–29 °C with a 16 h/8 h photoperiod. Higher temperatures (28–29 °C) further speed up plant growth. (**e**) In vitro plants coming out from liquid medium, 12–18 days after propagation. Plants should be fully rooted, and minimum have 2–4 leaves for transfer to solid medium. (**f**) Young plants that were transferred from liquid to solid culture medium. Assess growth and contamination on a weekly basis. After 3–6 weeks on solid culture medium, plants are ready for a new cycle

or completely renewed. Based on our experience with hundreds of genetically diverse sweetpotato accessions, the response was uniform and favorable in classic suspension liquid medium under rotation (Fig. 7.1), with only a very limited number of genotypes showing signs of hyperhydration.

The simple suspension method described here permits laboratories to propagate, with a limited input of resources up to 80–100 accessions in a physical space of ~0.7 m^2. The cross-contamination risk is minimal (as explants are always contained within the same sealed container), and the use of uni-nodal explants results in a high multiplication rate. The described protocol was originally developed to speed up the propagation process of the initial phase of cryopreservation at CIP (years 2017–2018), and subsequently underwent additional improvements in the framework of a phytosanitary project (years 2019–2020).

7.2 General Considerations, Equipment, Materials, and Preparation of Culture Media

All culture media should be prepared with distilled deionized water and commercial, pre-prepared, full-strength Murashige & Skoog medium (MS). Analytical grade reagents and calibrated/validated equipment should be used during all protocol steps (e.g. pH-meter, autoclave, laminar flow chamber, electronic dispensing pump, etc.). Chemicals are stored at room temperature (22 ± 3 °C), under refrigeration (5 ± 3 °C) or in the freezer (−15 ± 5 °C), according to specifications in the Material Safety Data Sheet of each product. Reagents should be dissolved using a magnetic stirrer at 600–1000 rpm, depending on the volume of culture media that are being prepared. Label stock solutions with the following minimum information: type and concentration of solution, name of person who prepared the solution, known hazards (included in a NFPA diamond), and date of preparation.

7.2.1 Equipment and Materials

1. Equipment: laminar flow chamber; analytical balance; autoclave; pH-meter; magnetic stirrer; microwave oven; water distiller; orbital shaker; dispensing pump; heat sealing machine; vacuum pump (for filter sterilization system); refrigerator/freezer; incubation chamber/room.
2. Materials: aluminum foil; autoclave indicator tape; scalpel holder; forceps for tissue culture (23 cm, fine point); scalpel blades (#10); Sterifil® Aseptic System (Merck Millipore); prefilter (Type: AP15, Ø 47 mm); filter (Type: GSWP, pore size: 0.22 μm, Ø 47 mm); glass beaker (100 mL); glass measuring cylinder (100 mL); glass test tubes (25 × 150 mm) with plastic caps; weight spatulas; pipette pump or bulb (for 10 mL pipette); plastic beakers (1 and 2 L); plastic racks for 25 × 150 mm test tubes (7-way tray); plastic weighing dishes (small and large size); Pyrex glass bottles (autoclavable, 1 and 2 L); saran wrap; parafilm; sterile Erlenmeyer glass flasks (50 mL); screw cap plastic flasks (20 mL); sterile paper sheets (A5); sterile plastic pipettes (10 mL); sterile tool holder; sterilization sleeves (width: 30 cm); stir bars (1/8″ and 3/8″); wash bottle with alcohol (70%).

7.2.2 Media Preparation

7.2.2.1 Preparation of Solid Culture Medium (1 L)

1. Components: 4.33 g/L of MS salts; 5 mL/L vitamin stock solution (see **Note 1**); 1 mL/L thiamine stock solution (at 100 ppm) (see **Note 2**); 1 mL/L gibberellic acid stock solution (at 1000 ppm) (see **Note 3**); 30 g/L sucrose; 3.0 g/L phytagel.
2. Pour ~600 mL distilled-deionized water into a plastic beaker (1 or 2 L). Place beaker on a magnetic stirrer and let stir at medium speed.
3. Dissolve 4.33 g MS salts without vitamins.
4. Dissolve 30 g sucrose.

5. Add 5 mL previously thawed vitamin stock solution and 1 mL gibberellic acid stock solution.
6. Add 1 mL previously thawed thiamine stock solution.
7. Pour culture medium into a plastic measuring cylinder (1 L) and bring it up to a final volume of 1000 mL with distilled-deionized water. Pour culture medium back into the beaker.
8. Measure pH and adjust with HCl (1 M) or NaOH (1 M) as needed to a value of 5.70 ± 0.02 (see **Note 4**).
9. Place 25 × 150 mm test tubes in autoclavable plastic racks (36 tubes per rack; 3 racks per liter of culture medium [= 108 tubes]).
10. Add 3.0 g of phytagel to the culture medium.
11. Dissolve phytagel in microwave at 100% intensity for 10 min, pause after 7 min and mix the medium on a magnetic stirrer (see **Note 5**).
12. Using an electronic dispensing pump, dispense 9 mL of culture medium per 25 × 150 mm test tube. Cap all test tubes. Label one of the tests tubes with type of media, date, and initials of the person that prepared it. Place an autoclave indicator tape onto one of the tube caps of each rack.
13. Autoclave the culture medium for 20 min (at 121 °C, 15 psi). Let the culture medium cool down to room temperature (23 ± 3 °C).
14. Store culture medium at 5 ± 3 °C for a maximum period of 2 weeks.

7.2.2.2 Preparation of Liquid Culture Medium (1 L)

1. Components: 4.33 g/L MS salts; 5 mL/L vitamin stock solution (see **Note 1**); 1 mL/L thiamine stock solution (at 100 ppm) (see **Note 2**); 1 mL/L sterile gibberellic acid stock solution (at 1000 ppm) (see **Note 3**); 30 g/L sucrose.
2. Pour ~600 mL of distilled-deionized water into a plastic beaker (1 or 2 L). Place beaker on a magnetic stirrer and let stir at medium speed.
3. Dissolve 4.33 g MS salts without vitamins.
4. Dissolve 30 g sucrose.
5. Add 5 mL previously thawed vitamin stock solution.
6. Add 1 mL previously thawed thiamine stock solution.
7. Pour culture medium into a plastic measuring cylinder (1 L) and bring the solution up to a final volume of 1 L with distilled-deionized water. Pour culture medium back into the beaker.
8. Measure pH and adjust with HCl (1 M) or NaOH (1 M) as needed to a value of 5.70 ± 0.02 (see **Note 4**).
9. Pour liquid medium into a Pyrex glass bottle (1 L), close the bottle loosely to allow pressure to escape during the autoclaving process. Label bottle (type of medium, date and preparer's initials). Place autoclave indicator tape on the bottle cap.
10. Autoclave the culture medium for 20 min (at 121 °C, 15 psi). Let the culture medium cool down to a temperature of ~35–40 °C (in a laminar flow chamber, LFC).
11. Using a sterile disposable pipette (10 mL) (see **Note 6**), add 1 mL of filter-sterilized gibberellic acid (GA3) stock solution. Shake gently.

12. Place sterile 50 mL Erlenmeyer flasks in LFC (see **Note 7**). Carefully remove the aluminum foil cap from the Erlenmeyer flask top and place it onto a sterile paper sheet (see **Note 8**). Using a separate sterile disposable pipette (10 mL), dispense 22 mL of culture medium per Erlenmeyer flask.
13. Cover flasks with sterile aluminum foil cap. Take care that the internal surface of the cap does not come in contact with any nonsterile object. Seal with Parafilm between aluminum foil cap and flask.
14. Randomly select 2–3 flasks from each batch and incubate at 25 ± 2 °C. Assess the flasks after 4–5 days for fungal/bacterial contamination. Place the rest of the sealed flasks in plastic containers (~20 flasks/container) and store at 5 ± 3 °C for a maximum period of 2 weeks.

7.3 Methods

7.3.1 In Vitro Propagation in Liquid and Solid Culture Medium

1. Turn on LFC and a glass bead sterilizer. Clean workspace of LFC (see **Note 9**). Set up tool holder and tissue culture forceps (23 cm) (see **Note 10**). Attach scalpel blades to scalpel holders (see **Note 11**). Apply best practices when working in the LFC (see **Note 12**).
2. Place sterile Erlenmeyer flasks in LFC (see **Note 7**). Set up sterile paper sheets (A5) in LFC (Fig. 7.2) (see **Note 8**). Place test tubes containing the in vitro mother plants in a test tube rack on a lab cart next to the LFC.
3. Read barcode labels of test tubes and print out new labels for Erlenmeyer flasks (see **Note 13**).
4. Using forceps, place a small pile of 3–4 sterile A5 paper sheets in the central part of the LFC workspace. The pile serves as a cutting surface during propagation (see **Note 14**).
5. Remove saran wrap from test tube cap and open cap. Using plant tissue culture forceps (23 cm) remove plants from the tube and place them in the central section of the paper sheet pile (see **Note 15**).
6. Using sterilized scalpel with a No. 10 blade and tissue culture forceps, remove leaves and roots of the plantlets and cut stem in several segments (or explants) of approximately 10–15 mm, with one single bud per segment (Fig. 7.3). If the distance between nodes is very short (<3 mm), cut stem segments with two buds. The internodal distance varies depending on genotype and plant age.
7. Uncap an Erlenmeyer flask (see **Note 16**). Transfer up to 15 stem segments per Erlenmeyer flask (size: 50 mL) (see **Note 17**). Close flask with aluminum foil cap (Fig. 7.4). Seal and label (see **Notes 18** and **19**).
8. Incubate Erlenmeyer flasks on an orbital shaker at 80–90 rpm at 25–29 °C (equipped with clamps or a rack) [see **Note 20**], light intensity of 100 ± 20 μmol/m^2/s and photoperiod of 16 h light/8 h darkness (Fig. 7.5) [see **Note 21**].

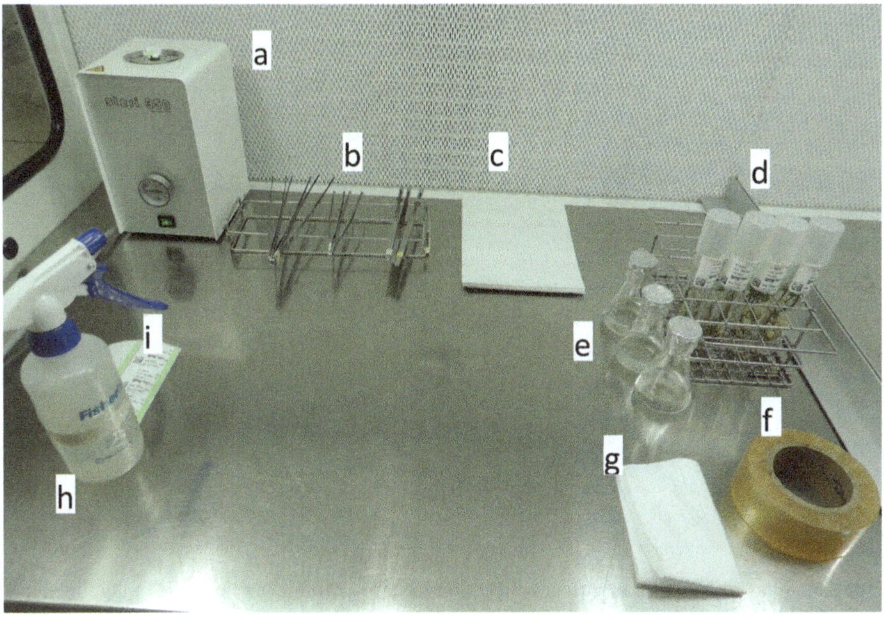

Fig. 7.2 Set up of supplies in Laminar Flow Chamber (LFC), required for in vitro propagation. (**a**) Glass bead sterilizer (at 250 °C). (**b**) Forceps for tissue culture (23 cm) and scalpel holders with blade No. 10, supported on sterile tool holder (stainless steel). It's recommended to work with various sets of forceps and scalpels, and to ensure that tools have cooled before use. (**c**) Pile of sterile paper sheets. For each propagation step locate a pile of 2–3 sheets in the central section of the LFC's working table and replace them by a new sheet after each accession. (**d**) In vitro sweetpotato plants used as starting material for propagation in liquid culture medium. (**e**) Sterile liquid culture medium (22 mL) contained in Erlenmeyer flasks (50 mL). (**f**) Roll of saran wrap used for sealing the joint between cap and vessel (Erlenmeyer flask or test tube). (**g**) Antibacterial wipe moistened with alcohol (70%). After each propagation cycle clean table surface with wipe. (**h**) Alcohol spray (disinfection of hands and working surface). (**i**) Barcode labels which are sticked on the vessels to identify the propagated accession

9. After 10–18 days in liquid medium (genotype-dependent), transfer rooted plantlets from Erlenmeyer flasks to glass test tubes (25 × 150 mm) with sterile solid culture medium (see **Note 22**). Use long tissue culture forceps for plant transfer. Place 2 plants per test tube (Fig 7.6) [see **Note 23**]. Explants that did not develop into complete and rooted in vitro plants are transferred to Erlenmeyer flasks with fresh liquid medium and incubated for 10–18 additional days. Occasionally, although rare, accessions may not grow well in liquid medium, in which case it is recommended to propagate them in deep petri dishes on solid culture medium, previous transfer to test tubes (Fig. 7.7) [see **Note 24**].
10. Close test tubes with sterile cap. Seal cap with saran wrap (2 layers) and label with barcode label (see **Note 19**).
11. Incubate test tubes for 4–5 weeks (genotype-dependent) at 24 ± 1 °C, light intensity of 100 ± 20 μmol/m^2/s and photoperiod of 16 h light/8 h darkness.

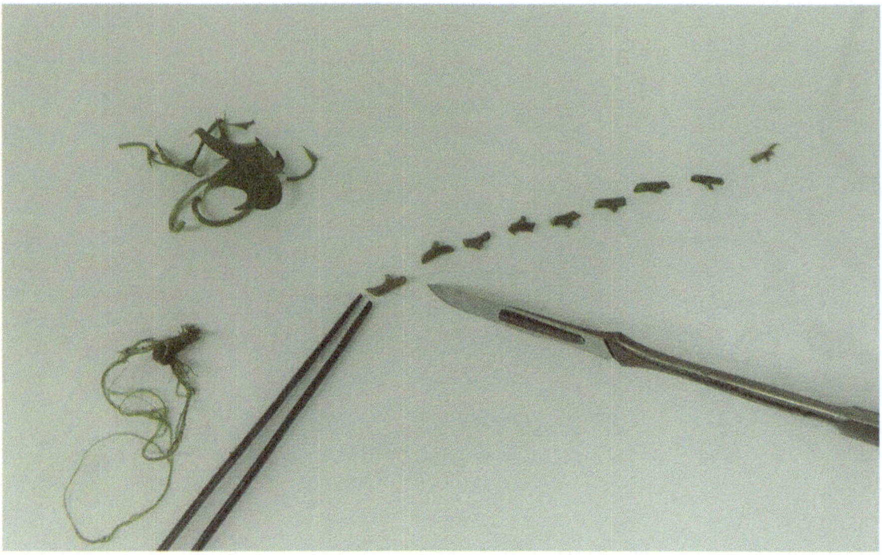

Fig. 7.3 Stem segments of in vitro sweetpotato plants prior to transfer to liquid culture medium. Leaves and roots were removed carefully, and the stem was cut in uni-nodal segments of ~0.5–1.5 cm (one bud per segment)

Fig. 7.4 (**a**) Transfer of uni-nodal stem segments to Erlenmeyer flask. Transfer 5–15 segments per flask (50 mL). (**b**) Erlenmeyer flask containing stem segments. The flask is closed with a sterile aluminum foil cap that was previously sterilized together with the empty vessel (the liquid culture medium was dispensed afterwards in the laminar flow chamber). The vessel is labeled with a bi-dimensional barcode and sealed with saran wrap

Fig. 7.5 (**a**) Vertical view of orbital shaker equipped with extended aluminum platform and special stainless-steel rack for Erlenmeyer support. The use of a rack (instead of clamps) permits to place more flasks per area. The shown model has a capacity of 81 Erlenmeyer flasks (50 mL). (**b**) Horizontal view of modified orbital shaker

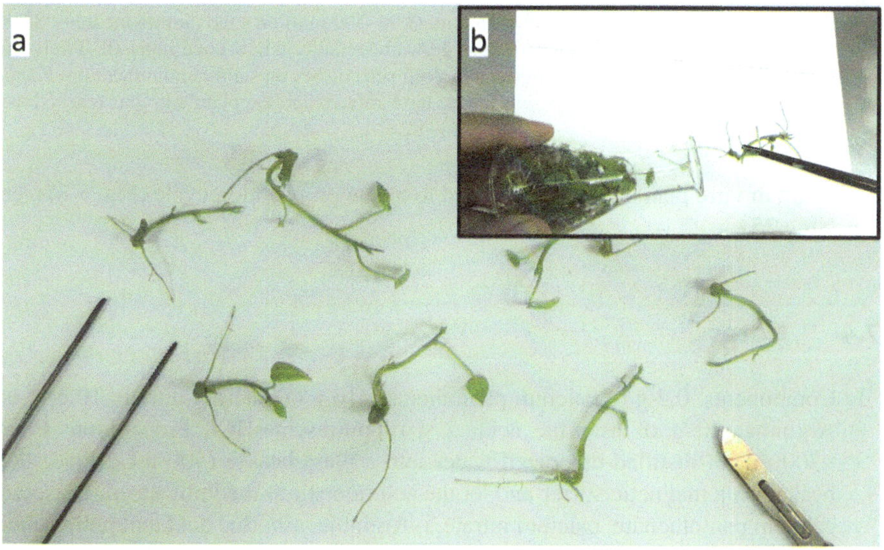

Fig. 7.6 (**a**) Aspect of in vitro sweetpotato plants after ~14 days in liquid culture medium, before its transfer to solid culture medium. In vitro plants were grown from uni-nodal stem segments. (**b**) Removal of in vitro plants from Erlenmeyer flask (liquid culture medium) before its transfer to solid medium. During transfer, ensure that roots, plants, and tools do not get in contact with unsterile surfaces (e.g. external surfaces of Erlenmeyer flask or test tubes)

Fig. 7.7 (**a**) An alternative propagation method for uni-nodal stem segments of sweeetpoato in vitro plants in deep petri dishes (depth: 25 mm, Ø 90–100 mm) on solid culture medium. Stick label on the upper part of the lid to ensure plants grow under uniform light conditions. (**b**) Twelve to 20-day-old in vitro sweetpotato plants grown in deep petri dishes on solid culture medium. Plants are ready for its transfer to test tubes, when they have rooted, and the plant's tip has reached the dish's lid

Adult in vitro plants can be submitted to a new propagation cycle (Fig. 7.8) (see **Note 25**).

7.4 Notes

1. Components: 0.2 g of calcium pantothenate; 10 g of calcium nitrate; 10 g of L-Arginine; 20 g of ascorbic acid; 2 g of putrescine-HCl. Preparation: Pour ~300 mL of distilled-deionized water into a glass beaker (500 mL). Place the beaker on a magnetic stirrer and let the solution stir at medium speed. Dissolve calcium pantothenate, calcium nitrate, L-Arginine, ascorbic acid and putrescine-HCl. Pour stock solution into a glass measuring cylinder (500 mL) and bring it to a final volume of 500 mL with distilled-deionized water. Pour the solution back into the beaker and thoroughly mix with a magnetic stirrer. Using a disposable pipette (10 mL), dispense 20 mL of vitamin stock solution per 20 mL plastic vial. Cap vials and label. Store vitamin stock solution at -20 ± 5 °C for a maximum period of 2 months.

Fig. 7.8 (**a**) Sweetpotato in vitro plants, ~1 week after its transfer from liquid to solid culture medium. (**b**) Aspect of sweetpotato in vitro plants after 12–18 days in liquid culture medium, followed by 3–4 weeks on solid medium. After this period, plants can be submitted to a new propagation cycle in liquid medium

2. Component: 10 mg of Thiamine-HCl. Preparation: Pour ~50 mL of distilled-deionized water into a glass beaker (100 mL). Place beaker on a magnetic stirrer and let stir at medium speed. Dissolve Thiamine-HCl. Pour stock solution into a glass measuring cylinder (100 mL) and bring it to a final volume of 100 mL with distilled-deionized water. Pour solution back into the beaker and homogenize it on a magnetic stirrer. Using a disposable pipette (10 mL), dispense 20 mL of thiamine stock solution per 20 mL plastic vial. Cap vials and label. Store Thiamine-HCl stock solution at -20 ± 5 °C for a maximum period of 2 months.
3. Component: 0.5 g of gibberellic acid (GA3). Preparation: Dissolve gibberellic acid in 2–3 drops of NaOH (1 M) on a small plastic weighing dish. Use a laboratory spoon for dissolving and mixing the solution. Empty dissolved GA3 into a 500 mL glass beaker. Rinse off weighing dish with distilled-deionized water over the glass beaker (all dissolved GA3 should end up in the beaker). Add ~200 mL of distilled-deionized water. Place beaker on magnetic stirrer and let the solution stir at medium speed. Pour solution into a glass measuring cylinder (500 mL) and bring it to a final volume of 500 mL with distilled-deionized water. Pour stock solution back into the beaker. Measure pH and adjust to 5.70 ± 0.02 (see **Note 4**). Filter sterilize GA3 through the Sterifil®

Aseptic System in the LFC. Using a sterile disposable pipette (10 mL), dispense 10–30 mL of GA3 stock solution per sterile Eppendorf flask (30 mL) in LFC. Close container with sterile cap and seal with parafilm. Store GA3 stock solution at 5 ± 3 °C for a maximum period of 2 weeks.

4. To increase and decrease the pH of any solution, add drop(s) of NaOH and HCl to a solution, respectively, slowly while stirring the solution and monitoring the change in pH. Solutions which have a final volume of less than 500 mL, it is recommended to use more diluted solutions of NaOH and HCl for pH adjustment (e.g. 0.5 M). Start the pH adjustment by adding three to five drops of HCl or NaOH at once, while stirring the solution. Wait 2–3 min for the pH to stabilize and then continue adding smaller volumes as needed. When the pH adjustment gets close to the required value (e.g. 5.70), dilute some drops of HCl (1 M) or NaOH (1 M) in distilled water and use this solution for the final pH adjustment. Patience is key aspect in changing the pH of a solution.
5. Use a microwave oven to dissolve the gelling agents for solid culture media that do not contain any thermolabile compounds. When the gelling agent has started to dissolve (before boiling), take the beaker out of the microwave and thoroughly mix the solution by stirring with a magnetic stirrer to avoid formation of agglomerates. Next, place it back in the microwave and heat until boiling and mix. The gelling agent is dissolved completely when the solution changes from cloudy (milky) to a transparent/clear solution. For instance, to prepare 1 L of culture media, heat the culture medium for 7–8 min in the microwave, homogenize the solution on a magnetic stirrer, and then continue heating for 3–4 min more. The heating time will be approximately doubled for a 2 L solution. When a dispensing pump is used for pouring the culture media, the beaker containing the media is placed on a magnetic stirrer (to ensure a homogeneous composition in all vessels).
6. Before opening the sterile sleeve of the disposable pipette, check that neither the wrapper nor pipette have damages or have been opened. Open the wrapper and remove the pipette aseptically and insert the top into a pipette-aid (pump or bulb). Fill the pipette a bit above the desired volume line and then slowly lower the meniscus to that line. Remove the pipette from the vessel, allowing the outside of the pipette to gently touch the inner lip of the vessel to remove any adherent liquid. Aseptically move the pipette to the receiving vessel and empty the pipette content. If you are emptying the entire content of the pipette you will have to "blow-out" the remaining liquid in the tip with a firm puff of air from the pipette-aid. If the tip or tip-close part of the pipette gets in touch with any nonsterile solution or object, change the pipette by a new sterile.
7. Set up of Erlenmeyer flasks: Fill a sink ¾ full of water. Add neutral detergent. Preferably, use a biodegradable, phosphate-free and nonionic detergent (e.g. Scienceware® Aquet®). Submerge 50 mL Erlenmeyer flasks in detergent solution and let them soak for about 15 min. Brush internal surfaces of the flask using a test tube brush (Ø 20 mm). Clean the external surfaces with a scour pad. Rinse flasks 3 times with deionized water to remove remaining detergent. Let flasks dry upside down on a plastic tray overnight (min. 18 h). Rinse flask

internally with alcohol (70%) and remove remaining label glue from the external surface (using alcohol [70%] and scour pad). Let flask dry for ~15 min until the alcohol has completely evaporated. Only use completely dry flasks. Using scissors or paper cutter, cut aluminum foil squares of ~6 × 6 cm. Place two centrally aligned squares onto the neck of each flask (one single layer could break). Pull down the four sides of the aluminum foil, forming a uniform flask cap. Firmly press aluminum foil along the opening of the flask and the neck. Place a maximum of 12 Erlenmeyer flasks per Kraft paper bag (#8). Seal bag with masking tape. Stick a piece of autoclave indicator tape on the bag. Dry heat-sterilize flasks for 20 min at 121 °C.

8. Cut A4 Bond paper in half using scissors or paper cutter. Pile up the paper sheets. Place a pile of ~1 cm of A5 paper sheets per sterilization sleeve. Seal sleeve with a heat-sealing machine. The sleeves are equipped with an internal autoclave control strip that changes color after successful autoclaving (no additional autoclave tape is required). Alternatively, a pile of A5 paper sheets (1–1.5 cm) can be placed in a kraft paper bag (#8). Seal bag with masking tape and stick a piece of autoclave indicator tape on the bag. Place 2 bags per autoclavable plastic bag and seal the plastic bag with masking tape. Autoclave sleeves or bags at 121 °C for 15 min. Dry sleeves/bags in a glassware dryer to ensure the paper sheets are completely dry. Before opening the sleeve in the LFC, check that the autoclave control strip has changed its color. Also verify that the sleeve has no signs of damage (cracks or holes). The sleeve is opened in the LFC by carefully pulling the two sides of the sleeve in opposite direction (at the sealed extreme of the sleeve). Kraft paper bags are opened in the LFC with sterile scalpel or scissors (at the extreme that was sealed with masking tape). Remove the pile of paper sheets using sterile forceps and place it in the upper left or right corner of your workspace in the LFC. The work surface of the LFC was previously disinfected with alcohol (70%). Do not to get in contact with nonsterile surfaces during the setup of the sterile paper sheets (i.e. opening of sleeve/bag, removal and colocation of paper sheets with sterile forceps). Cutting of stem segments for in vitro propagation is performed on a pile of 3–4 sterile paper sheets. Also, sterile items like aluminum foil caps of Erlenmeyer vessels are supported on a pile of 2–3 sterile papers to avoid bacterial or fungal contaminations.
9. After turning on the LFC check for stable air flow. Modern LFCs are equipped with a manometer to measure the velocity of the sterile air flow. A minimum air flow velocity of 0.35 m/s is required to ensure adequate sterile conditions within the chamber. Before sterilizing tools, check the glass bead sterilizer has reached its operative temperature of 250 °C. In the case no sterilizer is available, tools can also be flame sterilized using an alcohol burner. Disinfect working surface with alcohol (70%) using antibacterial wipes.
10. Tool holders are generally made of stainless steel or glass (= petri dish). Heat-sterilize tool holders in sterilization sleeves for 20 min at 121 °C. Alternatively, sterilize tool holders in sealed autoclavable bags. Before removing tool holder from the sleeve/bag, disinfect hands or gloves with alcohol (70%). Take care not

to touch those surfaces of the holder that will support the tools. Place tool holder in the upper right/left corner of the workspace, next to the glass bead sterilizer. Distance between holder and back wall should be ~10 cm. Clean forceps and scalpel holders using an antibacterial wipe moistened with alcohol (70%). It is recommended to work with a set of 4–5 forceps and 4–5 scalpel holders. Heat-sterilize forceps for 20–30 s. Insert forceps deeply into the sterilizer's glass beat vessel, as only those surfaces are adequately sterilized that get into contact with the glass beads. Place forceps on tool holder.

11. Open wrap of scalpel blade. Using forceps carefully remove blade from wrap. "*Grip blade with forceps, or similar, avoiding contact with cutting edge. Hold handle in left hand with tool's bayonet fitting uppermost. Place blade partway over handle fitting and engage slots. Slide blade until it clicks into position. To improve assembly, flex blade slightly upwards when sliding onto the handle. For removal or change of blades, grip blade with forceps making sure that the cutting edge is facing away from hand and body. Ensure the blade is pointing downwards and towards the trolley and NOT towards you or another member of your team. Whilst holding the handle firmly lift the back edge of the blade with the forceps or needle holders and slide away the handle*" (Swann-Morton 2021).
12. Maintenance of sterile conditions within the LFC requires some basic rules and guidelines for best practice: (a) Minimize clutter, (b) Wash your hands and arms properly and thoroughly before working in the LFC, (c) Arrange objects in such a way that the work area is directly bathed with clean air from the HEPA filter, (d) Avoid spraying or squirting solutions onto the HEPA filter, (e) Remove outer pouches and wraps on the edge of the work area, (f) Keep large objects away from the back of the hood to not disrupt the laminar airflow, (g) Avoid coughing, quick movements, and talking, (h) Remove jewelry around your hands and wrists when working in the LFC, (i) Place only sterile materials inside the clean work area, and (j) Propagate plants in the central part of the work space, and not close to the outer edge of the LFC.
13. Use of barcode labels for reading and printing reduces the risk of human mistakes and mix-up of accessions. Barcode labels should contain basic information identifying the accession and the process stage, such as accession identifier, phytosanitary status, propagation date, lot of culture medium and name (or initials) of the person who has propagated the plants. Bidimensional or Quick Response (QR) barcodes can contain larger quantity of alphanumeric data than unidimensional barcodes.
14. The pile should be thick enough to avoid scalpels coming in contact with the LFC's workspace surface during cutting. If any nonsterile object gets in contact with the upper surface of the paper pile, replace it by a new pile of three to four sterile sheets (A5). Replace the paper pile by a new one after each processed accession.
15. Cut sealing saran wrap vertically, using a discarded scalpel blade. Cut smoothly and away from the fingers holding the tube. Opening of test tube and removal of plants is best performed inclining the test tube to a nearly horizontal position, which avoids potential contaminants (bacteria, fungi, remnants of sealing tape)

falling into the test tube. Additionally, it is easier to remove plants from the tube in a close to horizontal position. Firmly grip the plants at the basal zone of the stem during removal and carefully remove them from the tube, together or without the block of solid culture medium. Take care that long roots or culture medium do not get into contact with external surfaces of the test tube.

16. Loosen the aluminum paper around the neck of the flask without touching the internal part of the aluminum foil nor the external surface of the flask which was covered by the foil. Grip the base of the foil cap, carefully remove the cap from the flask and place it upside up on a pile of 2–3 sterile paper sheets.
17. Stem segments coming from different test tubes should be transferred into separate flasks and well labeled. When required during handling, hold the Erlenmeyer flask in its base, but never touch the sterile neck zone of the flask. During transfer, grip stem segments smoothly with forceps without damaging the bud. Segments can be transferred in groups, gripping 2–5 segments at once.
18. Grip aluminum foil cap in its base during closing. After placing the cap on the flask, grasp around the neck-covering section of the foil and pull it smoothly downward until the aluminum foil has semihermetically closed with the mouth section of the flask. Press the aluminum foil around the neck-zone of the flask to ensure the cap is tightly closed. Using saran wrap or parafilm seal junction between cap and flask, as well as the flask's neck and mouth section. Wrap bottom up, and then bottom down (two layers).
19. When labeling test tubes or Erlenmeyer flask, first stick the label onto a piece of saran wrap and then label the container by wrapping the plastic wrap around the neck of the Erlenmeyer flask or test tube. This prevents the label from sticking directly on the container and its subsequent removal and washing is simpler. Place the label from the original container onto the new Erlenmeyer flask or test tube to which the material is transferred to, as a control label, along with a new printed label. This helps to track correct labeling in each multiplication cycle.
20. Check maximum capacity (weight) of orbital shaker and avoid overloading. Distribute flasks equidistantly on the shaker platform to ensure smooth rotation without misbalancing the shaker. Periodically, e.g. every 6 months, perform preventive maintenance of the shaker, which should include replacement and/or adjustment of motor belt, assessment of motor performance, check soldering points and components of mainboard and lubrication of axis when required. At CIP, we have manufactured and installed a larger and lighter aluminum platform to the orbital shaker (60 × 60 cm), which permits placing up to 80–100 Erlenmeyer flasks (50 mL) per shaker. High incubation temperatures of 28–29 °C speed up growth and development of in vitro plants in liquid culture medium.
21. Temperature, light intensity, photoperiod, and relative humidity can be controlled and recorded using an environmental measuring/control system or device. An easy, efficient, and cheap way to measure environmental variables is using mobile data loggers with external temperature, humidity, and light intensity probes. The interval for recording the data points can be adjusted individually and downloaded remotely (wireless) or through direct cable

connection with a desktop PC. Advanced systems can send automatic alarms by email or text message, when the predefined maximum and minimum values of the environmental conditions are overpassed.

22. When in vitro plants have developed minimum one root and three to four new nodes (or leaves), they can be transferred to solid culture medium. Stem should show robusticity and not be excessively elongated (avoid long distances between nodes). Transfer plants as soon as possible to solid medium (genotype-dependent).
23. Before removing the aluminum foil cap from the Erlenmeyer flask, first remove the sealing, and then carefully loosen the basal and central part of the cap to facilitate easy opening. Grab the foil cap at its basal part during opening, and don't touch the neck or opening of the flask. During transfer, hold roots and stem together and avoid loose roots and leaves that could come into contact with external vessel surfaces. Introduce basal section of the stem ~1.5 cm into solid culture medium (roots should completely be submerged in the medium). Plants should be located equidistantly in the tube, i.e. the distance between the two plants and the test tube wall should be the same.
24. Before transferring explants to fresh liquid medium, perform 2–3 small cuts in the base of the explant to stimulate root formation. Sweetpotato accessions that do not grow well in liquid medium can be successfully propagated in deep petri dishes (Ø: 90–100 mm, height: 25 mm) on solid culture medium (50 mL/petri dish). Place 10–36 uni-nodal explants per deep petri dish and incubate for 10–18 days (genotype-dependent) at 24 ± 1 °C, photoperiod of 16 h/8 h of light/darkness and a light intensity of 100 ± 20 $\mu mol/m^2/s$. After 14–20 days, transfer rooted plants to 25 × 150 mm glass test tubes (2 plants/tube) and incubate at the same environmental conditions.

    Ensure in vitro plants submitted to a new propagation cycle in liquid culture medium are fully rooted and have developed a robust stem and minimum 5–6 nodes or leaves. Do not use in vitro plants with anormal aspect (yellowing, bad growth, deformation, hyperhydration, etc.) for a new propagation cycle.

Acknowledgement The authors gratefully express thanks to the Global Crop Diversity Trust (GCDT) and the CGIAR Genebank Platform for funding our work and the publication of this book chapter.

References

Aitken-Christie J, Davies HE (1988) Development of a semiautomated micropropagation system. Acta Hortic 230:81–87

Alvard D, Cote F, Teisson C (1993) Comparison of methods of liquid medium culture for banana micropropagation. Effects of temporary immersion of explants. Plant Cell Tissue Organ Cult 32: 55–60

Haberlandt G (1902) Culturversuche mit isolierten Pflanzenzellen. Sitz-Ber. Mat Nat Kl Kais Akad Wiss Wien 111:69–92

Harris RE, Mason EB (1983) Two machines for in vitro propagation of plants in liquid media. Can J Plant Sci 63:311–316

Kotte W (1922) Kulturversuch mit isolierten Wurzelspitzen. Beitr Allg Bot 2:413–434

Murashige T, Skoog F (1962) A revised medium for rapid growth and bioassays with tobacco tissue cultures. Physiol Plant 15:473–497

Robbins WJ (1922) Cultivation of excised root tips and stem tips under sterile conditions. Bot Gaz 73:376–390

Simonton W, RobacKer C, Krueger S (1991) A programmable micropropagation apparatus using cycled medium. Plant Cell Tissue Organ Cult 27:211–218

Skoog F, Miller CO (1957) Chemical regulation of growth and organ formation in plant tissues cultured *in vitro*. Symp Soc Exp Biol 11:118–131

Swann-Morton (2021) Safe handling of Swann-Morton products. https://www.swann-morton.com/view_reading.php?reading_id=1

Tisserat B, Vandercook CE (1985) Development of an automated plant culture system. Plant Cell Tissue Organ Cult 5:107–117

Went FW (1926) On growth-accelerating substances in the coleoptile of *Avena sativa*. Proc Kon Ned Akad Wet 30:10–19

White PR (1933) Liquid media as substrata for the culturing of isolated root tips. Biol Zentralbl 53:359–364

White PR (1934) Potentially unlimited growth of excised tomato root tips in a liquid medium. Plant Physiol 9:585–600

White PR (1939) Controlled differentiation in a plant tissue culture. Bull Torrey Bot Club 66:507–513

Emerging Technologies in Plant Breeding for Fibre Crops, Cotton, and Sunn Hemp

8

Anu, Suresh Yadav, Vivek K. Singh, Pratik Istari Bhoyar, Vijay Sharma, Rizwana Rehsawla, and Rahul Kumar

Abstract

The fibre crops play a key role in the economy of a country. The major fibre crops include Cotton, Jute, Sunn hemp, etc. The major breeding approaches in fibre crops are to breed the cultivars with increased yield in the form of fibre quantity with improved quality. Apart from the fibre yield and quality, the composition of fibre, length of vegetative cycle, and resistance to various abiotic and biotic stresses are also considered as important. Enhancing our ability to broadly interview nucleic-acid-dependent knowledge in the cell is critical for genomics-assisted breeding. In this sense, molecular markers have proved to be extremely useful genetic instruments for generating genetic, physical, and trait maps in a variety of fibre crop organisms. With the advancement in technology, mutation identification has become possible by making it easier to scan vast numbers of plants for unusual, caused, or normal genetic variation in particular target genes proven to be involved in important traits. In this chapter, we have compiled a wealth of knowledge in order to emphasize the importance of numerous plant breeding techniques, such as conventional plant breeding tools, marker-assisted selection, genetic engineering, and next-generation sequencing technologies, in order to breed or develop fibre cultivars in a more suitable manner in light of

Anu · V. K. Singh · R. Rehsawla
CCS Haryana Agricultural University, Hisar, Haryana, India

S. Yadav (✉) · R. Kumar
Division of Genetics, Indian Agricultural Research Institute, New Delhi, India

P. I. Bhoyar
Indian Agricultural Research Institute, Regional Station, Indore, Madhya Pradesh, India

V. Sharma
Banda University of Agriculture and Technology, Banda, Uttar Pradesh, India

Kamaluddin et al. (eds.), *Technologies in Plant Biotechnology and Breeding of Field Crops*, https://doi.org/10.1007/978-981-16-5767-2_8

current concerns. With the advancement in technology, mutation identification has become possible by making it easier to scan vast numbers of plants for unusual, caused, or normal genetic variation in particular target genes proven to be involved in important traits. Targeted mutations disable or change the functioning of chromosomes, and chosen mutant genotypes may be used as breeding stock.

Keywords

Fibre crops · Cotton · Sunn hemp · Conventional breeding · Marker-assisted selection · Next-generation sequencing

8.1 Introduction

In the tribe of Gossypieae, Gossypium is a genus of flowering plants. It comes from the coastal areas of Old and New Worlds and subtropical regions. Cotton is cultivated as "cotton seed," and is then "ginned" into which the seed is separated from "lint." By spinning, long "lint" fibres are recycled to make wool knit or tissue into fabrics. While various varieties of fabric (source synthetic and natural) are used all over the world, cotton is still the most popular material for natural wear because it can be made into various types and weights, such as batiste, jersey, flannel, terry, corduroy, twill, denim, and canard. It can also be used for most garment styles, from underwear and sweaters to linens for homes, such as sheeting and curtains, due to its flexible nature. Cotton is very nice to wear because of the easy and convenient manner in which to walk. It's particularly in hot weather. They are the perfect option for people with asthma or allergies or allergic skin vulnerable to scratching because of the hypoallergenic and dust-miss resistance they have. In addition, it improves the circulation of air which helps extract body moisture and absorb it, draws heat from the skin, and keeps the body cool and dry while breathing better. Cotton is the best fabric for washing and dressing, and has always been used by human beings. For example, cotton history was traced to the fourth millennium BC in the Indian subcontinent (Santhanam and Sundaran 1997). In the rig Veda hymn is the first cotton reference. Stoffs dating back about 3000 BC, recovered from Mohenjodaro excavation in the Sind, were found originating from cotton, which was closely connected to *Gossyppium arboreum* (Gulati and Turner 1929), verifying therewith that even before 3000 BC cotton was spun into cloth and woven into cloth. Cotton fibre is usually rated on three points: colour or grade of whiteness, the quantity in cotton of external or plant content, and fibre gin preparation. Another important thing is that it is staple length. Egyptian cotton is renowned for its long staple and used in finer fabric styles. The yarn also contains extra-long staple cotton. Cotton is easy to paint and mix with other fibres when it comes to processing. In addition, cotton is the only fiver to be reinforced in wet conditions and is the most valuable option in hospitals as it can be safely sterilized at high temperatures. It is also the favourite choice for unit firefighter's uniforms. Since flame-retardant surfaces can be quickly coated. Cotton clothing makes a lot of sense in terms of environmental issues. Biodegradable and renewable, cotton is less than 10% wasted or considered unusable during cultivation. In addition, it is cheap, long lasting, and convenient to treat. Thus, cotton fibre production and use have effects in several ways: physical, social, and environmental benefits.

8.1.1 Cotton

8.1.1.1 History and Origin

In the Bible as in ancient and other Mediterranean literature, the plant is mentioned as nourishment, but its use as textile fibres is unclear. Kundu (1964) discovered that the main centre of origin of *C. olitorius* is Africa and secondly in India. However, in Africa and Australia, *C. capsularis* is not found; Indo-Burma is supposed to be its core. It is part of the *Malvaceae* family and is grown in tropical and subtropical condition. The crop is cross-pollinated by bees and cross-pollination varies from minimum 5–30% to an excess of 50%. The genus Gossypium contains a total of 50 species in which 45 are diploid with chromosomes. *Gossypium hirusutum* is the representative of cotton with 90% of total cotton production and 8% by Gossypium B. Species diploid was classified as genome (A, B, C, D, E, F, and G). In Africa or Asia, A, B, E, and F are found. In Australia C or G, D genome in America is found. The distribution in Mexico is wild. *G. hirusutum* is tetraploid cotton which is now grown in India. Crossing A and then Amphidiploid of two Diploid Species resulted in tetraploid cotton. G. raimondii is also known as a progenitor of tetraploid cotton and gene D (D5). There are five species of *G. herbacea: africanum, kuljianum, and wighitianum*. Likewise, *G. arboreum, Burmanecum, Sinense, Sudense,* and *Cernum* are the five breeds (Fig. 8.1) (Wendel et al. 2009).

8.1.2 Geographical Distribution

This species is native to Africa, Central and South America, Asia, Australia, Galapagos, and Hawaii. The term "cotton" means the four plants in the genus Gossypium (family *Malvaceae*) that have been domesticated as cloth fibre suppliers independent of each other, namely *Gossypium hirsutum L., G. barbadense L., G. arboreum L.,* and *G. herbaceum L.* The Gossypium genus includes about 50 species worldwide (Fryxell 1992). In fact, the genus *Gossypium* is not actually known; however, Western Central and Southern Mexico (18 species), Northeastern Africa, and Arabia are the main centres of diversity (14 species) and Australia (17 species). The current *Gossypium spp.* evidence on DNA sequences indicates that the species originated 10–20 million years ago (Seelanan et al. 1997). The cotton plants flourish in the warm climates of Africa and Australia, South America, the Indian Subcontinent, Arabia, the Galapagos, the Galaxies, and Hawaii (Fryxell 1992). The average worldwide production of cotton is between 33 million and 35 million hectares a year (approx. 2.5% of the world's arable land) (Townsend

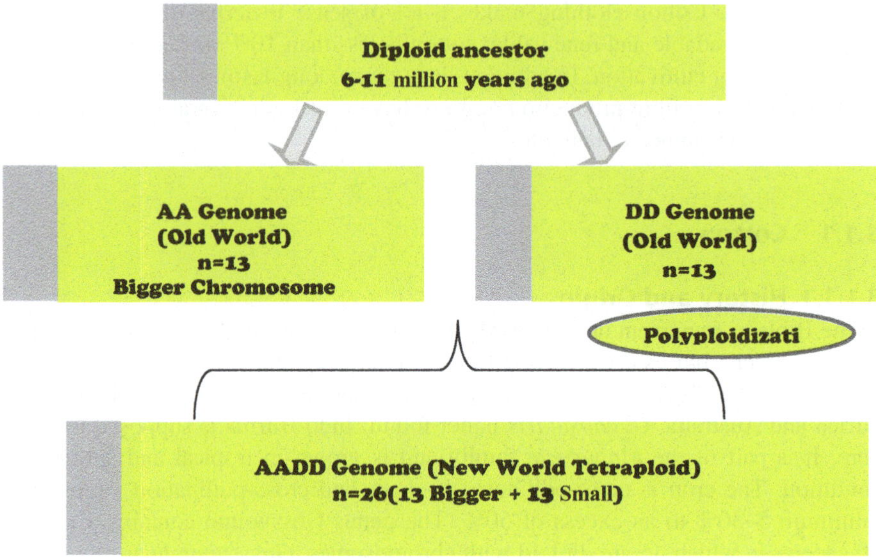

Fig. 8.1 Evolutionary framework of *Gossypium allotetraploids*

2010). It is said to be grown as a significant cash crop in more than 100 countries and hence to have a major effect on their economies. About 120 million family units are actively involved in worldwide cotton processing and about 350 million workers in the cotton industry are scheduled annually. The leading cotton growers are the countries like (China, India, US, Pakistan, Uzbekstan, Tajikistan, Mexico, Brasilia, Turkey, Egypt, Sudan, Australia, others Afro-States, and Israel) with the highest to the lowest overall output of cotton lint (Gruère and Sengupta 2011). China, India, and Pakistan remain the major raw cotton consumer: about 65% of these countries' global raw cotton consumption. With 34% worldwide and 21% worldwide production, India continued to retain the largest cotton field and the second-biggest cotton producer alongside China. In India, due to irratic weather conditions cotton growing areas were reduced by 0.5 to 0.6 million hectares during 2012–2013. More than 90% of the field consists of transgenic Bt cotton, which is interesting.

8.1.3 Taxonomy

Kingdom	Division	Class	Order	Family	Genus
Plantae	Magnoliophyta	Magnoliopsida	Malvales	Malvaceae	*Gossypium*

8.1.4 Emasculation and Pollination in Cotton

Select those buds in which anthesis is to occur the next day. Usually, white buds are selected for emasculation. Remove epicalyx, corolla, and staminal column with an emasculating instrument. Great care should be taken that there is no injury to stigma. The application of 100 ppm gibberellin acid solution at the base of anther at the time of the emasculation improves the seed set. To protect the stigma of the emasculated bud from contamination of foreign pollen, a soda straw of optimum length or a paper bag may be used. Proper tagging is also done. Emasculation and pollination should be completed simultaneously in the morning between 8.30 and 10.30 a.m. The extent of cross-pollination in cotton ranges from 5–30%.

In some cases, cross-pollination up to 50% has been reported. That is why cotton is considered as an often-cross-pollinated crop. The out crossing is mainly carried by insects particularly bees. The flowers which are expected to open next morning are bagged with butter bags to prevent cross-pollination. The pollen grains are collected from the ripe anthers of the flowers of the male parent with small sections of the soda straw. The anthers are slipped down or the stigma from the soda straw. The bracts are pulled up around the soda straw and are tied firmly to hold it. It is desirable to do both emasculation and pollination on the same day to avoid the drying up of the stigma.

8.1.5 Plant Characteristics

8.1.5.1 Characteristics of Cotton

- Size Short (90–120 cm).
- Pyramid-shaped compact and pleasant plant habit.
- Determination of the habit of fruiting with unimodal bolling delivery.
- Short length (150–165 days).
- Respond to a high dose of fertilizer.
- Competitiveness of high levels of interface.
- Strong standard of insect pest and disease tolerance.
- High performance in physiology.

8.1.5.2 Characteristics of Cotton Fibres and Products

- Fibre strength, yield, and performance.
- Strong strength and length of spinning fibre.
- Brilliant fibre.
- Power and elasticity.
- Nice to wear.
- Cellulosic natural fibre.
- Crafted of bollard cotton.
- Water absorption and "breath".
- Little to dry.
- Resists the build-up of stagnant power.
- Easily shrink.

- Can resist fire, bleach, and detergents.
- 20% better than dry when wet
- May suffer from mildew.
- Long exposure to sunlight affects fibre.
- Smooth, almost silky fabrics could be made of long staples (such as the Supima, Pima, Egyptians, and Sea Islands).

8.1.6 Cotton Breeding History

As a crop, cotton does not have to be introduced. Fibre is the most widely used natural fibre in the garment industry today for the production of a soft breathable textile. The Arabic word "Qutn" comes from the English name, which began to use about 1400 AD. Cotton has a long history; in the fifth and fourth millennium BC, it was cultivated in the Indus Valley Civilization. Cotton textiles were twined in India hundreds of years before the Christian era, and they have been used in Mediterranean countries. It's been grown from Peru to Mexico to Persia worldwide as natural fibre. Cotton was known as an imported fibre in northern Europe during the late mediaeval period. With no knowledge of what it was like, the people of the area can only believe that cotton should be manufactured from plant-born sheep with its similarities to wool. In many European languages, such as German cotton, this aspect is retained in the name of cotton, which translates as "tree wool" (Baum means "tree"; Wolle signifies wool). Cotton was cultivated in the warmer areas as in Asia and the United States by the end of the sixteenth century. A decisive moment in Indian independence history has sparked Cotton. The British Indian policy of exporting low-cost raw cotton and importing high-cost lines has enraged and damaged the economy. India has become a major cotton producer in the world today. After the US and China, it ranks third in global cotton production. India accounts for about 25% of the total cotton area in the world and 16% of world cotton production. In India, most cotton is cultivated under rain and about a third is irrigated. In India however, cotton yields are small and the global average yield is 300 kg/ha, compared to 580 kg/ha. Bollworms are the most devastating and require great effort to save the crop. Insecticides are used every year, with approximately half used only on cotton on every crop in India. *Armigera helicoverpa, Earias vittella*, *Gossypiella* pink bollworm, *Spodoptera litura*, which is mainly a leaf feeder, damages cotton balls as well. The problems of direct injury to plant and virus transmission are also sucking pests like aphids Aphis gossyii, jassids, *Amrasca bigutulla* and White flies, and *Bemisia tabaci*. *Helicoverpa armigera* are known as a resistant to most recommended insecticides (Ramasubramanian and Regupathy 2004). The problem of solving the problem and creating interest for farmers in the cultivation of bollworms leads to the development of GM cotton.

8.1.6.1 Cotton Conventional Breeding

Conventional breeding is the basis for improving traits and modern molecular breeding techniques can certainly result in expensive errors without a proper understanding of conventional breeding techniques. Conventional breeding was the only form of genetic improvement before the introduction of molecular marker technology. The invention of certain forms of growth was some of the common traditional cotton interventions. Early maturing varieties are developed (immune to boll weevils), morphologically displayed features such as frego bracts, bell leaves, nectarines, high gossypol, and bollworm resistant features. In Australian conventional breeding, the main issues are American bollworms, bacterial plague, and verticillium wilt (Constable et al. 2001). C T Patel in the 1970s introduced the breeding of standard *G. hirsutum* and production of the first intra-*hirsutum* hybrid cotton (H4) in India. B.H. Katarki in 1972 develop first interspecific hybrid by crossing (*G. hirsutum* x *G. barbadense*) in cotton (Varalaxmi) that led for higher productivity in the Indian subcontinent. Novel varieties like Badnawar-1, Khandwa-1, and Khandwa-2 from *G. hirsutum* x *G. tomentosum* cross, Arogya and PKV081 using *G. hirsutum* x *G. anomalum* cross, Devitej using *G. hirsutum* x *G. herbaceum* cross, SRT-1, Deviraj, and Gujarat 67 using *G. hirsutum* x *G. arboreum* cross were used through introgression breeding. Varieties like MCU2 and MCU5 are developed from *G. hirsutum* x *G. barbadense* cross. The production of early maturing short-staple varieties was one of the key strategies in Bollweevils as a major challenge to cotton cultivation and elimination. Some private enterprises, such as Delta Pine Plant Breeding – Current and Future Views 12, have made huge progress in improving cotton productivity. They also introduced a suitable mechanical processing variety known as Delta pine smooth leaves, which had approximately 25% US cotton by 1963, and a stronger resistance to Deltapine-16, which had approximately 28% US fibre content by 1972, and Delta pine Acala 90 premium-quality cotton, released as parental sprouts for several other varieties around the world. Their diligent analysis, which was developed early on, is invaluable and extremely important, had a big influence on cotton productivity changes and the preservation of germplasm.

8.1.6.2 Molecular Breeding

New technological developments are a major benefit for new young breeders so they can use new technologies, but it is important to keep an eye on their shortcomings until they have a full understanding of traditional breeding, its limitations, and advantages of advanced molecular breeding. An initial small error in the early molecular reproduction could ultimately make us pay a high price. Data such as quantitative characteristics loci (QTLs), Transgenic events, and the already evolved gene-edited lines can be carefully used in crop enhancement programs as well as in advanced technological analysis, such as genomic selection, fine mapping, and genetic editing, must be the focus field of sustainable study. Plant genomic science has created new molecular methods to enhance breeding performance and precision for agricultural scientists. Extremely promising for many attempts to boost crops, for example, is the use of DNA markers for the selection of marker (Tanksley et al. 1989). In potential cotton genetic reproduction, molecular markers have important significance. It is a relatively straightforward means of tracking genetic heritage. Special chromosome areas of major QTLs may be identified and used for effective selection techniques. Genetic uniformity among cotton cultivars that do not make a substantial genetic improvement to yield-related traits, caused by biotic and abiotic stresses, are the major concerns of the decrease in cotton productivity. Introducing and using new molecular technology for increasing the genetic advantage of economic characteristics will accomplish this aim. The influential genetic markers for aided molecular selection are DNA markers. SSRs have most of the attractive features relatively speaking, making them the most common marker for many crops. The use of SNP markers in MAS programming has increased faster, which makes it necessary for many plants to build the technology and platforms to discover SNP. Sequence-driven genotyping is an essential move towards the future with a whole spectrum of diversity and genomic research.

8.1.7 DNA Markers in Cotton

DNA-based markers are non-phenotypical, free of epistatic effects, and are clearly Mendelian. For features influenced by several QTLs, DNA markers are especially useful when introducing important genes from exotic germplasm and breeding (Edwards et al. 1987; Paterson et al. 1988). Established and currently used effectively in cotton breeding are new technologies such as high-performance marker systems and marker-based selection methods. As determined by fibre bundles, cotton fibre content has an additive inheritance that has enabled a steadfast genetic advancement in cotton (Kohel 1999a, b). However, it was difficult to identify the genetic factors responsible for this change, which limited breeding effort effectiveness.

The improved consistency of fibres can be caused by either the introduction of new genes or by new variations of existing genes (Kohel 1999b). Genetic knowledge and methods of breeding available to cotton breeders have not only allowed them to precisely satisfy textile industry needs for desired fibre-property combinations, while retaining high returns. Emerging genomic innovations have changed the genetics of plants and breeding (Tanksley et al. 1989). DNA cotton markers offer unparalleled opportunities to improve the leading fibre crop. These molecular markers may be used to analyze genes that regulate both basic and complex cotton characteristics. A collection of DNA markers are identified with fibre consistency QTLs in cotton 3–79 in ELS with the building of a molecular genetic map based on the cross between two better cottons *G. hirsutum* acc. TM-1 and *G. barbadense* acc. 3–79. There are currently approximately 145 morphologic markers in cotton, but they are of low value in variant production owing to the inability to bring together various markers in a genotype, some of them are listed in Table 8.1. Isozymes produced by allelic variants are called authentic, but not commonly utilized in different stages of growth because of their differential expression. Cotton genetic markers are more important than morphological or isozyme markers in order to improve efficiency and other primary quantitative characteristics (Percy and Kohel 1999). A major breakthrough has been made in "omics," which has enabled the use of these signals in various ways for genetic research, not only for phylogenetic studies, in order for molecular markers to maximize their benefits (Joshi et al. 2011). The abstention of pure DNA is of great significance for the discovery of cotton molecular markers, and genetic

Table 8.1 A summary of cotton QTL study and use of molecular markers for genetic improvement of cotton

Traits	Descriptor	Population	Size	Marker (number and type)	QTLS number	Reference
Fibre quality	FS, FL, and FF	F_2	171	RFLPs and 85 RAPDs	13	Kohel et al. 2001
	LY, LP, SW, NS, UQ, SF, FL, FE, FT, FF, and IF	F_2	120	144 AFLPs, RFLPs, and 150 SSRs	28	Mei et al. 2004
	FS, FE, FL, FU, LP, and FF	F_2	117	290 SSRs and 9 AFLPs	16	Zhang et al. 2005
	FF	BC_3F_2	3662	262 RFLPs	41	Draye et al. 2005
	FL, FLU, and SFC	BC_3F_2	3662	262 RFLPs	45	Chee et al. 2005
	FS, FE, FU, FL, and FF	RIL's	270	7508 SSRs, 384 SRAPs, and 740 IT-ISJs	13	Zhang et al. 2009
	FL, FS, FF, and FE	F_2	–	1378 SSRs	39	Shen et al. 2005
	FS, FL, FF, FMT, FE, and SFI	RIL's	180	4106 SSRs, AFLPs, RAPDs, and SRAPs	48	Wang et al. 2006a, b
	FE, FL, FS, FF, and FU	CP	172	16,052 SSRs	63	Zhang et al. 2012
Morphological	LBNO, SL1, L1, W1, L2, W2, L3 and W3	F_2	180	261 RFLPs	62	Jiang et al. 2000
	EM	F_2 and F_3	–	4083 SSRs	54	Li et al. 2013
	NFB	F_2	251	1165 SSRs	5	Guo et al. 2008
Plant architectural	PH, FBL, FBN, FBA, FBL/PH, and NMUB	RIL's	180	2130 SSRs, 2 RAPDs, and 1 SRAP	16	Wang et al. 2006a, b
Yield and fibre	NB, BW, SI, LP, LI, SCY, LY, FL, FS, FF, FE, and FU	4WC and inbred lines	280	6123 SSRs and EST-SSRs	31	Qin et al. 2008

research is adversely affecting the consistency of DNA and protein in tissue grinding because of phenolic compounds (Aslam et al. 2013).

8.1.8 Linkage Maps

Due to their close connection with essential agronomic characters, genetic linkage maps are useful in introgression, genome structure analysis, and MAS in plant improvement studies (Bolek 2003). A genetic linkage map is often used to display genetic information from a crop genome.

Quantitative trait loci, or QTLs, are regions in genomes that have genes linked to a quantitative trait (Collard et al. 2005), and QTL mapping is used to create linkage maps and perform QTL research. Crossing over principals was used to trace QTLs, allowing for the study of genes and markers in the progeny (Paterson et al. 1988). In nature, these traits are often inherited as oligogenic traits. Although for certain quality traits, a limited number of significant QTLs or genes may account for a large proportion of the trait's phenotypic variance.

Cotton molecular aided selection technique has been extensively explored in genomics (Zhang et al. 2008; Paterson et al. 1988) to achieve effective and rapid cotton improvement at the global level with high seed cotton yield and improved fibre consistency. The donor parent can express markers that are tagged to the target gene, and a high saturated map can be generated using markers that are polymorphic between close isogenic lines. By assuming hyponeuploid-based backcross, Stelly et al. 2005 created alien chromosome replacement lines in a close isogenic genetic history of TM-1. Furthermore, scientists have looked into the impact of chromosomes on lint yield and fibre consistency traits by using CS-B lines (Saha et al. 2006).

Reinisch et al. created the first cotton genetic map in Reinisch et al. 1994. Despite the fact that a vast number of linkage maps have been created since then due to the abundance of many DNA markers, there is still a need to identify accurate QTLs from a breeding standpoint. Yu et al. 2013 used simple sequence repeats (SSRs) to scan genotypes in a backcross inbred line for loci linked to fibre consistency and lint yield, and established a pioneer genetic map using BIL inside allotetraploid cultivated cotton organisms.

Latest developments in next-generation sequencing technology also provided low-cost tools for direct identification of high-quality single nucleotide polymorphism (SNP) markers for genotyping mapping populations (Schuster 2008) In a backcross inbred line [(TM1 Hai7124) TM1], ESTSSRs from *G. arboreum* were used to create a connection diagram. With these 99 ESTSSRs inserted into the backbone map, 111 loci were observed in total, including 511 SSR loci. These ESTSSRs can aid in the improvement of fibre quality in MAS.

In a RIL population, wang et al. (2011) used microsatellites to find QTLs relevant to fibre consistency. Two typical QTLs for lint percentage and fibre length were published with the genetic map. The pioneer linkage map was published by Park et al. (2005) using ESTSSRs in a RIL population derived from (*G. hirsutum* TM1

G. barbadense Pima) for fibre. The linkage has about 27% genome coverage, spanning 1277 cM of the genome, and 193 loci, with 121 of them newly mapped for fibre traits. Using SSRs and ESTSSRs, Saleem et al. (2015) identified two QTLs linked to drought resistance in F2 progeny formed from diverse parents. Using hydroponic culture, the progeny were tested for osmotic pressure with their kin Using SSRs and ESTSSRs in a RIL population, Abdurakhmonov et al. (2007) discovered QTLs regulating lint percentage on chromosomes 12, 18, 23, and 26. In an F2 population (*G. hirsutum* L. cv. Handan 208 vs. *G. barbadense* L. cv. Pima 90), 4 QTLs for lint index, 8 for seed index, 11 for lint yield, 4 for seed cotton yield, 9 for number of seeds per boll, 3 for fibre strength, 5 for fibre volume, and 8 for fibre fineness were discovered (He et al. 2007). SSRs were used to screen F2 progeny for nematode resistance (Ynturi et al. 2006), and researchers discovered the gene "GB713," which regulates resistance and could be used to fight reniform nematodes. Disease infestation is a serious problem in cotton, as evidenced by *Xanthomonas oxysporum* (Rungis et al. 2005), root-knot nematode (Niu et al. 2007), *Verticillium*, and cotton leaf curl disease (CLCuD) (Rahman et al. 2006), all of which alert cotton scientists to locate natural resistance sources and exploit them as soon as possible utilizing MAS. It's also likely that MAS will develop some economically essential traits including fibre consistency, yield, *Verticillium* wilt resistance, cotton leaf curl virus resistance, drought tolerance, and nematode resistance. It is becoming clear that attempts should be made to improve cotton production by the use of molecular breeding methodologies, which can be aided by recent advances in NGS. Furthermore, heavily saturated maps are helpful for determining genetic manipulations from a heredity standpoint, with SNPs being the greatest for this. In the cotton genomics age, these markers, along with QTLs, provide revolutionary instruments.

8.1.9 Genotyping by Sequencing

This technique has proved to be swift across a wide number of species and capable of evolving huge markers (Elshire et al. 2011; Poland et al. 2012). The ultimate aim of functional genomics is to use phenotypic information from phenotype to genotype to screen improved plant types for crop improvement. Genotyping by sequencing will progress from capturing further gene variations to whole-genome resequencing (Poland and Rife 2012). Understanding the practical and agronomic value of polyploidy and genome size heterogeneity within the Gossypium genus can be improved by decoding cotton genomes. Cotton-genome sequencing will ultimately stimulate fundamental research on genome evolution, polyploidization and related diploidization, gene expression, cell differentiation and development, cellulose synthesis, cell proliferation, molecular determinants of cell wall biogenesis, and epigenomics, all of which will be useful in the long-term processing of high-yielding and high-quality fibre, crop, and biomaterials (Chen et al. 2007). The cotton-genome sequence should be characterized and used in the future by integrating functional and structural genomic resources at the molecular and in silico levels, sequencing full-length cDNAs for genome annotation and expression assays, performing detailed annotation of the cotton-genome sequence to support gene discovery and map-based cloning in this species, and putting in place a large-network for detecting DNA-sequence variation (single nucleotide polymorphisms and genome-specific polymorphisms), enabling high-resolution whole-genome interaction studies, and designing genomic tilling arrays for gene expression and epigenomic research Sorghum bicolour (*G. hirsutum* L.), with minimal modifications to the basic protocol (Poland et al. 2012) of biological and agronomic characteristics, and sequence and annotate small RNAs and microRNAs, as well as classify their targets (Chen et al. 2007). This technique has been applied to a variety of cotton species.

8.1.10 Genome Wide Association (GWAS)

By using segregating populations by DNA marker techniques (Chen et al. 2007; Zhang et al. 2008), which are important for fastening marker-assisted selection, researchers have been able to explore the genetic variation available in germplasm, create genetic maps, and map QTLs for economic and agronomic traits. Due to restricted crossing over, it is difficult for biparental populations to detect closely related markers for molecular breeding. Furthermore, since certain small QTLs are not observed, the density of polymorphism in the bi-parental population is restricted. There is an alternative approach for QTL mapping called "association mapping," which uses cultivars with distinct traits and depends on linkage disequilibrium (Zhao et al. 2014). Association mapping is based on allele association between marker and phenotypic loci. This technique can be caused by mutation, genetic drift, population selection, and other factors, particularly in plants where hybridization has resulted in substantial inbreeding (Hart and Clark 1997). Due to providing improved precision to the locus and using highly studied populations of genetic heterogeneity consistent with phenotype, this approach overcomes many of the challenges of conventional genetic mapping. This method is based on linkage disequilibrium between the loci. It is required in LD mapping to define the magnitude and trend of LD in the population under observation in order to achieve the desired goals. The magnitude of the relationship, the degree of parental recombination, and the linkage disequilibrium in the gene pool both help to determine the collection is best for connection mapping (Lu et al. 2011).

Many scientists from all over the world have researched seed cotton production, yield elements, and fibre consistency traits in cotton using association mapping (Abdurakhmonov et al. 2009). Scientists have been able to research the variance present in germplasm materials thanks to association mapping. Through the detection of single nucleotide polymorphisms, it is now possible to investigate the entire genome's relationship with desired quantitative trait loci in plants, allowing for the creation of heavily saturated mapping populations (Waqas et al. 2014).

8.1.11 Tilling

Variation of DNA bases causes phenotypic variation in plant genomes, which can be caused spontaneously or by the use of various chemicals (Till et al. 2007). Tilling (targeting mediated local lesions in genomes) is a technique that helps researchers to determine an allelic variant in a single base pair for a specific gene. SNP mutations have been developed using chemical therapies. Tilling and eco-tilling techniques can detect point mutations, which are helpful from a breeder's viewpoint (Simsek and Kacar 2010). In tilling, the mutagens used to induce point mutations are highly selective, and at optimum concentrations, single base alternations may occur spontaneously at a high frequency. The knockout population is created by chemically processing the seed and causing a difference in the DNA sequence (Mehboob et al. 2012). Tilling was used in *G. arboreum* by Auld et al. 2009, who showed the technique's applicability in cotton. The ability to create a sufficient number of sequence variants in the target genome is determined by the length of application, the relative capability of ethyl methane sulfonate (EMS), and rays. Aslam et al. created a kill curve by screening three Gossypium species (*G. hirsutum, G. barbadense, and G. arboreum*). They looked at the effects of various mutagens (EMS and rays) at eight different rates of EMS (0.1–0.8%) and two different amounts of rays (10%). Every species' genotypes were assessed using morphological parameters such as emergence and plant height, as well as yield traits (number of bolls per plant, boll weight, lint yield, and lint percentage). From the perspective of reverse and forward genetics, viable accessions were chosen from mutagenized genotypes. They discovered that EMS had a much higher mutation rate than rays. Many software instruments exist to assist in the observation of base variation; for example, conservation-based SIFT (sorting intolerant from tolerant) is a mechanism for deciding when a difference in an amino acid hampering codon exists. PARSESNP (for Project Aligned Related Sequences and Evaluate SNPs, http: wwwproweborgparsesnp) can detect any gene alternation, according to Taylor and Greene 2003; graphs display the changes in sequence by using specific co-segregating material, positioning of coding/and non-coding regions, and reference DNA sequence.

8.1.12 Genetically Modified Cotton

While biotechnology is not a pest management technique in and of itself, it has the ability to unleash a torrent of new pest-control products. In reality, biotechnology offers a range of innovative, healthy, and efficient tools that could effectively tackle pest management issues. Significant developments in chemistry, biochemistry, behaviour, neurophysiology, molecular genetics, and genetic engineering have resulted in a range of bio-rational chemicals and materials that are less toxic and harmful to humans and the environment than traditional pesticides. Genetically modified (GM) refers to the conversion of genes between species using a range of laboratory methods such as cloning genes, splicing DNA fragments together, and introducing genes into cells. Recombinant DNA technology applies to both of these approaches together. There is no major technical advance to drive the sluggish yield plateau to higher peaks since the age of transgenic introductions. There is a need for solid, focused, and organized cotton research among the world cotton research community to fulfil future demand with restricted cultivable land and unpredictable climatic vulnerabilities. Coker genotypes are commonly used in genetic engineering studies since they can be regenerated in vitro using somatic embryogenesis (Jadhav and Katageri 2017). The two main aims of cotton genetic modification in the beginning were to confer pest resistance and immunity to more environmentally friendly herbicides (John and Stewart 1992). To date, India and other countries have certified more than 65 transgenic cotton cases. Bollworms that have been exposed to BT cotton on a daily basis have developed resistance, diminishing the effectiveness of regulation. The P450 monooxygenase gene (CYP6AE14) in cotton bollworms was silenced to reduce larval resistance to gossypol through the plant-mediated path. (Mao et al. 2007). Genetic innovation has made a major contribution to modern crop improvement. Bt cotton arrived at the perfect moment, when bollworms were wreaking havoc on the cotton crop and leaving farmers helpless. Since its introduction in the United States in 1995, China in 1997, and India in 2002, Bt. technology has had a major effect on bollworm control, resulting in a decline in pesticide use. The acceptance of GM cotton in different parts of the world has opened up new possibilities for cotton yield and quality improvement.

GhUGP1 (Cotton uridine diphosphate glucose pyrophosphorylase) overexpression increases fibre consistency and lowers fibre sugar content in upland cotton (Li et al. 2015). Overexpression of the novel sucrose synthase gene GhSusA1 results in a significant increase in biomass and fibre weight, as well as a modest increase in fibre power (Jiang et al. 2012). To increase the shape and consistency of the fibre, a silkworm fibroin gene was used (Li et al. 2009). The fibre lengths and micronaire values of transgenic cotton plants expressing the fibre expansin gene (GhEXPA8) increased dramatically. To investigate the biological functions of PHYA1 and (indirectly) other phytochrome genes in fibre, RNAi was used to target the fibre content QTL-associated phytochrome PHYA1 (Abdurakhmonov et al. 2014). Geneticists have been working to exclude gossypol from cottonseed for a long time. Cotton with lysigenous glands was created using antisense technology against the (+)-delta-cadinene gene to inhibit terpenoid aldehydes (gossypol). Plants that developed ultra-low gossypol cottonseed (ULGCS) were bred using RNAi-knockdown of delta-cadinene synthase gene(s) (Benedict et al. 2004). Increased research expenditure in biotic and abiotic stresses using a transgenic approach is expected in the future. Exploiting and optimizing cotton fibre and yield traits with the aid of alien gene integration requires a lot of attention. In terms of public acceptance and concerns, a major public awareness programme, including benefits, biosafety, and risk evaluation, is needed.

8.1.13 CRISPR-CAS System for Crop Improvement in Cotton

In the last decade, there has been a breakthrough in the field of genome alteration, with selective genome modification technologies continuing to progress. Until CRISPR Cas9 technology, genome editing tools such as zinc-finger nucleases (ZFN) and transcription activator-like effector nucleases (TALENs) were commonly utilized. To date, the CRISPR/Cas9 method has been used to successfully edit genomes of both model plants and crop plants. Via targeted mutagenesis of two endogenous genes [Cloroplastos alterados 1 (GhCLA1) and vacuolar H + pyrophosphatase (GhVP)], Chen et al. 2017 demonstrated for the first time that CRISPR/Cas9 can be used for advanced functional genomic research in cotton. Many aspects affect the CRISPR/Cas9 system's ability to achieve high mutation rates. Long et al. 2018 examined sgRNA expression and mutagenesis performance by replacing the current promoter with the endogenous U6 promoter (*Arabidopsis* AtU6–29 promoter). The use of an endogenous U6 promoter to push sgRNA expression improved mutagenesis efficiency by 4–6 times. This research used CRISPR/Cas9 to verify sgRNA mutagenesis efficacy in cotton in a quick and efficient manner. Gao et al. 2017 used a transient expression analysis of two genes in cotton, Translation elongation factor 1 (GhEF1) and Phytoene desaturase (GhPDS), to examine the existence of mutations caused by the CRISPR/Cas9 system. Wang et al. 2018, used the CRISPR/Cas9 system to successfully modify several sites of the genome in allotetraploid cotton by targeting the exogenously transformed gene Discosoma red fluorescent protein2 (DsRed2) and an endogenous gene Cloroplastos alterados 1. (GhCLA1). Table 8.2 lists some of the improved varieties of cotton through biotechnological interventions.

8.2 Sunn Hemp

Sunn hemp (*Crotalaria juncea L.*) is one of India's oldest and most important economically. Sunn hemp belongs to the family of legumes (Fabaceae) and has great potential as a multi-purpose fibre crop. *C. juncea* is the most common species of genus *Crotalaria*, which consists of more than 350 species. In the tropics, Sunn hemp is mainly grown as one of the most commonly grown green manure crops in rotation with many different crops (Kundu 1964; White and Haun 1965; Mascarenhas et al. 1980; Rotar and Joy 1983). Smooth and lignified fibres obtained from Sunn hemp stem could be used in the manufacture of pulp and paper. The additional features of Sunn hemp that increase its value potential as a non-wood fibre crop include low nitrogen fertilization requirements, N fixation, drought tolerance, resistance to root-knot nematodes, etc. Sunn hemp is more resistant to root-knot nematodes in comparison with kenaf (*Hibiscus cannabinus* L.) that is another potential non-wood fibre crop which has high yield potential, less vulnerable to lodging, and the stalks dry out more easily after killing frost before harvesting. Past and current research has demonstrated Sunn hemp as a potential source of non-wood fibre that can be used in the manufacture of newspaper, specialty papers, part of commercial nursery potting media, and many other commercial products.

Table 8.2 Improved varieties of cotton

Varieties	Year	Yield	Description
F 1861	2002	6.5 q/ acre	Cotton leaf curl virus resistant and medium staple of 2.5% span. It gives 33.5% ginning.
FDK 124 (SVAC		9.28 q per acre	It is an early maturing, short staple, with coarse fibre, high yielding, resistant to jassid, and whitefly variety of desi cotton.
LD 327	1987	11.5 q/acre	High yielding, high ginning, tolerant to fusarium wilt variety of cotton.
LD 694	2001	7 q/acre	It is short staple, coarse fibre variety with ginning, and resistant to jassid, fusarium wilt, and bacterial blight
LH 1556	1995	8.5 q/acre	It is early maturing, short duration variety of cotton with round bolls with good fluffy opening.
LH 2076	2008	7.8 q/acre	It has green broad lobed leaves and green stem and resistant to cotton leaf curl virus.
LHH 144	1998	7.6 q/acre	It is intra-*hirsutum* leaf curl virus resistant cotton with superior medium staple fibre that is ideal for spinning at 40s counts.
MRC 6301 Bt	2006	10.0 q/acre	It is resistant to spotted bollworm and American bollworm with good boll size (4.3 g) and fluffy opening.
MRC 6304 Bt	2006	10.1 q/acre	It resists bollworm and American bollworm. It has good fibre characteristics
MRC 7017 BG-II	2010	10.4q/ acre	It is tolerant to Para wilt, spotted and pink bowl bugs, cotton leaf curls, and tobacco caterpillars. It has semi-sympodious plants with thick green lobed leaves and light-yellow cream flowers. The fibre properties are very strong. With 29.7 mm fibre length and a 33.6% ginning output.
MRC 7031BG-II	2010	9.8 q/acre	It is tolerant to Para wilt, spotted and pink bowl bugs, cotton leaf curls, and tobacco caterpillars. This hybrid's fibre length is 29.4 mm and its ginning outcome is 33.4%.
PAU 626 H	2007	9.8 q/acre	Its genetic hybrid cotton focused on early maturation and high yield. It has medium-sized bollards with a fluffy aperture. Fusarium and bacterial blight are best tolerated.
RCH 308 Bt	2008	8.0 q/acre	This variety of cotton is resistant to bollworm and has good boll size (3.7 g) with fluffy opening
RCH 314 Bt	2008	8.2 q/acre	This variety of cotton is resistant to boll worms with 2.5% span length of 28.2 mm.
RCH 317 Bt	2006	10.5 q/acre	This variety is resistant to spotted bollworm and American bollworm with 28.6 mm 2.5% span length and 33.9% ginning result.

8.2.1 History and Origin

C. juncea is thought to have been discovered in India (Montgomery 1954). The genus *Crotalaria* gets its name from the rattle made when the seeds in mature pods are shaken (White and Haun 1965). Sunn hemp (*C. juncea L.*) was given the *juncea* name because of its resemblance to Spanish broom (*Spartium junceaum* L.) with its green rushlike, sparsely leaved leaves (Kundu 1964). Sunn hemp is typically thought of as a tropical or subtropical plant. It is drought tolerant and adaptable to a wide range of soil types. Fibre production thrives in a light loam soil that is relatively fertile and well-drained (Montgomery 1954; Kundu 1964). On clay or low-lying soils, robust growth may be achieved; however, the bast fibre can be coarser and may lead to lower yields. During the crop-growing season, the major Sunn hemp-growing areas of India and Brazil are characterized by average temperatures varying from 23.0 °C to 29.4 °C, high humidity, and rainfall of about 170–200 mm (Dempsey 1975). Sunn hemp is a photoperiod-sensitive weed, which means that it flowers in reaction to short days (White and Haun 1965).

8.2.2 Taxonomy

Sunn hemp is an erect, short-day, annual shrub, usually 1 to 4 m in height. The stems are ribbed and cylindrical. Thick plantings of Sunn hemp lead to reduced branching in the upper portion. The simple, elliptic to oblong-shaped leaves are arranged spirally on the stem. A long, solid taproot, well-formed lateral roots, and several branched and lobed nodules, up to 2.5 cm in length, characterize the root system. Sunn hemp inflorescence is a terminal open raceme that consists of deep yellow flowers up to 25 cm in length. Flowering of Sunn hemp is indeterminate in nature. Sunn hemp has widespread cross-pollination, and self-pollination happens only after insect or mechanical stimulation of the stigmatic surface (Purseglove 1968). Seeds are small, kidney-shaped, flattened, and have a protein content of about 35%. Sunn hemp seed weight varies greatly from 18,000 to 30,000 seeds per kg, owing to the diverse environmental requirements for cultivation (Dempsey 1975). "Tropic Sun," a Hawaiian cultivar, is said to have 30,000 to 35,000 seeds per kilogramme (Rotar and Joy 1983).

8.2.3 Emasculation and Pollination in Sunn Hemp

Sunn hemp stems are made up of two fibres: bast and woody root. The bast fibres in the outer bark are much longer than those in the woody heart, but their fibre widths are very similar (Cunningham et al. 1978). The proportion of bark in the overall stalk ranges between 15% and 20% by dry weight. According to Kundu (1964), the percentage of bast fibre in dry stalks ranges between 6.4% and 10.5%. The flower is zygomorphic, full, and pentamerous. Five fused (gamosepalous) sepals, five free (polypetalous) petals, ten free stamens in two whorls, and a single pistil represented by gynoecium make up the flower. Androecium is known for its anther dimorphism. The second whorl is represented by five globose, basifixed anthers on slender and longer filaments, and is located in the centre of the flower. Anther dimorphism has also been observed in *C. retusa* (Tidke and Patil 2000) and C. digitata (Tidke and Patil 2000). (Muthu and Ganesan 2012). The ripening of the first whorl of anther almost correlates with the stigma's receptivity. In pollen dispersal, the second whorl of anther is late and supplies pollen to stigma upon elongation until the first whorl of anthers withers. The fruit is an inflated pod with 10–15 seeds inside it. The seeds are small, kidney-shaped, and mostly black. Exceptions occur in the K-12 yellow variety with yellow seeds. Different views on the pollination behaviour of *C. juncea* were expressed, some reports indicate it to be obligatory outcrossing species (Mitra 1934). By contrast, it was also classified as a crop that was self-pollinated. Sunn hemp is described as a mainly self-pollinated crop benefiting from insect visits. Self-pollinated flowers do not result in the formation of pods. The failure of self-forming pod formation indicates the role of self-incompatibility. Sunn hemp crops are reported to be visited by a large number of insect species. However, pollination is provided only by the species powerful enough to perform tripping. Several species of honeybees *like Apis florae*, *A. cerana,* and *A. indica* are reported to be ineffective tripper. Several species of bumble bees are efficient pollinators effecting pollination in *Crotalaria* like *Xylocopa fenistroides* Fabr. in *C. juncea*, *X. frontalis*, and *X. grisescens* in *C. retusa*. Species of Xylocopa prefer zyomorphic, yellow flowers present in Sunn hemp. Bees visit flowers that are fully open and tend to gather pollen in the afternoon. Xylocopa species have been reported to prefer medium to large flowers that are sufficiently strong to withstand insects' rough foot work. In most species of *Crotalaria*, the yellow-coloured flower attracts xylocopa bees. Plants attract insects as they provide all kinds of food sources including pollen or nectar, or both. It is found that certain species of *Crotalaria* have nectar and attract big carpenter bees like *C. laburnifolia*. Other species, such as *Apis mellifera* and ants, are also involved in pollination of related species in *C. micans*. Mohan established self-incompatibility in Sunn hemp and suggested that it led to outcrossing. He attributed any variation from 100% in out-crossing to the breakdown of self-incompatibility due to the age of the flowers.

Self-incompatibility in Sunn hemp was gametophytically controlled. The stigmatic surface, typical of the tribe of the crotalariea, is characterized by the presence of the membrane.

The absence of such a membrane in *C. retusa* makes it autonomous. Another factor that works against self-incompatibility may be responsible for the screening of pollen with pistil. The self-interest is widely reported as in *C. sericeae* in several types of *Crotalaria*, on the other hand. Since the 1930s extensive study in the United States on Sunn hemp has been conducted where it has proven to be an excellent crop for soil growth. It has led to higher organic matter intake; increased nitrogen fastening and suppressed the growth of populations of nematodes (Breitenbach 1958; Dempsey 1975). Because of its difficulties in seed production, many farmers have left Sunn hemp cultivation. During World War II, interest in growing Sunn hemp was revived. In 1942, *C. Juncea* was listed on the list of the necessary war materials for its potential use as cordage fibre. The subject was once again on Sunn hemp from the late 1950s to the 1960s. This has been done by attempts to identify renewable annual suppliers of non-wood fibres for paper and pulp processing. Since studies have shown that Sunn hemp has excellent pulping properties and that it can achieve high maceration yields (Nieschlag et al. 1960; Nelson et al. 1961).

Because of its ability to reliably deliver better returns and take less accommodation, Kenaf received greater attention. A more recent Cunningham et al. (1978) research revealed that sun hemp, including the length and the width of bast fibre (3.79 vs.2.62 mm) (24.3 vs.19.7 μm), could be superior to kenaf for specific fibre properties. They argued further that Sunn hemp is an outstanding paper candidate with the following characteristics:

1. High yields of sulphate pulp bleachable.
2. Pulp strength higher than or equal to the blended hardwood pulp in the south.
3. The fibre bast is longer than that of wood fibre in the longitude to width ratio.

Sunn hemp has a greater possible benefit in soils infesting the southern root knot than kenaf, as seen by Scott Jr. and Cook (1994). Tests in increasing conditions in South Texas, Kansas, and South Carolina reported total stalk yields ranging from 14,800 to 18,450 kg/ha (Scott Jr. and Cook 1994, Scott Jr. et al. 1991). In Indian, Bangladesh, and Brazil, the main cultivation areas of Sunn hemp are grown as green manure crop, a feed crop, or a bast fibre. High-quality paper and cordage are used for the treatment of bast fibre. In order to assess the suitability of shorter core fibres of Sunn hemp for commercial nursery use in the cultivation of less potting soil, research is also being carried out in Texas on its uses as a green manure crop and a source of fibre.

8.2.4 Plant Characteristics

Sunn hemp is a member of the genus *Crotalaria*, containing approximately 600 species. Except for the case of *C. Juncea*, for fibre, no species is cultivated. Other plants are used either as green manure or as fodder. Sunn hemp, an upright herbaceous shrub, is a short-day plant. The plants consist of tap root system with root surface nodules. Generally, plants reach up to 2.5–3.0 m in height. Leaves are simple in form, stipulated, whole, and elliptical to oblong. Sunn hemp bears an indeterminate growth habit of the terminal raceme inflorescence. Flowers with a broad ovate standard petal with a strong midrib at the back of the petal are a traditional standard form. Petals, because of their orientation, serve as a point of attraction for various types of insects. Wings are medium in size and they are pointed and slightly twisted with keel petals. Sunn hemp, 2n = 16, is commonly stated to be a plant that is self-incompatible. Cross-pollination is extensive and only after insects or any other means have activated the stigmatic surface (Kundu 1964) does self-pollination. More recently, successful self-compatibility attempts have been recorded in breeding (Ribeiro et al. 1977). The growth of self-compatible germplasm can speed up the production of real, stable, pure lines of breeding. Most of the Sunn hemp cultivars, according to Purseglove (1968) and Dempsey (1975), originated from the selection of improved types appropriate for particular locations rather than through breeding procedures. In general, these choices focused on early maturity, enhanced fibre yield, and pest resistance. Dempsey (1975) offers lists of improved Sunn hemp cultivars selected for India, Brazil, and Taiwan's growing conditions. Past attempts to establish strains between *Crotalaria* from interspecific crosses have not been successful (Kundu 1964).

8.2.5 Genetics and Breeding of Sunn Hemp

The nature and magnitude of genetic diversity is of vital significance to any crop's genetic improvement. The lack of genetic diversity in phenotypic characteristics in the genus *Crotalaria*, especially at the intra-specific stage, has hampered crop improvement so far. Most of the plants in the genus are unifoliated, with green sepals and yellow showy petals bearing zygomorphic flowers. Floral symmetry exceptions (actinomorphic flowers in *C. graminicola*), sepal colour (blue in *C. incana*) and petal colours (blue in *C. verrucosa*) are found. *C. retusa* is unusual in having the longest and widest pods. Studies have been performed to explain genetic heterogeneity for various characteristics, such as green manure characteristics and fibre yield characteristics. Variability between the lines for various fibre yield traits was discovered by examination of germplasm obtained from different sources. Several Sunn hemp germplasm lines collected from different parts of India and major variations were identified for most of the characters, and they further suggested the presence of genotypic diversity. Restricted genetic variation at intra-specific and moderate genetic diversity at inter-specific levels was discovered by genetic diversity measured using various techniques. Studies of genetic variation using the seed protein profile by SDS-PAGE in various species of the genus *Crotalaria* showed strikingly different patterns of protein. Random amplified polymorphic DNA (RAPD) study was used by Jayanthi and Mandal (2001) in two populations of *C. longipes* reported minimal polymorphism (0 to 33%). Wang et al. (2006a, b) analyzed the genetic variation of the sample of *Crotalaria* germplasm using medicago and soybean-derived expressed sequence tag-simple sequence repeat (EST-SSR) markers. In general, phylogenetic research grouped accessions along species lines into four major classes. To create a more satisfactory classification and understanding of evolutionary relationships, a systematic analysis including some additional economically significant species of *Crotalaria* seems to be appropriate. The intraspecific differences between the two accessions of either species can only be overcome through DNA-based approaches. For the genus *Crotalaria*, sequence analysis of chloroplast genes could produce a far more resolved phylogeny. There are few studies available concerning the genetics of various crop traits. The lack of classical genetic research may be due to the lack of characters that are morphologically distinct. The only distinctiveness found is for seed colour with two derivatives (i.e. black and yellow). Another feature is hypocotyl pigmentation, where two types can be distinguished. The inheritance pattern of pigmentation of anthocyanin in flowers and hypocotyls in *C. juncea* has been reported to be dominated by a single gene with dominant anthocyanin pigmentation and recessive type without pigmentation (de Miranda et al. 1989). Similarly, the seed coat colour was also reported to be a monogenic trait. For pigmentation on the seeds, petal bases, and hypocotyls, pleiotropy was noted. There was an important positive association between plant height, green and dry biomass, and root nodule characters in the research on the green manuring feature of Sunn hemp. The fibre yield was also strongly associated with plant height and overall green weight at harvest. In all the component characteristics, such as plant height, basal diameter, etc., with the exception of fibre percentage, predominance of specific combining capacity (SCA) was reported. The general combining ability (GCA) was only important in the case of plant height.

8.2.6 Breeding Methodologies

Systematic study on the breeding of Sunn hemp in India began at the beginning of the twentieth century with the development of new strains of Sunn hemp, viz. Jabalpur, Black K-12, etc. As a result of experiments performed at Kanpur, the earliest Sunn hemp strain officially published was K-12 earlier identified as "Cownpore-12" established in 1926 (Sabnis 1931). The variety "K-12 gold" was produced from this variety (K-12) with a distinctive yellow seed coat colour in 1971. The Sunn hemp breeding programme subsequently gained momentum with the assistance of the Indian Council of Agricultural Research (ICAR, New Delhi, India). This resulted in new varieties such as M-18, M-35, etc. being identified with novel characteristics such as earliness and higher yield. In the following paragraphs, different methodologies used for genetic improvement are listed.

8.2.6.1 Conventional

Very less consideration has been given for the breeding of successful varieties of Sunn hemp. Any of the varieties that have been published so far are primarily picked from germplasm. Some crop characteristics such as complex breeding behaviour have impeded the production of superior varieties, including the prevalence of self-incompatibility and widespread cross-pollination almost to the exclusion of self-pollination, loss of genetic heterogeneity, and incongruence of distant crosses. The prevalent use of green manure, where the field is either ploughed at the vegetative stage or at the start of flowering, is another fundamental factor. However, when the crops are to be grown for commercial benefit, such as yield or quality, successful varieties are needed. The mass collection results in most of the varieties published so far in Sunn hemp. The production of the variety focuses on greater yield (K-12 Black, K-12 Yellow), early maturity (M-18, ST-112), shoot borer resistance (M-19 and M-35), wilting resistance (ST-95, D-IX), and drought resistance (M-18). It was confirmed that the T-6 variety was day-neutral. Until recently, K-12 Black and K-12 Yellow were the most common strains. The K-12 yellow type, however, has been vulnerable to viral infection over time. The introduction of superior varieties, including Shailesh, Swastika, Ankur, and Prankur, has been demonstrated to be resistant to viral infections during the current decade. Some of the improved varieties are mentioned in Table 8.3.

Table 8.3 Improved varieties of Sunn hemp

Varieties	Yield (q/ha)	Area Recommended
K-12 yellow	15	It is suitable for all Sunn hemp-growing areas with wider adaptability. It also shows some resistance to wilt disease.
Shailesh (SH-4)	10–11	It is suitable for all Sunn hemp-growing areas.
Swastik (SUIN-053)	11–12	It is suitable for all Sunn hemp-growing areas.
Ankur (SUIN-037)	11–12	It has wider adaptability and recommended for all Sunn hemp-growing areas.
Prankur (JRJ-610)	11–12	It is suitable for all Sunn hemp-growing areas.
K-12 (black)	10–12	It is recommended for U.P. it also shows some resistance to shoot borer and wilt disease.
M-18	10–11	It is recommended for light soils with low rainfall areas of M.P. it is a drought-tolerant variety.
M-35	11–13	It is similar to M-18 variety in characteristics. It also has some resistance to shoot borer.
BE - 1	9–11	It is mainly cultivated in Bihar and Orissa.
Belgoan	10–12	It is suitable for heavy soil and recommended for M.P.
Chindwara	11–13	It is mainly cultivated in M.P.
Jabalpur	10–13	It is a selection from a type collected from Jabalpur region of M.P. it is recommended for the cultivation in M.P. and U.P.
Bellary	9–11	It is a high-yielding variety and mainly cultivated in Tamil Nadu.
D - IX	11–12	It is recommended for the cultivation in Maharashtra. It has resistance to wilt disease.
T - 6	9–11	It is used as a green manure crop and mainly cultivated in W.B.
ST - 95	10–13	It has slight resistance to wilt and shoot-borer. It is a selection from a type introduced from Taiwan.
ST-55	14–15	It has moderate resistance to wilt diseases.

8.2.6.2 Mutation

Mutation is assumed to improve genetic variation as an established phenomenon (Yilmaz and Boydak 2006). In order to maximize heterogeneity and isolate novel lines with better characteristics, induced mutagenesis was attempted in Sunn hemp. Due to their effect on various quantitative characteristics, translocation lines were induced in *C. juncea* as a result of irradiation with gamma rays followed by treatment with EMS or MMS (Verma and Raina 1991). Irradiated seeds can be used as a source of generating genetic diversity. For their usefulness in developing successful genotypes for increased fibre yield and improved green manuring efficiency, mutations may be investigated.

8.2.6.3 Polyploidy

In multiple crop varieties, polyploidy has been attempted to maximize the forage value of the crop. The findings have not, however, been positive in *C. juncea*

(Dyansagar 1983) and in *C. Sericea*. Polyploidy induction has been attempted in *C. linofolia* with the goal of increasing its importance as forage. Autotetraploid plants have been produced with colchicine by seedling therapy. The tetraploid plant leaves were dense, dark green, and broad with a coarse texture. In tetraploids, the large size of leaves increases the performance of the forage. In other species of *Crotalaria*, however, contrary findings have been recorded in this regard. Other results of tetraploids include increased size of stomata, delayed flowering, extended flowering time, reduced flower frequency per inflorescence, increased pollen size, sterility of pollen, low fruit set, rudimentary pod appearance, smaller pods, and wrinkled seeds. These may be due to genetic and chromosomal imbalances. Seeds derived from polyploid plants were significantly larger than seeds derived from diploid plants. The sterility of the pollen was due to the high frequency of laggards, micronuclei during anaphase and telophase. It is more economically important because of the increased size of the leaves. Most of the induced polyploids became sterile within a few generations and reverted to diploidy. Selection among the segregating generation of polyploid can be required to produce enhanced genotype.

8.2.6.4 Heterosis Breeding

In many self-pollinated and cross-pollinated crops, heterosis has been successfully used with high success resulting in improved yield. The underlying conditions for using the process of heterosis are male sterility. The identification and maintenance of male sterile lines in germplasm lines can necessarily break the yield barrier of Sunn hemp. In various species of the genus, male sterile plants have been recorded. In comparison to the large, yellow, and plumpy anthers with ample yellow pollen. Kempanna and Sastry noted a single plant in *C. striata* with poorly formed, hollow, small anthers that failed to dehisce. They confirmed shrivelled microspores devoid of any material. The plant did not set any fruit when bagged. Male-sterile plants revealing male-sterile factor superiority were also formed by the progeny from outcrossing such a male-sterile line. Male-sterility was also subsequently found in *C. mucronata Desv.*, *C. juncea L.*, and *C. pallida* Ait. Male sterility in *C.* pallida was ascribed due to some biochemical instead of morphological changes in the tapetum and due to lack of cytokinesis following meiosis in mother cells of the microspore. By almost absence of seed setting under selfing, the development of inbreds in Sunn hemp is limited. Male sterility chemical induction was attempted in *C. juncea* using Maleic Hydrazide and Mendok. Most of the chemicals capable of causing male sterility, however, had unwanted effects, including harmful effects on female gametes, decreased pod set, and other harmful effects. It is strongly justified to perform an exhaustive survey of several new chemicals with the properties of ideal male gametocides.

8.2.6.5 Non-Traditional/Modern/Biotechnological

In vitro tissue culture and transgenics technologies provide the scope for seed genetic enhancement and complement conventional crop improvement instruments. The Micropropagation Experiments in *C. juncea* is missing. However, reports on related species like *C. retusa* are available, where multiple firing, supplemented with growth regulators, was induced from leaf cuttings on MS medium (Devendra et al. 2011). With 90% success, the rooted shoots were successfully transplanted. Micropropagation has been accomplished in *C. laburnifolia* in 2012 by Rajender et al., *C. verrucosa* in 2008 by Hussain et al. and in *C. lutescens* by Naomita and Rai in 2000. In inducing desirable plants by somaclonal variance and in the mass replication of desired plants, micropropagation may be used. Under in vitro settings, inbred lines can be achieved from doubled haploids. Under regulated conditions such as glasshouse/poly house, genetic integrity can be preserved by sib-mating (with the assistance of insects) of clones. Genetic engineering has arisen as an important instrument for breeding effective varieties or causing diversity in existing types of plants. Transgenesis can be utilized to obtain ideal plants by adding new genes in the kingdom and broadening the genetic base of the seed. Sunn hemp should predict a lot of genetic advancement by the use of transgenic technologies, which is challenging due to floral limitations through traditional means. Sunn hemp has been developed as a transformation mechanism free of in vitro plant regeneration following Agrobacterium infection. The gene of the two main serotypes of the foot and mouth disease virus (FMDV) was developed to harbour transgenics. Other in vitro approaches, such as somaclonal variation and somatic hybridization, can help to induce heterogeneity across organisms or genera and can generate successful variability. Table 8.4 enlists some of the milestones in the genetic improvement of Sunn hemp over time.

8.3 Achievements

The world is facing an increased demand for fibre. Therefore, Sunn hemp and cotton, non-wood fibre crops, have the ability to be developed on a broad industrial scale and to meet out the current fibre requirements. The processing of pulp, specialty papers, and as a part of a soil-less potting medium are existing established industrial uses for both fibre crops. Sunn hemp bast fibre has been used in the past for cordage and for specialized paper production.

8.4 Summary and Future

Cotton, the most important fibre used in large-scale garment manufacturing, continues to dominate global fibre production (18 million tonnes). However, pesticide use is heavy – cotton accounts for almost half of all pesticides used in the United States – and cotton harvesting in developed countries is labour-intensive. Because of concerns regarding pesticide tolerance in cotton pests, organic and integrated pest control of cotton is becoming more common. However, as new markets for fibres emerge, many cotton farmers, especially in the United States, are considering the cultivation of pesticide-free fibre crops such as hemp or kenaf. Sunn hemp is also cultivated in crop rotations in countries outside the United States as a green manure crop in order to improve soil health and suppress root-knot nematode infestations.

Table 8.4 Sunn hemp's genetical improvement over time

S. No.	Year	Scientist	Study
1.	1967	Kaul and Singh	• Male sterility chemical induction was attempted in *C. juncea* using maleic hydrazide and Mendok
2.	1989	Miranda et al.	• Studied the inheritance pattern of plant height and seed yield in Sunn hemp • Reported gametophytically regulated self-incompatibility in Sunn hemp • Studied the inheritance pattern of pigmentation of anthocyanin in flowers and hypocotyls in Sunn hemp
3.	1991	Verma et al.	• Translocation lines were induced in *C. juncea* as a result of irradiation with gamma rays followed by treatment with EMS or MMS
4.	1991	Mangotra and Koul	• Reported that most of the induced polyploids become sterile within a few generations and revert to diploidy
5.	1992	B.C.Patil	• Polyploidy induction was attempted in C. Linofolia
6.	1994	Islam et al.	• Documented morphological anomalies of *C. juncea* due to gamma-ray treatment and temperature for 12 to 36 hours
7.	2001	Jayanthi and Mandal	• Reported minimal polymorphism (0 to 33%) using random amplified polymorphic DNA (RAPD) study in two populations of *C. longipes*
8.	2004	Kumar et al.	• Study of impact of gamma irradiation on growth response viz. function of active mitotic index, germination percentage, and survivability percentage in *C. juncea*
9.	2004	Virdi et al.	• Reported positive association between plant heights, green and dry biomass, and root nodule characters with the green manuring feature of Sunn hemp
10.	2005	Tripathi et al.	• Reported positive association of basal diameter of the plant with the fibre yield in *C. juncea*
11.	2006	Wang et al.	• Study of genetic variation of *crotalaria* germplasm using medicago and soybean-derived expressed sequence tag-simple sequence repeat (EST-SSR) markers.
12.	2008	Shah et al.	• Reported post-irradiation ageing effects on seed germinability and various morphological characteristics of *C. saltiana*
13.	2009	Okeke et al.	• Reported floral symmetry exceptions (actinomorphic flowers in *C. graminicola*), sepal colour (blue in *C. incana*) and petal colours (blue in *C. verrucosa*)
14.	2011	Raj et al.	• Study of genetic variation using the seed protein profile by SDS-PAGE in various species of the genus *crotalaria*.
15.	2012	Rao et al.	• Sunn hemp was developed as a transformation mechanism free of in vitro plant regeneration following agrobacterium infection.
16.	2012	Rajender et al.	• Micropropagation has been accomplished in C. laburnifolia
17.	2014	Kumar et al.	• Reported mitotic disruptions and impaired germination of the seed due to gamma-ray irradiation
18.	2018	Naresh Kumar et al.	• Studies the genetic divergence of selected genotypes in Sunn hemp (*C. juncea*)

The extension of product research and production, the development of harvesting and processing equipments, and the discovery of more successful crop management techniques should be the focus of future efforts. The future of Sunn hemp looks very exciting, provided that worldwide demand for fibre is growing and existing stocks of wood fibre will not afford to satisfy the demands of the future industrial industry. Analysis has shown that Sunn hemp can be successfully grown as a crop of fibre and that much of the existing fibre industry requirements can be replaced by these fibres.

References

Abdurakhmonov IY, Buriev ZT, Saha S, Jenkins JN, Abdukarimov A, Pepper AE (2014) Phytochrome RNAi enhances major fibre quality and agronomic traits of the cotton *Gossypium hirsutum L.* Nat Commun 5(1):1–10

Abdurakhmonov IY, Buriev ZT, Saha S, Pepper AE, Musaev JA, Almatov A, Shermatov SE, Kushanov FN, Mavlonov GT, Reddy UK, Yu JZ, Jenkins JN, Kohel RJ, Abdukari, mov. A. (2007) Microsatellite markers associated with lint percentage trait in cottonǰ *Gossypium hirsutum*. Euphytica 156:141–156. https://doi.org/10.1007/s10681-007-9361-2

Abdurakhmonov IY, Saha S, Jenkins JN, Buriev ZT, Shermatov SE, Scheffler BE, Pepper AE, Yu JZ, Kohel RJ (2009) Linkage disequilibrium-based association mapping of fibre quality traits in *G. hirsutum L.* variety germplasm. Genetica 136(3):401–417. https://doi.org/10.1007/s10709-008-9337-8

Aslam U, Khan AA, Cheema HMN, Imtiaz F, Malik W (2013) Kill curve analysis and response of ethyl methanesulfonate and γ-rays in diploid and tetraploid cotton. Int J Agric Biol Eng 15:11–18

Auld D, Light G, Fokar M, Bechere E, Allen RD (2009) Mutagenesis system for genetic analysis of Gossypium. In: Paterson AH (ed) Genetics and genomics of cotton, vol 3. Springer Inteurnational Publishing, New York, pp 209–226. https://doi.org/10.1007/978-0-387-70810-2_9

Benedict CR, Martin GS, Liu J, Puckhaber L, Magill CW (2004) Terpenoid aldehyde formation and lysigenous gland storage sites in cotton: variant with mature glands but suppressed levels of terpenoid aldehydes. Phytochemistry 65(10):1351–1359

Bolek Y (2003) Status of genome mapping and use in cotton improvement. KSÜ Fen ve Mühendislik Dergisi 6(2):72–79

Breitenbach CA (1958) Land management practices for kenaf production. In: Proc. World Conf. on Kenaf. ICA, Havana, pp 97–101

Chee PW, Draye XC, Jiang et al (2005) Molecular dissection of phenotypic variation between *Gossypium hirsutum* and *Gossypium barbadense* (cotton) by a backcross-self approach: III. Fibre length. Theor Appl Genet 111(4):772–781

Chen X, Lu X, Shu N, Wang S, Wang J, Wang D et al (2017) Targeted plant breeding - current and future views 36 mutagenesis in cotton (*Gossypium hirsutum L.*) using the CRISPR/Cas9 system. Sci Rep 7:44304

Chen ZJ, Scheffler BE, Dennis E, Triplett BA, Zhang T, Guo W, Chen X, Stelly DM, Rabinowicz PD, Town CD, Arioli T, Brubaker C, Cantrell RG, Lacape JM, Ulloa M, Chee P, Gingle AR, Haigler CH, Percy R, Saha S, Wilkins T, Wright RJ, Van Deynze A, Zhu Y, Yu S, Abdurakhmonov I, Katageri I, Kumar PA, Ur. Rahman M, Zafar Y, Yu JZ, Kohel RJ, Wendel JF, Paterson AH (2007) Towards sequencing cotton (Gossypium) genomes. Plant Physiol 145(4):1303–1310

Collard BCY, Jahufer MZZ, Brouwer JB, Pang ECK (2005) An introduction to markers, quantitative trait loci (QTL) mapping and marker-assisted selection for crop improvement: the basic concepts. Euphytica 142(1–2):169–196

Constable, G, Reid, P. & Thomson, N. (2001), Approaches utilized in breeding and development of cotton cultivars in Australia

Cunningham RL, Clark TF, Bagby MO (1978) *Crotalaria juncea*-annual source of papermaking fibre. TAPPI 61:37–39

de Miranda MAC, Bulisani EA, Teixeira JPF, Mascarenhas HAA (1989) Haranca da pigmentacao com antocianina em *Crotalaria juncea* L. Bragantia 48:87–94

Dempsey JM (1975) Fibre Crops. The University Presses of Florida, Gainesville

Devendra BN, Srinivas N, Naik GR (2011) Direct somatic embryogenesis and synthetic seed production from Tylophora indica (Burm. f.) Merrill an endangered, medicinally important plant. Int J Bot 7(3):216–222

Draye X, Chee P, Jiang CX et al (2005) Molecular dissection of interspecific variation between *Gossypium hirsutum* and *G. barbadense* (cotton) by a backcross-self approach: II. Fibre fineness. Theor Appl Genet 111(4):764–771

Dyansagar VR (1983) Induced tetraploidy in Crotalaria juncea L. Cytologia 48:483–489

Elshire RJ, Glaubitz JC, Sun Q, Poland JA, Kawamoto K, Buckler ES, Mitchell SE (2011) A robust, simple genotyping-by sequencing (GBS) approach for high diversity species. PLoS One 6(5): e19379

Fryxell PA (1992) A revised taxonomic interpretation of Gossypium L. (Malvaceae). Rheedea 2 (2):108–165

Gao W, Long L, Tian X, Xu F, Liu J, Singh PK et al (2017) Genome editing in cotton with the CRISPR/Cas9 system. Front Plant Sci 8:1364

Gruère G, Sengupta D (2011) Bt cotton and farmer suicides in India: an evidence-based assessment. J Dev Stud 47(2):316–337

Gulati AN, Turner AJ (1929) 1—A note on the early history of cotton. J Tex Inst Trans 20(1): T1–T9

Guo YJ, McCarty CJ, Jenkins N, Saha S (2008) QTLs for node of first fruiting branch in a cross of an upland cotton, *Gossypium hirsutum L.*, cultivar with primitive accession Texas 701. Euphytica 163(1):113–122

Hart DL, Clark AG (eds) (1997) Principles of population genetics. Sinauer Associates, Sunderland, Massachusetts, pp 1–635

He DH, Lin ZX, Zhang XL, Nie YC, Guo XP, Zhang YX, Li W (2007) QTL mapping for economic traits based on a dense genetic map of cotton with PCR-based markers using the interspecific cross of *Gossypium hirsutum* vs *Gossypium barbadense*. Euphytica 153:181–197. https://doi.org/10.1007/s10681-006-9254-9

Jadhav M, Katageri I (2017) Agrobacterium tumefaciens mediated genetic transformation in Coker312 (*Gossypium hirsutum L.*) using hypocotyls explants. Int J Curr Microbiol App Sci 6(12):2771–2779

Jayanthi M, Mandal PK (2001) Low genetic polymorphism in natural populations of Crotalaria longipes. Biol Plant 44(3):455–457

Jiang C, Wright RJS, Woo ST, DelMonte A, Paterson AH (2000) QTL analysis of leaf morphology in tetraploid Gossypium (cotton). Theor Appl Genet 100(3–4):409–418

Jiang Y, Guo W, Zhu H, Ruan YL, Zhang T (2012) Overexpression of GhSusA1 increases plant biomass and improves cotton fibre yield and quality. Plant Biotechnol J 10(3):301–312

John ME, Stewart JM (1992) Genes for jeans: biotechnological advances in cotton. Trends Biotechnol 10:165–170

Joshi SP, Prabhakar K, Ranjekar PK, Gupta VS (2011) Molecular markers in plant genome analysis. Science Publishers, Enfield, pp 1–19

Kohel, R. J., (1999a), Cotton improvement: A perspective. Cotton World 1: (in press)

Kohel RJ (1999b) Cotton germplasm resources and the potential for improved fibre production and quality. In: Basra AS (ed) Cotton Fibres. The Haworth Press, Inc, New York, pp 167–182

Kohel RJ, Yu J, Park YH, Lazo GR (2001) Molecular mapping and characterization of traits controlling fibre quality in cotton. Euphytica 121(2):163–172

Kundu BC (1964) Sunn-hemp in India. Proc Soil Crop Soc Fla 24:396–404

Li B, Yang Y, Hu WR, Li XD, Cao JQ, Fan L (2015) Over-expression of Gh UGP 1 in upland cotton improves fibre quality and reduces fibre sugar content. Plant Breed 134(2):197–202

Li C, Wang X, Dong N et al (2013) QTL analysis for early-maturing traits in cotton using two upland cotton (*Gossypium hirsutum L.*) crosses. Breed Sci 63(2):154–163

Li F, Wu S, Lü F, Chen T, Ju M, Wang H et al (2009) Modified fibre qualities of the transgenic cotton expressing a silkworm fibroin gene. Chin Sci Bull 54(7):1210–1216

Long L, Guo DD, Gao W, Yang WW, Hou LP, Ma XN et al (2018) (2018), optimization of CRISPR/Cas9 genome editing in cotton by improved sgRNA expression. Plant Methods 14(1): 85

Lu YL, Shah T, Hao ZF, Taba S, Zhang SH, Gao SB, Liu J, Cao MJ, Wang J, Prakash AB, Rong TZ, Xu YB (2011) Comparative SNP and haplotype analysis reveals a higher genetic diversity and rapider LD decay in tropical than temperate germplasm in maize. PLoS One 6(9):e24861

Mao YB, Cai WJ, Wang JW, Hong GJ, Tao XY, Wang L et al (2007) Silencing a cotton bollworm P450 monooxygenase gene by plant-mediated RNAi impairs larval tolerance of gossypol. Nat Biotechnol 25(11):1307–1313

Mascarenhas HAA, Braga NR, de Miranda MAC, Feitosa CT, Bataglia OC (1980) Efeito de adubos verdes e organicos na producao de soja. Instituto Agronomico, Campinas. *Boletim Tecnico* 63

Mehboob R, Shaheen T, Iqbal M, Ashraf et al (2012) Cotton genetic resources. A review. Agron Sustain Dev 32:419–432. https://doi.org/10.1007/s13593-011-0051-z

Mei M, Syed NHW, Gao et al (2004) Genetic mapping and QTL analysis of fibre-related traits in cotton (*Gossypium*). Theor Appl Genet 108(2):280–291

Mitra M (1934) Wilt disease of *Crotalaria juncea* Linn. (sunn-hemp). Indian J Agr Sci 4:701–714

Montgomery B (1954) Sunn fibre. In: Mauersberger HR (ed) Mathew's textile fibres, 6th edn. Wiley, New York, pp 323–327

Muthu RK, Ganesan R (2012) Rediscovery of Crotalaria digitata (Fabaceae) from Madurai district, Tamil Nadu, India. Rheedea 22(2):103–106

Nelson GH, Nieschlag HJ, Daxenbichler ME, Wolf IA, Perdue RE (1961) A search for new fibre crops. III Laboratory scale pulping studies. TAPPI 44:319–325

Nieschlag HJ, Nelson GH, Wolf IA, Perdue RE (1960) A search for new fibre crops. TAPPI 43: 193–201

Niu C, Hinchliffe DJ, Cantrell RG, Wang CL, Roberts PA, Zhang JF (2007) Identification of molecular markers associated with rootknot nematode resistance in upland cotton̄. Crop Sci 47: 951–960. https://doi.org/10.2135/cropsci2006.07.0499

Park YH, Alabady MS, Ulloa M, Sickler B, Wilkins TA, Yu J, Stelly D, Kohel RJ, El-Shiny OM, Cantrell RG (2005) Genetic mapping of new cotton fibre loci using EST-derived micro-molecular breeding of cotton 151 satellites in an interspecific recombinant inbred (RIL) cotton population. Mol Gen Genomics 274:428–441. https://doi.org/10.1007/s00438-005-0037-0

Paterson AH, Lander ES, Hewitt JD, Peterson S, Lincoln SE, Tanksley SD (1988) Resolution of quantitative traits into Mendelian factors by using a complete linkage map of restriction fragment length polymorphisms. Nature 335(6192):721–726

Percy RG, Kohel RJ (1999) Qualitative genetics. In: Smith CW, Cothren JT (eds) Cotton̄ origin, history, technology, and production. John Wiley & Sons, New York, pp 319–360

Poland JA, Brown PJ, Sorrells ME, Jannink JL (2012) Development of high-density genetic maps for barley and wheat using a novel two enzyme genotyping-by-sequencing approach. PLoS One 7(2):e32253

Poland JA, Rife TW (2012) Genotyping-by-sequencing for plant breeding and genetics, *the plant.* Genome 5(3):92–102. https://doi.org/10.3835/plantgenome2012.05.0005

Purseglove JW (1968) Tropical crops. Wiley, New York

Qin H, Guo W, Zhang YM, Zhang T (2008) QTL mapping of yield and fibre traits based on a four-way cross population in *Gossypium hirsutum L.* Theor Appl Genet 117(6):883–894

Rahman M, Ahmed N, Asif M, Zafar Y (2006) Identification of DNA markers linked with cotton leaf curl disease (CLCD). In: International cotton genome initiative (ICGI) work- shop, 2006; Brasilia, Brazil, pp 77–78

Ramasubramanian T, Regupathy A (2004) Pattern of cross-resistance in pyrethroid-selected populations of Helicoverpa armigera Hübner (Lep., Noctuidae) from India. J Appl Entomol 128(9-10):583–587

Reinisch AJ, Dong J, Brubaker CL, Stelly DM, Wendel JF, Paterson AH (1994) A detailed RFLP map of cotton. *Gossypium hirsutum* × *G. barbadense*: chromosome organization and evolution in a disomic polyploid genome. Genetics 138:129–147

Ribeiro IJA, de Miranda MCA, Bulisani EAL, Almeida DA, Lovadini LAC, Sugimori MH, Filho OP (1977) Melhoramento da crotalaria I-autocompatibilidade e resistancia a murcha de *Ceratocystis fimbriata*. Bragantia 36:291–295

Rotar PP, Joy RJ (1983) Tropic Sun' sunn hemp, *Crotalaria juncea* L. Univ. of Hawaii, College of Tropical Agr. and Human Resources, Institute of Tropical Agr. and Human Resources Research Extension, Honolulu. Series 36

Rungis D, Llewellyn D, Dennis ES, Lyon BR (2005) Investigation of the chromosomal location of the bacterial blight resistance gene present in Australian cotton (Gossypium hirsutum L.) cultivar. Australian J Agric Res 53:551–560. https://doi.org/10.1071/ar04190

Sabnis TS (1931) Report on Hemp marketing in India

Saha S, Jenkins J, Wu J, McCarty J, Gutierrez O, Percy R, Cantrell R, Stelly D (2006) Effect of chromosome specific introgression in upland cotton on fibre and agronomic traits. Genetics 172 (3):1927–1938

Saleem MA, Malik TA, Shakeel A, Ashraf M (2015) QTL mapping for some important drought tolerant traits in upland cotton3 *the journal of Animal & Plant*. Science 25:502–509

Santhanam V, Sundaram V (1997) Agri-history of cotton in India—an overview. Asian Agrihist 1:235–251

Schuster SC (2008) Next-generation sequencing transforms today's biology. Nat Methods 5:16–18. https://doi.org/10.1038/nmeth1156

Scott AW Jr, Cook CG (1994) Kenaf, crotalaria, roselle--two-year comparison in South Texas. In: Proc. Int. Kenaf Assoc. Conf, pp 97–101

Scott AW Jr, Namken LN, Garza J (1991) Kenaf/crotalaria variety study--1989. In: 50th Anniversary Research Report. Rio Farms, Inc., Monte Alto, pp 44–46

Seelanan T, Schnabel A, Wendel JF (1997) Congruence and consensus in the cotton tribe (Malvaceae). Syst Bot:259–290

Shen X, Guo W, Zhu X et al (2005) Molecular mapping of QTLs for fibre qualities in three diverse lines in upland cotton using SSR markers. Mol Breed 15(2):169–181

Simsek O, Kacar Y (2010) A. (2010), discovery of mutations with TILLING and ECOTILLING in plant genomes. Sci Res Essays 5:3799–3802

Stelly D, Saha S, Raska D, Jenkins J, McCarty J, Gutierrez O (2005) Registration of 17 upland (*Gossypium hirsutum*) germplasm lines disomic for different *G. barbadense* chromosome or arm substitutions. Crop Sci 45(6):1435–0653

Tanksley SD, Young ND, Paterson AH, Bonierbale MW (1989) RFLP mapping in plant breeding: new tool for an old science. Biotechnology 7:257–264

Taylor NE, Greene EA (2003) PARSESNP: a tool for the analysis of nucleotide polymorphisms. Nucleic Acids Res 31:3808–3811. https://doi.org/10.1093/nar/gkg574

Tidke JA, Patil GV (2000) Floral biology and pollination of Crotalaria sericea Retz. Environ Ecol 18(4):1020–1022

Till BJ, Comai L, Henikoff S (2007) TILLING and ECOTILLING for crop improvement. Genomics-assisted crop improvement. Genom Approach Platforms 1:333–349

Townsend DJ (2010) Stretching the dispute settlement understanding: US-Cotton's relaxed interpretation of cross-retaliation in the World Trade Organization. Rich J Global L & Bus 9:135

Verma RC, Raina SN (1991) Cytogenetics of Crotalaria. VII. Induced cytochimeral and asynaptic mutants of Crotalaria juncea (Sunnhemp). Indian J Genet Plant Breed 51:315–319

Wang B, Guo W, Zhu X, Wu Y, Huang N, Zhang T (2006b) QTL mapping of fibre quality in an elite hybrid derived-RIL population of upland cotton. Euphytica 152(3):367–378

Wang BH, Wu YT, Huang NT, Zhu XFW, Guo Z, Zhang TZ (2006a) QTL mapping for plant architecture traits in upland cotton using RILs and SSR markers. Acta Genet Sin 33(2):161–170

Wang FR, Gong YC, Zhang CY, Liu GD et al (2011) Genetic effects of introgression genomic components from Sea Island cotton (*Gossypium barbadense L.)* on fibre related traits in upland cotton (*G. hirsutum L.)*. Euphytica 181:41–53. https://doi.org/10.1007/s10681-011-0378-1

Wang P, Zhang J, Sun L, Ma Y, Xu J, Liang S et al (2018) High efficient multisites genome editing in allotetraploid cotton (*Gossypium hirsutum*) using CRISPR/ Cas9 system. Plant Biotechnol J 16(1):137–150

Waqas M, Khan AA, Ashraf J, Qayyum A, Ahmad MQ, Iqbal MZ et al (2014) Molecular markers and cotton genetic improvement: current status and future prospects. Sci World J 2014., Article ID 607091:15

Wendel JF, Brubaker C, Alvarez I, Cronn R, Stewart JM (2009) Evolution and natural history of the cotton genus. In: Genetics and genomics of cotton. Springer, New York, pp 3–22

White GA, Haun JR (1965) Growing *Crotalaria juncea*, a multi-purpose fibre legume, for paper pulp. Econ Bot 19:175–183

Yilmaz A, Boydak E (2006) The effects of cobalt-60 applications on yield and yield components of cotton (Gossypium barbadense L.). Pak J Biol Sci 9(15):2761–2769

Ynturi P, Jenkins JN, McCarty JC, Gutierrez OA, Saha S (2006) Association of root- knot nematode resistance genes with simple sequence repeat markers on two chromosomes in cotton. Crop Sci 46:2670–2674. https://doi.org/10.2135/cropsci2006.05.0319

Yu J, Zhang K, Li S, Yu S, Zhai M, Wu X, Li S et al (2013) Mapping quantitative trait loci for lint yield and fibre quality across environments in *a Gossypium hirsutum* × *Gossypium barbadense* backcross inbred line population. Theor Appl Genet 126:275–287. https://doi.org/10.1007/s00122-012-1980-x

Zhang HB, Li Y, Wang B, Chee PW (2008) Recent advances in cotton genomics. Int J Plant Genom. https://doi.org/10.1155/2008/742304

Zhang K, Zhang J, Ma J et al (2012) Genetic mapping and quantitative trait locus analysis of fibre quality traits using a three-parent composite population in upland cotton (*Gossypium hirsutum L.*). Mol Breed 29(2):335–348

Zhang ZS, Hu MC, Zhang J et al (2009) Construction of a comprehensive PCR-based marker linkage map and QTL mapping for fibre quality traits in upland cotton (Gossypium hirsutum L.). Mol Breed 24(1):49–61

Zhang ZS, Xiao YH, Luo M et al (2005) Construction of a genetic linkage map and QTL analysis of fibre-related traits in upland cotton (*Gossypium hirsutum L.*). Euphytica 144(1–2):91–99

Zhao Y, Wang H, Chen W, Li Y (2014) Genetic structure, linkage disequilibrium and association mapping of Verticillium wilt resistance in elite cotton (*Gossypium hirsutum L.*) germplasm population. PLoS One 9(1):e86308

Recent Advances in Omics Approaches for Mungbean Improvement

9

Chandra Mohan Singh, Aditya Pratap, Hitesh Kumar, Smita Singh, Bhupendra Kumar Singh, Durga Prasad, Indrapreet Dhaliwal, and Mukul Kumar

Abstract

Mungbean is the most important legume crop among the genus *Vigna*. It is widely cultivated in South-East Asian continents. It is grown throughout the year in different warm temperatures and tropical regions of India and other countries as well. It has easily digestible proteins along with the other micro-nutrients. Therefore, it is used as food, feed, forage, green manure and cover crop. It also improves soil health through the symbiotic association of *Rhizobia*. Keeping the importance of mungbean, it is still affected by various biotic and abiotic stresses. A modest effort has been made by several researchers and some extent of success was also noticed. Now, molecular breeding is more sound after decoding the whole genome sequence data of mungbean. Various genes/QTLs were also tagged and mapped for many important traits like yellow mosaic disease, powdery mildew, *Cercospora* leaf spot, and leaf shape. A very recently, the

C. M. Singh (✉) · H. Kumar · S. Singh · M. Kumar (✉)
Department of Genetics and Plant Breeding, Banda University of Agriculture and Technology, Banda, Uttar Pradesh, India

A. Pratap
Division of Crop Improvement, ICAR-Indian Institute of Pulses Research, Kanpur, Uttar Pradesh, India

B. K. Singh
Department of Entomology, Banda University of Agriculture and Technology, Banda, Uttar Pradesh, India

D. Prasad
Department of Plant Pathology, Banda University of Agriculture and Technology, Banda, Uttar Pradesh, India

I. Dhaliwal
Department of Plant Breeding and Genetics, Punjab Agricultural University, Ludhiana, Punjab, India

Kamaluddin et al. (eds.), *Technologies in Plant Biotechnology and Breeding of Field Crops*, https://doi.org/10.1007/978-981-16-5767-2_9

bruchid-resistant gene was also mapped. Besides the genomics, the transcriptome data for various traits like desiccation tolerance and root traits are also available. Some of the linked markers are reported by researchers based on mungbean genome sequence and other related legumes, which will be useful for mungbean improvement. Nevertheless, the use of molecular breeding tools still lag in mungbean due to unavailability of the robust marker validation, and therefore need special attention. The recent advancements in omics approaches after six years of decoding the genome sequence and the significant achievements made in mungbean have been discussed in this chapter.

Keywords

Double haploid · Marker-assisted recurrent selection · Marker-assisted selection · Mungbean · Proteomics · Recombinant inbred line

9.1 Introduction

The genus *Vigna* is an immensely variable and large genus comprising several species. It includes mungbean (*Vigna radiata* L. Wilczek), urdbean (*Vigna mungo* L. Hepper), cowpea (*Vigna unguiculata* L. Walp), adzukibean (*Vigna angularis* Willd. Ohwi and Ohashi), bambara groundnut (*Vigna subterranea* L. Verdn.), mothbean (*Vigna aconitifolia* Jacq.), rice bean (*Vigna umbellata* Thunb. Ohwi & Ohashi), and beach pea (*Vigna marina*) (Pratap et al. 2015). Of these, mungbean having chromosome number $2n = 2X = 22$ is the most important among the grain legumes in developing countries of Asia, Africa and Latin America, where it is consumed as food and feed and also used as green manuring crop (Fig. 9.1).

The dried seeds of mungbean and urdbean contain about 25–28% protein, 1.0–1.5% oil, 3.5–4.5% fiber, 4.5–5.5% ash and 62–65% carbohydrate on dry weight basis (Singh et al. 2017). Besides dried seeds, sprouts and fresh green pods

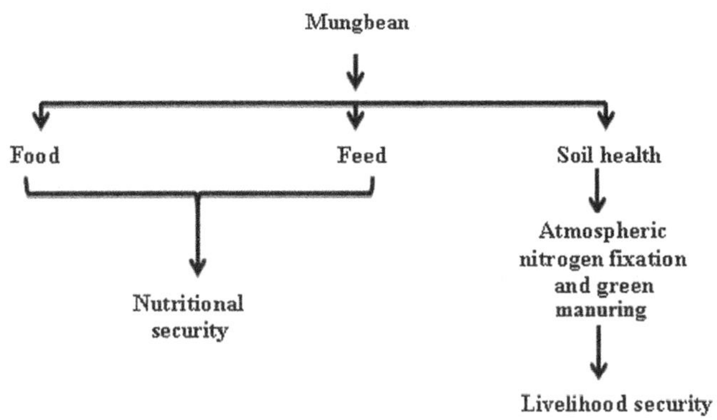

Fig. 9.1 The importance of mungbean as multipurpose crops

are also consumed (Karuppanapandian et al. 2006). The shorter life cycle of mungbean makes it more suitable for cultivation in spring/summer season and fits well in the wheat-rice cropping system on the vast plains of central and northern India, providing additional yield benefit to growers. Their wider adaptability to extreme environmental conditions, short life cycle, low input requirements and nitrogen-fixing ability highly augment their importance (Hoorman et al. 2009; Jat et al. 2012). It also improves the soil texture and enhance the potential of the subsequent crops (Yaqub et al. 2010). Owing to these features, there is ample scope of vertical as well as horizontal expansion of mungbean worldwide. However, this crop is highly affected by various biotic and abiotic stresses (Singh et al. 2015; Kumar et al. 2016, 2021; Nair et al. 2019; Mishra et al. 2020). Therefore, the development of high-yielding genotypes with enhanced resistance to diseases and insects/pests remains the most prioritized goal in their genetic improvement. To this end, molecular breeding in mungbean is now on a sounder footing after the availability of the whole genome sequence of mungbean (Kang et al. 2014; Jiao et al. 2016). The genome size of *Vigna* ranged from 416–1394 Mb (Parida et al. 1990; Lakhanpaul and Babu 1991) in which mungbean has a genome size of 512 Mb (Kang et al. 2014). A relatively recent approach, genotyping by sequencing is conducive towards the identification and introgression of useful genes from wild relatives into cultivated species. Several new cultivars with desirable agronomic traits and improved characteristics have already been developed in mungbean. Noteworthy progress has also been made in mungbean through genetic transformation. This chapter thus aims to elaborate such novel efforts and provide an insight into the improvement of mungbean.

9.2 Origin and Evolution

The origin of a species is generally concluded on the basis of information retrievable from archeological remains as well as on the basis of the existence of wild species or progenitors. Natural forces such as spontaneous mutation, migration, natural hybridization and genetic drift alter these wild species/progenitors resulting in their evolution into cultivated species (Pratap and Kumar 2011). Mungbean (*V. radiata* var. *radiata*) is the most important crops among *Vigna* crops, is believed to have originated in the Indian subcontinent (de Candolle 1884; Vavilov 1926; Zukovskij 1962). Further, a wide range of genetic diversity of cultivated as well as weedy and wild types of *Vigna* in India also make it a strong candidate as the region of its first domestication (Baudoin and Marechal 1988). Mention of mungbean has also been made in Vedic texts such as in Kautilyas' *Arthashastra* and *Charak Samhita* point to their origin further beyond the pre-Christian era (Jain and Mehra 1980). It is believed that mungbean (*V. radiata*) has been domesticated from *Vigna sublobata,* considered as wild progenitor of mungbean. It is found in abundance as weeds in cultivated and wasteland areas of India (Singh et al. 1974; Chandel et al. 1984; Lawn and Cottell 1988) (Fig. 9.2) and in subtropical wetland regions of Northern and Eastern Australia (Lawn and Cottell 1988). A considerable amount

Vigna radiata *Vigna sublobata*

Fig. 9.2 Cultivated and progenitor species of mungbean

of genetic variability has also been observed in the Western Ghats (Aditya Pratap and Joseph K. John, unpublished data).

9.3 Taxonomy and Plant Characteristics

The genus *Vigna*, earlier known as *Phaseolus*, belongs to family Leguminoceae and sub-family Papilionaceae. It is a large tropical genus consisting of six subgenera: *Ceratotropis*, *Haydonia*, *Lasiospron*, *Plectotropis*, *Sigmoidotropis* and *Vigna*. *Ceratotopis* has its centre of diversity in Asia and consists of 21 species, eight of which are domesticated; *viz*., moth bean [*V. aconitifolia* (Jacq.) Maréchal], adzuki bean [*V. angularis* (Willd.) Ohwi and Ohashi], urdbean [*V. mungo* (L.) Hepper], mungbean [*V. radiata* (L.) Wilczek], creole bean [*V. reflexo-pilosa* Hayata var. *glabra*, Maréchal, Mascherpa and Stainer N. Tomooka and Maxted], jungli bean [*V. trilobata* L. Verdc Toapée], Thai [*V. trinervia*, Heyne and Wall Tateishi and Maxted] and rice bean [*V. umbellata*, Thunb. Ohwi and Ohashi]. The most important among these are moth bean, adzuki bean, urdbean and mungbean (Tomooka et al. 2006). Mungbean has been specially named as *radiata* due to the terminal position of the pods. The key morphological features of mungbean are summarized in Table 9.1.

9.4 Floral Biology

Mungbean is predominantly self-pollinated crop due to the cleistogamous nature of flowers. The flower is complete pentamerous, small, yellow-coloured, with bracts and hermaphrodite. It has five sepals (three large and two small, fused), five petals (one standard, two wings and two keels), diadelphous androecium (9+1) and one monocarpellary gynoecium. The stigma is hairy and oblique (Fig. 9.3).

Table 9.1 Major characteristics of mungbean

Trait	Key features
Plant type	Erect, semi erect, spreading
Growth habit	Non-determinate (some are determinate)
Leaf shape	Regular trifoliate (lobed, in some cases)
Stipule	Broad
Hairiness	Sparse
Pod position	Terminal
Pod wall thickness	Thin
Number of seeds	10–13
Maturity	Mostly non-synchronous
Seed coat colour	Green, yellow
Seed size	Small, medium bold, bold
Seed lustre	Shiny, dull
Hilum	Flat

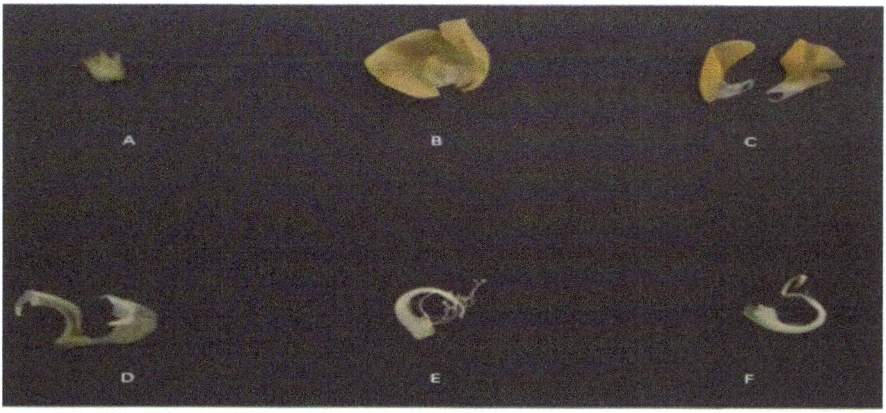

Fig. 9.3 Presentation of flower parts of mungbean. (**a**) Sepals. (**b**) Petals. (**c**). Wing. (**d**) Keel. (**e**) Androecium. (**f**) Gynoecium

9.5 Priorities and Thrust Areas

The development of high-yielding varieties is the prime objective of any crop improvement programme. Due to the high weather-susceptible nature of mungbean, development of short duration varieties with distinct vegetative and reproductive phases for fitting into the narrow windows of cereal-cereal cropping system is our basic need for summer cultivation. Besides, the areas where heavy rain is received during the early monsoon, development of waterlogging tolerance cultivars at seedling stage should be our focus. Breeding of erect plant type and synchronous maturing, wider adaptability and stable varieties make it convenient for enhancing the yield by the incorporation of more number of plants. Incorporation of photo-thermo-insensitivity traits from wild to cultivars should help in the expansion of mungbean throughout the years. Pyramiding of genes for resistance to major insect pests like thrips, jassids, pod borer and hairy caterpillar; and diseases such as YMD, PM and CLS, for which resistance level is not so high in presently cultivated germplasm, make it more suitable (Singh et al. 2020a, b; Pratap et al. 2020). Along with the biotic stresses, the abiotic stresses are also prevalent in mungbean (Singh et al. 2021a). This crop is highly sensitive to salt, drought and heat stress, so the focus on developing abiotic stress-tolerant varieties are also the aim of breeders. Identification of diverse germplasm sources including wild and weedy species for incorporating important economic traits and breeding new plant types. Marker-assisted selection (MAS) fastens the breeding cycles and help in pyramiding the desirable genes from wild and weedy sources to cultivated background.

9.6 Breeding Strategies

It includes both conventional and recent molecular breeding. The effective breeding approaches for mungbean improvement are given in Fig. 9.4.

9.6.1 Conventional Breeding Methods

9.6.1.1 Introduction and Selection

Introduction, transposition of crop plants from one place to another place, is the quickest method of varietal development. In this method, only we need to acclimatize the introduced variety and release for its general cultivation. In selection, the genotypes with differential reproduction rates are tagged and superior plants from the same genotypes are bulked or multiply the superior individual plant and develop a variety.

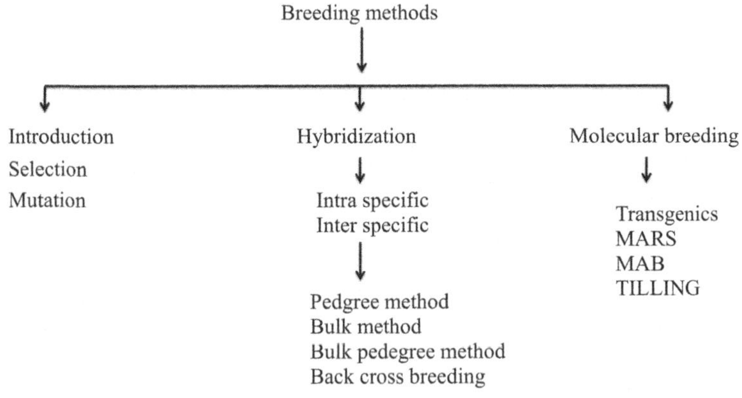

Fig. 9.4 The breeding methods applicable to mungbean improvement

9.6.1.2 Mutation Breeding

Induced mutations have been found useful in creating useful variability for yield traits, plant type and resistance to various stresses. So far 52 varieties have been developed through mutation breeding in different pulse crops among which 8 are in mungbean. Most of these have been developed from already released and adapted varieties. In general, gamma-irradiation has been used extensively while rarely chemical mutagens have been used. The mutant varieties are an improvement over their maternal parent and/or standard check for character(s) such as plant type, seed size, seed colour, maturity duration and resistance to disease(s). Mutation breeding has resulted in the development of Pant Moong 2, MUM 2, Co 4, LGG 407, LGG 405 and BM 4 in mungbean.

9.6.1.3 Hybridization

The initial phase of varietal development had been aimed at improving locally adapted but genetically variable populations, mainly by pedigree breeding followed by selections with major emphasis on traits other than yield. This resulted in the release of a large number of purelines, few of which are still cultivated in some parts of the country. Lately, the focus has shifted towards development of short duration, photo- and thermo-insensitive varieties; coupled with resistance to major biotic stresses viz., yellow mosaic disease, *Cercospora* leaf spot, powdery mildew, etc. which have significantly contributed to the national mungbean production. Wild relatives offer a number of useful genes which are not found in cultivated germplasm (Pratap et al. 2021a). Distant hybridization has led to the development of a number of promising advanced breeding lines under *Vigna* improvement programme. Mungbean × urdbean crosses have been routinely attempted as the derivatives from these crosses exhibit many desirable features such as lodging resistance, synchronous podding, non-shattering pods and enhanced YMD resistance (Reddy and Singh 1990; Gill et al. 1983). Singh and Dikshit (2002) successfully introgressed yield genes in mungbean from urdbean imparting 15–60% yield advantage. Similarly, progenies from mungbean × rice bean and mungbean × *V. sublobata* crosses have also exhibited a higher extent of resistance to YMV (Verma and Brar 1996). Chaisan et al. (2013), Basavraja et al. (2018), Bhanu et al. (2018) generated the cross combinations between *V. umbellata* × *V. radiata* for transferring the MYMV resistance in mungbean. Bruchid (*Callosobruchus chinensis* L.) is major post-harvest constraint of grain legumes. One accession of wild mungbean (*Vigna sublobata*) exhibited complete resistance to adzuki bean weevils and cowpea weevils (Fujii et al. 1989) which has been successfully used in the breeding programme (Tomooka et al. 1992). An accession of *V. mungo* var. *silvestris* is also reported to be immune to bruchids (Fujii et al. 1989; Dongre et al. 1996). Likewise, Pandiyan et al. (2020) also transfered the bruchid resistance from cultivated rice bean into mungbean and identified three bruchid-resistant recombinant inbred lines (RILs).

9.6.2 Molecular Breeding

In current trends of plant breeding, high yield coupled with enhanced tolerance and traits relating to yield stability and sustainability should be a major focus for crop improvement. These major prime traits include durable resistance to diseases, insect pest and abiotic stress tolerance (Nair et al. 2019). Despite the continued yield improvement through conventional breeding, new biotechnological techniques will be needed to maximize the success probability (Mehandi et al. 2019; Bohra et al. 2020; Varshney et al. 2021; Pratap et al. 2021b; Singh et al. 2021c). Molecular marker technology and developing genomic resources, offers great promise for plant breeding (Prakash et al. 2016; Roorkiwal et al. 2020; Bohar et al. 2020; Singh et al. 2021b). Owing to genetic linkage, DNA markers can be used to detect the presence of allelic variation in the genes and increase the efficiency of breeders. The effective utilization of these markers as a tool in plant breeding accelerates MAS.

9.6.2.1 Mapping Populations

Various types of plant populations are used for mapping and tagging of quantitaive trait loci (QTLs)/genes governing various developmental traits, biotic and abiotic stresses. Nature of mapping population is very important for detecting its power for mapping. In general, F_2 segregating populations originated from extreme phenotype of a trait(s) are used for mapping and tagging. Besides this, backcross (BCF_2) populations and some of the fixed population such as double haploids (DH), RILs, near isogenic lines (NILs), nested association mapping (NAM), multiple advanced generations inter-cross (MAGIC) populations are frequently used for the purpose of mapping. However, these populations are developed by using parents with extreme phenotypes, therefore only bi-parental segregation occurs, which is the major limit of the linkage mapping (Jaiswal et al. 2012). In recent years, exploring QTLs by association analysis has been one of the effective approaches in quantitative genetics, which performs rapid and fine-mapping of the target locus (Singh et al. 2020a, b). Thomas (2017) developed a mapping panel of mungbean which consisted of 30 crosses including four interspecific crosses to *Vigna sublobata* and advanced to F5 generations for mapping complex traits like drought and heat. Association mapping has an ability to detect more QTLs because it uses diverse germplasm that has more allelic diversity and occurrence of several random events because of its parental evolution history than bi-parental population. Many researchers used marker-trait associated through association mapping studies for mapping various important traits, for example, fibre quality in cotton (Nie et al. 2016), MYMIV resistance in soybean (Kumar et al. 2014), seed weight in soybean (Zhao et al. 2019), agronomic and flowering traits in lentil (Kumar et al. 2017a, b), seed coat colour in mungbean (Noble et al. 2018) and *MYMIV* resistance in mungbean (Singh et al. 2020a, b).

9.6.2.2 Genome Sequence Information

The molecular breeding is accelerated after the availability of genome sequence data (Kang et al. 2014). Before the decoding of genome sequence information, the

Table 9.2 Genome sequence information of mungbean and related *Vigna* species

Vigna species	Genotype	Chromosome number	Genome size (Mb)	Reference
Vigna radiata	*VC1973A*	$2n = 2x = 22$	548	Kang et al. (2014)
Vigna angularis	Gyeongwon	$2n = 2x = 22$	591	Kang et al. (2015)
Vigna stipulacea	JP245503	$2n = 2x = 22$	445	Takahashi et al. (2019)
Vigna marina	ANBp-14-03	$2n = 2x = 22$	265	Singh et al. (2019)
Vigna umbellata	VRB3	$2n = 2x = 22$	414	Kaul et al. (2019)
Vigna unguiculata	IT97K-499-35	$2n = 2x = 22$	519	Lonardi et al. (2019)
Vigna mungo	PU-31	$2n = 2x = 22$	498	Souframanien et al. (2020)

researchers also used transferable markers from other *Vigna* species in the analysis of mungbean (Pratap et al. 2016, 2017) and other legumes (Choi et al. 2004; Choumane et al. 2004; Guohao et al. 2006; Gupta et al. 2013). Kang et al. (2014) sequenced a mungbean pure line namely VC1973A, its relative *V. reflexo-pilosa* var. *glabra*, *V. radiata* var. *sublobata* to construct draft genome sequence. The estimated genome size ranged from 579 Mb to is about 968 Mb (*Vigna reflexo-pilosa* var. *glabra*). In addition, they also sequenced a wild relative namely TC1966, a *V. radiata var. sublobata* accession with the estimated genome size of 501 Mb. The mungbean whole-genome sequence information will further boost genomics research in *Vigna* species and accelerate mungbean breeding programmes. Jiao et al. (2016) re-sequenced the two accessions of mungbean namely Salu and AL127 for mapping *lma* locus and 236,998 single nucleotide polymorphisms and 8896 insertion/deletions (InDels) were identified (Table 9.2).

9.6.2.3 QTL Identification

The genetic linkage analysis helps in detecting genes/QTLs controlling the traits, which is based on the recombination events during meiosis (Tanksley 1993). Using statistical methods such as single-marker analysis or interval mapping or composite mapping to detect associations between DNA markers and trait of interest, the genes or QTLs can be detected (Kearsey 1998). The identification of QTLs using molecular markers was a major breakthrough in characterizing quantitative traits (Paterson et al. 1988). Previously, the researchers believed that the identified markers associated with QTLs from preliminary mapping studies were directly used in MAS. However, in the recent past, it has become widely accepted that QTL confirmation and their validation are required (Langridge et al. 2001). However, limitations of linkage analysis such as the bi-parental segregation and determination of two alleles only at the same loci restricted to explore the many types of QTLs. Limited number of possible events occurring in gene loci leads to low resolution QTLs. Several researchers identified various QTLs for many traits using linkage mapping approach as listed in Table 9.3.

In recent years, exploring QTLs by association analysis is the most active approach in quantitative genetics to trace novel QTLs. Table 9.4 gives the list of gene(s)/QTLs governing various biotic and abiotic stress tolerance in mungbean. It

Table 9.3 List of gene(s)/QTLs governing various traits in mungbean

Trait[a]	Gene/QTL name	LG	Marker interval	Position (cM)	Variation covered (%)	References
DF	*Fld2*	2	VR0364	72.70	15.88	Kajonphol et al. (2012)
DF	*Fld4.1*	4	CEDG241-VR-SSR019	9.85	7.39	
DF	*Fld4.2*	4	DMB-SSR199-CEDG107	69.27	28.57	
DF	*Fld11*	11	VR0216-CEDG168	14.01	6.28	
DF	*Fld5.2.1*	2	GMES0477-CEDG026a	62.1	32.9	Isemura et al. (2012)
DF	*Fld5.4.1*	4	CEDC085b-MBSSR015a	75.7	24.0	
DF	*Fld5.6.1*	6	GMES0659-CEDG146	30.1	17.4	
DF	*Fld5.11.1*	11	CEDG281-CEDG072	15.1	6.6	
DFF	DFL4.1	4	MB213-VR057	5.77	33.38	Kajonphol et al. (2012)
DFF	DFL4.2	4	CEDG154-CEDG091	8.35	25.70	
DFF	DFL7.1	7	VR126-CEDG143	4.43	5.01	
DFPM	*Pddm2*	2	VR0364	72.70	12.58	
DFPM	*Pddm4.1*	4	CEDG241-VR-SSR019	9.85	8.43	
DFPM	*Pddm4.2*	4	DMB-SSR199-CEDG107	69.27	27.83	
DM	DMT4.1	4	CEDG086-VR057	4.89	25.67	
DM	DMT7.1	7	GBssr-MB7-CEDG143	2.64	7.24	
DM	*Pddh2*	2	VR0364	72.70	15.60	
DM	*Pddh4.1*	4	CEDG241-VR-SSR019	9.85	11.73	
DM	*Pddh4.2*	4	VR0313	17.03	11.63	
DM	*Pddm5.2.1*	2	cp00228a-GMES1156a	53.9	20.3	Isemura et al. (2012)
DM	*Pddm5.4.1*	4	CEDC085b-MBSSR015a	74.6	19.9	
DM	*Pddm5.6.1*	6	CEDG146-1584	30.8	8.3	
DM	*Pddm5.9.1*	9	CEDG166-GATS11	17.8	10.2	
100-SW	*Sd100wt2.1*	2	VR078-CEDG065	4.01	14.56	Kajonphol et al. (2012)
100-SW	*Sd100wt2.2*	2	VR17-VR0200	19.13	12.99	
100-SW	*Sd100wt4*	4	VR0366-VR035	35.08	11.96	

(continued)

Table 9.3 (continued)

Trait[a]	Gene/QTL name	LG	Marker interval	Position (cM)	Variation covered (%)	References
100-SW	*Sd100wt8*	8	VR-SSR031-VR0225	52.22	8.16	
100-SW	*Sd100wt9*	11	CEDG259-CEDG166	19.21	7.22	
100-SW	*Sd100wt11*	1	MB-SSR104-VR-SSR011	53.08	8.74	
100-SW	SD100WT1.1	2	CP1713-CP5137	6.52	18.41	
100-SW	SD100WT2.1	8	VR078-CEDG136	2.60	6.80	
100-SW	SD100WT8.1	9	CP2470-CLM871	2.51	11.70	
100-SW	SD100WT9.1	10	VRSSR010-MB36	2.90	17.88	
100-SW	SD100WT10.1	10	CEDG097-CP2142	3.02	12.38	
100-SW	*Sd100wt5.1.1*	1	CEDG053-CEDG001	25.7	4.4	Isemura et al. (2012)
100-SW	*Sd100wt5.1.2*	1	cp05137-CEDG074b	64.6	4.9	
100-SW	*Sd100wt5.2.1*	2	GMES0477-CEDG026a	58.8	16.6	
100-SW	*Sd100wt5.3.1*	3	GMES2040-GMES6175	29.2	7.0	
100-SW	*Sd100wt5.7.1*	7	GMES1424-MBSSR203	1.5	6.3	
100-SW	*Sd100wt5.8.1*	8	VM37-CEDG030	46.6	22.2	
SP	*Sdnppd1.1*	1	VR-SSR015-VR-SSR018	34.69	12.29	Kajonphol et al. (2012)
SP	*Sdnppd1.2*	1	VR0194-VR0198	46.18	12.08	
SP	*Sdnppd5.1.1*	1	CEDG220-GMES4400	69.8	9.1	Isemura et al. (2012)
SP	*Sdnppd5.9.1*	9	CEDG166-GATS11	16.7	7.0	
PL	*Pdl7*	7	CEDG111-VR0126	66.86	10.74	Kajonphol et al. (2012)
PL	*Pdl8*	8	VR-SSR005-VR-SSR031	50.43	9.26	
PL	*Pdl5.1.1*	1	CEDG256-cp04220	49.4	17.8	Isemura et al. (2012)
PL	*Pdl5.2.1*	2	CEDAAG002-GMES4137	80.2	15.0	
PL	*Pdl5.7.1*	7	CEDG064-CEDG174	6.5	20.5	

(continued)

Table 9.3 (continued)

Trait[a]	Gene/QTL name	LG	Marker interval	Position (cM)	Variation covered (%)	References
PL	*Pdl5.8.1*	8	CEDG257-CEDG059	50.9	5.9	
PL	*Pdl5.9.1*	9	CEDG166-GATS11	15.5	5.5	
PN	*Pdtn5.2.1*	2	CEDG096a-GMES0216b	51.6	6.5	
PN	*Pdtn5.4.1*	4	GMES1124-CEDG107	37.3	5.8	
PN	*Pdtn5.4.2*	4	GMES0216a-GMES1216a	64.4	12.0	
PN	*Pdtn5.7.1*	7	GMES0856b-GMES6625	19.4	6.2	
SPA	SDPAP4.1	4	CEDG139-MB-SSR179	3.78	11.24	Kajonphol et al. (2012)
SPA	SDPAP11.1	11	BM141-VR222	4.00	6.06	
STP	SDTP4.1	4	Bmd25-MB-SSR179	2.64	10.58	Isemura et al. (2012)
STP	–	7	GMES1424-cp07863	7.2	–	
STP	–	8	VM37-CEDG059	6.00–22.00	–	
SD	–	1	CE CEDCAA001a-cp04220	15.00	–	

[a]*DF* days to flowering, *DFF* days to 50% flowering, *DM* days to maturity, *SW* seed weight, *SP* seeds per pod, *PN* pod number, *SPA* seed phytic acid, *STP* seed total phosphorus, *SD* seed dormancy

is based on linkage disequilibrium and diversity analysis of trait of interest/locus. It offers the advantages of natural population and detecting multiple alleles on the same locus and targeting single genes.

9.6.2.4 Genome-Wide Association Studies (GWAS)

GWAS is an effective approach for detecting marker-trait association with higher precision in lesser time (Brachi et al. 2011). This approach is also effective in the condition of the non-availability of whole genome sequence data. Various researchers like Sai et al. (2017) prepared the genotyping by sequencing (GBS)-based linkage map for *MYMIV* resistance in mungbean. Mathivathana et al. (2019) prepared the GBS-based linkage map for *MYMV* resistance in inter-specific population of mungbean × rice bean and reported five QTLs. Likewise, Schafleitner et al. (2016) adopted GBS approach for detecting QTLs for bruchid resistance in mungbean. Noble et al. (2018) characterized a mungbean panel consisting of 466 cultivated and 16 wild accessions for tagging QTLs associated with seed coat colour.

Table 9.4 List of gene(s)/QTLs governing various biotic and abiotic stress tolerance in mungbean

Trait[a]	Marker/ gene/QTL name	LG	Marker interval	Position (cM)	PV (%)	References
MYMIV	*qYMIV1*	2	CEDG100-cp02662	38.8	9.33	Kitsanachandee et al. (2013)
MYMIV	*qYMIV2*	4	DMB-SSR008-VR113	51.8	10.67	
MYMIV	*qYMIV3*	9	CEDG166-CEDG304	50.2	12.55	
MYMIV	*qYMIV4*	2	CEDG100-cp02662	29.7	27.93	
MYMIV	*qYMIV5*	6	CEDG121-CEDG191	0.1	6.24	
MYMIV	*qYMIV2.1*	2	CEDG275-CEDG006	9.0/12.3	–	Alam et al. (2014)
MYMIV	*qYMIV7.1*	7	CEDG041-VES503	8.9/8.6	–	
PM	*qPMR-1*	–	CEDG282-CEDG191	–	20.10	Kasettranan et al. (2010)
PM	*qPMR-2*	–	MB-SSR238-CEDG166	–	57.81	
CLS	qCLS-1	3	CEDG117-VR393	26.91		Chankaew et al. (2014)
CLS	qCLS-2	3	CEDG117-VR393	31.67		
Drought	*qPH5A*	5	GMES5773 to MUS128	–	6.40–20.06	Liu et al. (2017)
Drought	*qMLA2A*	2	Mchr11-34 to the HAAS_VR_1812	–	6.97–7.94	

[a]*MYMIV* mungbean yellow mosaic India virus, *PM* powdery mildew, *CLS Cercospora* leaf spot, *PV* percent variation

9.6.2.5 Marker-Assisted Selection and Backcross Breeding

Marker-assisted breeding is now more sounds because of introgression of actual genes and genomic segments in lesser time. The main advantage of MAS is as a simpler and rapid method than phenotypic screening, which saves time, resources and effort. Selection can be carried out at the early plant growth stage such as seedling stage and single plants can be selected instead of many plant families of conventional screening methods. MAS allows selecting the individual plants based on their own genotype. It is also allowing to select the targeted gene for a trait of interest under transfer through foreground selection and recovery of recurrent parental background through background selection. Table 9.5 gives the list of DNA markers linked to gene(s) of interest governing various traits in mungbean.

Backcrossing is a widely used technique in plant breeding to incorporate the one or a few genes into an adapted or elite variety. The parent under transfer of the gene

Table 9.5 List of DNA markers linked to gene(s) of interest governing various traits in mungbean

Crop	Trait	Marker/gene name	Reference
Mungbean	*MYMIV*	CYR1	Maiti et al. (2011)
		YR4	
		CEDG293, DMB-SSR008, DMB-SSR059	Singh et al. (2020a, b)
	MYMV	MYMVR-583	Binyamin et al. (2015)
	Salt	YMV1	
		SSR9293	
		SSR3435	
		SSR6263	
	Bruchid	DMB-SSR-158	Chen et al. (2012)
		MB-87-COPU11	Hong et al. (2015)
		MB-RP-COPU06	
		CAPS3, CAPS4, CAPS12, CAPS13	Schafleitner et al. (2016)
		g779p	Liu et al. (2016)
		g34458p	
		g34480p	

MYMIV mungbean yellow mosaic India virus, *MYMV* mungbean yellow mosaic virus

trait is called non-recurrent parent, and parent used in repeated backcrossing to recover the background is called recurrent parent. Molecular markers help to increase the efficiency of the selection. Generally, three types of marker-assisted backcrossing are used in plant breeding i.e. foreground selection, recombinant selection and background selection. Screening for the target gene or QTLs in the population is referred as foreground selection (Hospital and Charcosset 1997). It helps in shortlisting the material quickly than the phenotypic screening. The second level of backcrossing involves, selecting BCF_2 progenies with the target gene or QTLs and recombination events between the target locus and linked markers, referred as recombinant selection. The purpose of recombinant selection is to reduce the size of the donor chromosome segment and to reduce the linkage drag (Hospital 2005). The third level of MABC involves selecting BCF_2 progenies with the greatest proportion of recurrent parent (RP) genome, using markers that are unlinked to the target locus, referred as background selection (Bertrand et al. 2008). It is also referred to for the use of tightly linked flanking markers for recombinant selection and unlinked markers to select the background of recurrent parent (Hospital and Charcosset 1997; Frisch et al. 1999b). Foreground markers should be tightly linked with the target gene/QTL, whereas background markers should be unlinked to the target gene/QTLs. In conventional backcrossing, it takes six to seven BCF_2 generations to recover the background but by using molecular markers, it can be achieved in early BCF_2 generations (Visscher et al. 1996; Hospital and Charcosset 1997; Frisch et al. 1999a, b).

9.7 Targeting Induced Local Lesion in Genome (TILLING)

Targeting induced local lesion in genome (TILLING) first began in the late 1990s (Henikoff and Comai 2003). TILLING is a reverse genetic approach, which aims to identify SNPs (single nucleotide polymorphisms), insertions or deletions in a gene (s) of interest from a mutagenized population. Therefore, the first step in TILLING is the creation of a mutagenic population, by treatment with a chemical mutagen such as EMS. Generally, EMS produces transition mutations (Henikoff and Comai 2003). The plant seeds are treated with EMS and allow to produce M_1 generation, which are subsequently self-fertilized to produce the M2 generation. The DNA from M_2 plants are collected and used for mutational screening (Colbert et al. 2001). EcoTILLING is also a molecular technique similar to TILLING, except with the objective to uncover natural genetic variation. It allows to rapid screening of several samples with a gene of interest to identify naturally mutagenic variations (SNPs). It also has the benefit of multiple polymorphisms detection in a single fragment.

9.8 Conclusion and Summary

A tremendous advancement is noticed in mungbean breeding after the release of whole genome sequence of mungbean (Kang et al. 2014). Now the days, various model legumes i.e. *Lotus japonicas*, *Medicago truncatula*, etc. sequences data are available for comparative genomics studies. After the whole genome sequencing of mungbean, the genome was re-sequenced also by several workers Schafleitner et al. (2016), Liu et al. (2016), Chen et al. (2012), Liu et al. (2017) for various traits. These sequences might be very much useful in developing markers for accelerating, marker-assisted breeding for various trait improvements in mungbean. The previously identified QTL regions need to be saturated with high density to help in identification of more closely related markers to the gene of interest using genomics and transcriptomics approaches. The tightly linked markers and donors for various traits may be effectively utilized in genomic-enabled breeding for mungbean and urdbean improvement.

Acknowledgement The authors acknowledge the contribution of authors cited. The authors declared that there is no conflict of interest among them. All the authors have read the manuscript and approved for its publication.

References

Alam AKMM, Somta P, Srinives P (2014) Identification and confirmation of quantitative trait loci controlling resistance to mungbean yellow mosaic disease in mungbean [*Vigna radiata* (L.) Wilczek]. Mol Breed. 34:1497–1506. https://doi.org/10.1007/s11032-014-0133-0

Basavraja T, Murthi N, Kumar LV et al (2018) Studies on cross compatibility in interspecific crosses of *Vigna radiata* × *Vigna umbellata* species. Legume Res 42:699–704. https://doi.org/10.18805/LR-3974

Baudoin JP, Marechal R (1988) Taxonomy and evolution of the genus *Vigna*: mungbean. In: Proceedings second international symposium. AVRDC, Bangkok

Bertrand C, Collard Y, Mackill DJ (2008) Marker-assisted selection: an approach for precision plant breeding in the twenty-first century. Philos Trans R Soc Lond B Biol Sci 12:557–572. https://doi.org/10.1098/rstb.2007.2170

Bhanu AN, Singh MN, Srivastava K (2018) Crossability studies of interspecific hybridization among *Vigna* Species. Biomed J Sci Technol Res 2:1–7. https://doi.org/10.26717/BJSTR.2018.02.000818

Binyamin R, Khan MA, Khan NA, Khan AI (2015) Application of SCAR markers linked with mungbean yellow mosaic virus disease-resistance gene in Pakistan mungbean germplasm. Genet Mol Res 14:2825–2830

Bohar R, Chitkineni A, Varshney RK (2020) Genetic molecular markers to accelerate genetic gains in crops. Biotechniques. https://doi.org/10.2144/btn-2020-0066

Bohra A, Jha UC, Godwin ID et al (2020) Genomic interventions for sustainable agriculture. Plant Biotech J 18(12):2388–2405. https://doi.org/10.1111/pbi.13472

Brachi B, Morris GP, Borevitz JO (2011) Genome-wide association studies in plants: the missing heritability is in the field. Genome Biol 10:232

Chaisan T, Somta P, Srinives P et al (2013) Development of tetraploid plants from an interspecific hybrid between mungbean (*Vigna radiata*) and rice bean (*Vigna umbellata*). J Crop Sci Biotech 16:45–51

Chandel KPS, Lester RN, Starling RJ (1984) The wild ancestors of urd and mungbeans [*Vigna mungo* (L.) Hepper and *V. radiata* (L.) Wilczek]. Bot J Linn Soc 89:85–96

Chankaew S, Isemura T, Isobe S et al (2014) Detection of genome donor species of neglected tetraploid crop *Vigna reflexo-pilosa* (Creole bean) and genetic structure of diploid species based on newly developed EST-SSR markers from azukibean. PLoS One 9:e104990. https://doi.org/10.1371/journal.pone.0104990

Chen H-M, Ku H-M, Schafleitner R et al (2012) The major quantitative trait locus for mungbean yellow mosaic Indian virus resistance is tightly linked in repulsion phase to the major bruchid resistance locus in a cross between mungbean [*Vigna radiata* (L.) Wilczek] and its wild relative *Vigna radiata* ssp. *sublobata*. Euphytica 192:205–216. https://doi.org/10.1007/s10681-012-0831-9

Choi HK, Mun JH, Kim DJ et al (2004) Estimating genome conservation between crop and model legume species. Proc Natl Acad Sci U S A 101:15389–15394

Choumane W, Winter P, Weigand F, Kahl G (2004) Conservation of microsatellite flanking sequences in different taxa of leguminosae. Euphytica 138:239–245

Colbert T, Till BJ, Tompa R et al (2001) High-throughput screening for induced point mutations. Plant Physiol 126:480–484

de Candolle A (1884) Origine des plantes cultives. Nabu Press, Charleston, p 424

Dongre TK, Pawar SE, Thakare RG et al (1996) Identification of resistant source to cowpea weevil [*Callosobruchus maculatus* (F.)] in *Vigna* sp. and inheritance of their resistance in black gram (*Vigna mungo* var. *mungo*). J Stored Prod Res 32:201–204

Frisch M, Bohn M, Melchinger AE (1999a) Comparison of selection strategies for marker-assisted backcrossing of a gene. Crop Sci. 39:1295–1301

Frisch M, Bohn M, Melchinger AE (1999b) Minimum sample size and optimal positioning of flanking markers in marker-assisted backcrossing for transfer of a target gene. Crop Sci 39:967–975

Fujii K, Ishimoto M, Kitamura K (1989) Pattern of resistance to bean weevil Bruchidae. in *Vigna radiata - mungo - sublobata* complex inform the breeding of new resistant varieties. Appl Ent Zool 24:126–132

Gill AS, Verma MM, Dhaliwal HS et al (1983) Interspecific transfer of resistance to mungbean yellow mosaic virus from *Vigna mungo* to *Vigna radiata*. Curr Sci 52:31–33

Guohao H, Woullard FE, Marong I et al (2006) Transferability of soybean SSR markers in peanut (*Arachis hypogaea* L.). Peanut Sci 33:22–28

Gupta S, Gupta DS, Anjum KT et al (2013) Transferability of simple sequence repeat markers in blackgram (Vigna mungo L. Hepper). Australian. J Crop Sci 7(3):345–353

Henikoff S, Comai L (2003) Single-nucleotide mutations for plant functional genomics. Annu Rev Plant Physiol Plant Mol Biol 54:375–401

Hong MG, Kim KH, Ku JH et al (2015) Inheritance and quantitative trait loci analysis of resistance genes to bruchid and bean bug in mungbean (*Vigna radiata* L. Wilczek). Plant Breed Biotechnol 3:39–46

Hoorman JJ, Islam R, Sundermeier A (2009) Sustainable crop rotation with cover crops. Ohino State University, Extension Factsheet. Agric. Natural Resour 9:1–6

Hospital F (2005) Selection in backcross programmes. Phil. Trans. R. Soc. B. 360:1503–1511. https://doi.org/10.1098/rstb.2005.1670

Hospital F, Charcosset A (1997) Marker-assisted introgression of quantitative trait loci. Genetics 147:1469–1485

Isemura T, Kaga A, Tabata S et al (2012) Construction of a genetic linkage map and genetic analysis of domestication related traits in mungbean. PLoS One 7:41304. https://doi.org/10.1371/journal.pone.0041304

Jain HK, Mehra KL (1980) Evaluation, adaptation, relationship and cases of the species of *Vigna* cultivation in Asia. In: Summerfield RJ, Butnting AH (eds) Advances in legume science. Royal Botanical Garden, London, pp 459–468

Jaiswal V, Mir RR, Mohan A et al (2012) Association mapping for pre-harvest sprouting tolerance in common wheat (*Triticum aestivum* L.). Euphytica 188:89–102. https://doi.org/10.1007/s10681-012-0713-1

Jat SK, Shivay YS, Parhar CM et al (2012) Evaluation of summer legumes for their economic feasibility, nutrient accumulation and soil fertility. J Food Legumes 25:239–242

Jiao K, Li X, Guo W et al (2016) Genome re-sequencing of two accessions and fine mapping the locus of lobed leaflet margins in mungbean. Mol Breeding 36:128. https://doi.org/10.1007/s11032-016-0552-1

Kajonphol T, Sangsiri C, Somta P et al (2012) SSR map construction and quantitative trait loci (QTL) identification of major agronomic traits in mungbean. Sabrao J Breed Genet 44:71–86

Kang JY, Kim SK, Kim MY et al (2014) Genome sequence of mungbean and insights into evolution within *Vigna* species. Nat Comm 5:5443. https://doi.org/10.1038/ncomms6443

Kang YJ, Satyawan D, Shim S et al (2015) Draft genome sequence of adzuki bean, *Vigna angularis*. Sci Rep 5:8069. https://doi.org/10.1038/srep08069

Karuppanapandian T, Karuppudurai T, Sinha PB et al (2006) Genetic diversity in green gram (*Vigna radiata* (L.)) landraces analyzed by using random amplified polymorphic DNA (RAPD). Afr J Biotechnol 5:1214–1219

Kasettranan W, Somta P, Srinives P (2010) Mapping of quantitative trait loci controlling powdery mildew resistance in mungbean (*Vigna radiata* (L.) Wilczek). J. Crop Sci. Biotechnol. 13:155–161. https://doi.org/10.1007/s12892-010-0052-z

Kaul T, Eswaran M, Thangaraj A et al (2019) Rice bean (*Vigna umbellata*) draft genome sequence: unravelling the late flowering and unpalatability related genomic resources for efficient domestication of this underutilized crop. bioRxiv 816595. https://doi.org/10.1101/816595.

Kearsey MJ (1998) The principles of QTL analysis (a minimal mathematics approach). J. Exp. Bot 49:1619–1623. https://doi.org/10.1093/jexbot/49.327.1619

Kitsanachandee R, Somta P, Chatchawankanphanich O et al (2013) Detection of quantitative trait loci for mungbean yellow mosaic India virus resistance in mungbean (*Vigna radiata* (L.) Wilczek) in India and Pakistan. Breed Sci 63:367–373. https://doi.org/10.1270/jsbbs.63.367

Kumar B, Talukdar A, Verma K et al (2014) Mapping of yellow mosaic virus resistance in soybean (*Glycine max* L. Merr.) through association mapping approach. Genetica. https://doi.org/10.1007/s10709-014-9801-6

Kumar C, Mishra SB, Nilanjaya et al (2016) Evaluation of selection indices for improving terminal heat tolerance in mungbean. Legume Res 18(2):216

Kumar J, Gupta DS, Gupta S et al (2017a) Quantitative trait loci from identification to exploitation for crop improvement. Plant Cell Rep 36:1187. https://doi.org/10.1007/s00299-017-2127-y

Kumar J, Gupta S, Biradar RS, Gupta P, Dubey S, Singh NP (2017b) Association of functional markers with flowering time in lentil. J Appl Genet. https://doi.org/10.1007/s13353-017-0419-0

Kumar R, Singh CM, Arya M et al (2021) Investigating selection indices to discrminate the physiologically efficient heat tolerant genotypes of mungbean. Legume Res 43(1):43–49

Lakhanpaul S, Babu C (1991) Symposium on grain legumes. New Delhi, pp 47–57

Langridge P, Lagudah E, Holton T et al (2001) Trends in genetic and genome analyses in wheat: a review. Aust J Agric Res 52:1043–1077. https://doi.org/10.1071/AR01082

Lawn RJ, Cottell A (1988) Wild mungbean and its relative in Australia. Biologist 35:267–273

Liu MS, Kuo TCY, Ko CY et al (2016) Genomic and transcriptomic comparison of nucleotide variations for insights into bruchid resistance of mungbean. BMC Plant Biol 16(1):1

Liu C, Wu J, Wang L et al (2017) Quantitative trait locus mapping under irrigated and drought treatments based on a novel genetic linkage map in mungbean. Theor Appl Genet. https://doi.org/10.1007/s00122-017-2965-6

Lonardi S, Munoz-Amatria M, Liang Q et al (2019) The genome of cowpea (*Vigna unguiculata* [L.] Walp.). The Plant Journal 98:767–782. https://doi.org/10.1111/tpj.14349

Maiti S, Basak J, Kundagrami S et al (2011) Molecular marker-assisted genotyping of *mungbean yellow mosaic India virus* resistant germplasms of mungbean and blackgram. Mol. Biotechnol. 47:95–104

Mathivathana MK, Murukarthick J, Karthikeyan A et al (2019) Detection of QTLs associated with mungbean yellow mosaic virus (MYMV) resistance using the interspecific cross of *Vigna radiata* × *Vigna umbellata*. J Appl Genetics 60:255–268. https://doi.org/10.1007/s13353-019-00506-x

Mehandi S, Quatadah SM, Mishra SP et al (2019) Mungbean: retrospect and prospects. Legume Crops. https://doi.org/10.5772/intechopen85657

Mishra GP, Dikshit HK, Ramesh SV et al (2020) Yellow mosaic disease (YMD) of mungbean (Vigna radiata (L.) Wilczek): current status and management opportunities. Front. Plant Sci. https://doi.org/10.3389/fpls.2020.00918

Nair RM, Pandey AK, War AR et al (2019) Biotic and abiotic constraints in mungbean production—progress in genetic improvement. Front. Plant Sci. https://doi.org/10.3389/fpls.2019.01340

Nie X, Huang C, You C, Li W, Zhao W, Shen C, Zhang B, Wang H, Yan Z, Dai B, Wang M, Zhang X, Lin Z (2016) Genome-wide SSR-based association mapping for fiber quality in nation-wide upland cotton inbreed cultivars in China. BMC Genom 17:352. https://doi.org/10.1186/s12864-016-2662-x

Noble TJ, Tao Y, Mace ES, Williams B, Jordan DR, Douglas CA, Mundree SG (2018) Characterization of linkage disequilibrium and population structure in a mungbean diversity panel. Front Plant Sci 8:2102. https://doi.org/10.3389/fpls.2017.02102

Pandiyan M, Krishnaveni A, Sivakumar C, Vaityalingam M, Sivakumar P, Jamuna E, Radhakrishnan V, Senthilkumar P (2020) Development of bruchid resistant genotypes in mungbean through introgression of wild genotypes. International J Chemical studies 8:62–66. https://doi.org/10.22271/chemi.2020.v8.i2b.9584

Parida A, Raina S, Narayan R (1990) Quantitative DNA variation between and within chromosome complements of *Vigna* species. Genetica 82:125–133

Paterson AH, Lander ES, Hewitt JD, Peterson S, Lincoln SE, Tanksley SD (1988) Resolution of quantitative traits into Mendelian factors by using a complete linkage map of restriction fragment length polymorphisms. Nature 335:721–726. https://doi.org/10.1038/335721a0

Prakash N, Kumar R, Choudhary VK, Singh CM (2016) Molecular assessment of genetic divergence in pea genotypes using microsatellite markers. Legume Res 39(2):183–188

Pratap A, Kumar J (2011) History origin and evolution. In: Pratap A, Kumar J (eds) Biology and breeding of food legumes. CABI, Oxfordshire, p 432

Pratap A, Gupta S, Malviya N, Tomar R, Maurya R, Johm KJ, Madhavan L, Singh NP (2015) Genome scanning of Asiatic *Vigna* species for discerning population genetic structure based on microsatellite variation. Mol Breed 35:178. https://doi.org/10.1007/s11032-015-0355-9

Pratap A, Gupta S, Tomar R, Malviya N, Maurya R, Pandey VR, Mehandi S, Singh NP (2016) Cross-genera amplification of informative microsatellite markers from common bean and scarlet runner bean for assessment of genetic diversity in mungbean (*Vigna radiata*). Plant Breed 135: 499–505

Pratap A, Chaturvedi SK, Tomar R, Rajan N, Malviya N, Thudi M, Saabale PR, Prajapati U, Varshney RK, Singh NP (2017) Marker-assisted introgression of resistance to *Fusarium* wilt race 2 in Pusa 256, an elite cultivar of *desi* chickpea. Mol Genet Genom 292:1237–1245

Pratap A, Gupta S, Rathore M, Basavraja T, Singh CM, Prajapati U, Singh P (2020) Mungbean. In: Pratap A, Gupta S (eds) The beans and peas. Elsevier, New York, pp 1–32

Pratap A, Das A, Kumar S, Gupta S (2021a) Current perspectives on introgression breeding in food legumes. Front Plant Sci 11:589189. https://doi.org/10.3389/fpls.2020.589189

Pratap A, Singh CM, Gupta S et al (2021b) Genetic enhancement in mungbean (*Vigna radiata*) as revealed by genome-wide mapped microsatellite markers. Agric Res. https://doi.org/10.1007/s40003-020-00507-x

Reddy KR, Singh DP (1990) The variation and transgressive segregation in the wide and varietal crosses of mungbean. Madras Agric J 77:12–14

Roorkiwal M, Bharadwaj C, Barmukh R et al (2020) Integrating genomics for chickpea improvement: achievements and opportunities. Theor Appl Genet 133:1703–1720. https://doi.org/10.1007/s00122-020-03584-2

Sai CB, Nagrajan P, Raveendran M, Rabindran R, Kannan BJ, Senthil N (2017) Understanding the inheritance of *mungbean yellow mosaic virus* resistance in mungbean. Mol Breed 37:63. https://doi.org/10.1007/s11032-017-0650-8

Schafleitner R, Shu-mei H, Shui-hui C, Jo-yi Y, Chen-yu L, Miao-rong Y, Krishnan B, Mao-sen L, Hsiao-feng L, Chien-yu C, Long-fang O, Chen W D-c, Thu-Giang TB, Ramasamy S, Chih-wei T, Nair R (2016) Identification of single nucleotide polymorphism markers associated with resistance to bruchids (*Callosobruchus* spp.) in wild mungbean (*Vigna radiata* var. *sublobata*) and cultivated V. radiate through genotyping by sequencing and quantitative trait locus analysis. BMC Plant Biology 16:159. https://doi.org/10.1186/s12870-016-0847-8

Singh BB, Dikshit HK (2002) Possibilities and limitations of interspecific hybridization involving green gram (*Phaseolus radiatus*) and black gram (*Phaseolus mungo*). Indian J Agric Sci 72: 676–678

Singh HB, Joshi BS, Chandel KPS, Pant KC, Saxena RK (1974) Genetic diversity in some Asiatic *Phaseolus* species and its conservation. Indian J Genet 34:52–57

Singh CM, Kumar R, Mishra SB, Pandey A, Arya M (2015) Characterization of mungbean genotypes against *Mungbean yellow mosaic virus* and cercospora leaf spot disease under north-east plain zone. International J Agric Environ Biotechnol 8(1):119–125. https://doi.org/10.5958/2230-732x.2015.00016.9

Singh DP, Singh BB, Pratap A (2017) Genetic improvement of mungbean and urdbean and their role in enhancing pulse production in India. Indian J Genet 76:550–567

Singh CM, Singh P, Pratap A, Pandey R, Purwar S, Vibha DCA, Baek K-H, Mishra AK (2019) Breeding for enhancing legumovirus resistance in mungbean: current understanding and future directions. Agronomy 9(10):622. https://doi.org/10.3390/agronomy9100622

Singh AK, Velmurugan A, Gupta DS, Kumar J, Kesari R, Konda A, Singh NP, Roy SD, Biswas U, Kumar RR, Singh S (2020a) Draft genome sequence of a less-known wild Vigna: beach pea (V. marina cv. ANBp-14-03). Crop J 7(5):660–666. https://doi.org/10.1016/j.cj.2019.05.007

Singh CM, Pratap A, Gupta S, Birder R, Singh NP (2020b) Association mapping for mungbean yellow mosaic India virus resistance in mungbean (*Vigna radiata* L. Wilczek). 3-Biotech 10:33. https://doi.org/10.1007/s13205-019-2035-7

Singh CM, Singh P, Tiwari C, Purwar S, Kumar M, Pratap A et al (2021a) Improving drought tolerance in mungbean (*Vigna radiata* L. Wilczek): morpho-physiological, biochemical and molecular perspectives. Agronomy 11:1534

Singh CM, Prajapati U, Gupta S, Pratap A (2021b) Microsatellite-based association mapping for agronomic traits in mungbean (*Vigna radiata* L. Wilczek). J Genet 100:87. https://doi.org/10.1007/s12041-021-01336-9

Singh P, Mishra AK, Singh CM (2021c) Genome-wide identification and characterization of Lectin receptor-like kinase (*LecRLK*) genes in mungbean (*Vigna radiata* L. Wilczek). J Appl Genetics 62:223–234. https://doi.org/10.1007/s13353-021-00613-8

Souframanien J, Raizada A, Dhanasekar P, Suprasanna P (2020) Draft genome sequence of the pulse crop blackgram reveals potential R-genes. BioRxiv. https://doi.org/10.1101/2020.06.21.163923

Takahashi Y, Sakai H, Yoshitsu Y, Muto C, Anai T, Pandiyan M, Senthil N, Tomooka N, Naito K (2019) Domesticating *Vigna stipulacea*: a potential legume crop with broad resistance to biotic stress. Front Plant Sci. https://doi.org/10.3389/fpls.2019.01607

Tanksley S (1993) Mapping polygenes. Ann. Rev. Genet. 27:205–233. https://doi.org/10.1146/annurev.ge.27.120193.001225

Thomas N (2017) Development of the mungbean nested association mapping (NAM) resource. *InterDrought V*, 2017-02-21-2017-02-25

Tomooka N, Lairungruang C, Nakeeraks P, Egawa Y, Thavarasook C (1992) Development of bruchid resistant mungbean using wild mungbean germplasm in Thailand. Plant Breed 109:60–66

Tomooka N, Kaga A, Vaughan DA (2006) The Asian *Vigna* (*Vigna* subgenus Ceratotropis) biodiversity and evolution. In: Sharma AK, Sharma A (eds) Plant genome: biodiversity and evolution. Part-C phanerogams (angiosperms- dicotyledons), vol 1. Science Publishers, Enfield, pp 87–126

Varshney RK, Bohra A, Yu J, Graner A, Zhang Q, Sorrells ME (2021) Designing future crops: gemic assisted breeding comes of age. Trends Plant Sci. https://doi.org/10.1016/j.tplants.2021.03.010

Vavilov NI (1926) Studies on the origin of cultivated plants. CABI, Leningrad, p 1951

Verma MM, Brar JS (1996) Breeding approaches for increasing yield potential of mungbean. In: Asthana AN, Kim DH (eds) Recent advances in mungbean research. Indian Society of Pulses Research and Development, Kanpur, pp 102–123

Visscher PM, Haley CS, Thompson R (1996) Marker-assisted introgression in backcross breeding programs. Genetics 144:1923–1932

Yaqub M, Mahmood T, Akhtar M, Iqbal MM, Ali S (2010) Induction of mungbean as a grain legume in the annual rice-wheat double cropping system. Pak J Bot 42:3125–3135

Zhao X, Dong H, Chang H et al (2019) Genome wide association mapping and candidate gene analysis for hundred seed weight in soybean [*Glycine max* (L.) Merrill]. BMC Genomics 20:648. https://doi.org/10.1186/s12864-019-6009-2

Zukovskij PM (1962) Cultivated plants and their wild relatives. Commonwealth Agriculture Bureau, London

Genomics-Assisted Breeding Approaches in Lentil (*Lens culinaris* Medik)

10

D. K. Janghel and Vijay Sharma

Abstract

Lentil is an important *Rabi* season food legume crop, commonly known as "poor man's meat", globally cultivated as rainfed or in residual moisture condition. Conventional breeding techniques are suffering from narrow genetic base, lack of genomic information and accurate selection procedure which restricts the lentil breeders in achieving the major breeding objectives. The accessibility of genomic tools and technologies, i.e. gene-specific markers (simple sequence repeat (SSR) or single nucleotide polymorphism (SNP)), expressed sequenced tags (EST), genome sequences, transcriptome sequences, Quantitative trait locus (QTL) maps and next generation sequencing (NGS), have unlocked a novel means to assist the lentil breeding programs by deploying diverse genomic resources which eventually changed the legume status from orphan to rich in genomic information. The transcriptome analysis identifies the ESTs derived functional markers and intron-targeted primers (ITP) in lentil. There is urgent need of high-throughput phenotyping (HTP) technologies for rapid, highly efficient, high quality and accurate prediction and measurement of complex quantitative traits in modern plant breeding. Functional genomics approaches help in tagging the unknown or targeted genes controlling the traits of interest, whereas comparative genomics approaches improved the cross-species transferability of genetic information among legume species. Marker-assisted selection (MAS) and DNA-based

D. K. Janghel (✉)
Department of Genetics and Plant Breeding, CCS Haryana Agricultural University, Hisar, Haryana, India
e-mail: jangheld1515@hau.ac.in

V. Sharma
Banda University of Agriculture and Technology, Banda, Uttar Pradesh, India

Kamaluddin et al. (eds.), *Technologies in Plant Biotechnology and Breeding of Field Crops*, https://doi.org/10.1007/978-981-16-5767-2_10

selection of desirable plant are remarkable ways to introduce QTLs/genes associated with yield enhancement, biotic or abiotic traits.

Keywords

Lentil · NGS · HTP · MAS functional · Comparative genomics

10.1 Introduction

Lentil commonly called as "poor man's meat" is an important *Rabi* season food legume crop, globally cultivated as rainfed or in residual moisture after monsoon season (Erskine et al. 2011). It is the earliest domesticated pulse crop around 8000 years ago at the same time as wheat, barley and pea (Dhuppar et al. 2012; Zohary and Hopf 1973); indigenous to south western Asia and Mediterranean region (Cokkizgin and Shtaya 2013) and known as by several names around the world namely *masser* in India, *messer* in Ethiopia, das in Arabic, *mercimek* in Turkey and *heramame* in Japan (Muehlbauer and McPhee 2005). It is cultivated in major parts of Africa, America, Asia and Australia continents, and India is the world's largest lentil producer, the other major producers of world includes Canada, Turkey, Australia and the United States (FAOSTAT 2019). In India, two types of lentil cultivars/ varieties is growing commonly known as *masur* or *malk masur* (bold seeded cultivars) in Bundelkhand region of Uttar Pradesh (UP), Madhya Pradesh (MP) and in Maharashtra (MH); and *masuri* (small seeded cultivars) are popular in Bihar, Eastern Uttar Pradesh, West Bengal and Assam.

Lentil is a self-pollinated crop belongs to pea family (Leguminosae or Fabaceae) with diploid chromosome number of $2n = 2x = 14$ and haploid genome size of about 4 Gbp (Ogutcen et al. 2018). Table 10.1 lists the taxonomic details of lentils-based USDA, Plants Database 2019.

The genus *Lens* comprised of five annual species out of them four were wild species *viz.*, *L. ervoides*, *L. orientalis*, *L. montbretti* (Ladizinsky 1979; Zohary 1972) and *L. nigricans*, and fifth species was only cultivated *L. culinaris* Medik (Ferguson 2000). However, latest classification of lentil included four species which consisted of seven taxa (Ferguson and Erskine 2001; Sarker and Erskine 2006). *Lens culinaris*

Table 10.1 Taxonomic classification of lentil (USDA 2019)

Kingdom	Plantae
Sub-kingdom	*Tracheobionta*
Super-division	*Spermatophyta*
Division	*Magnoliophyta*
Class	*Magnoliopsida*
Subclass	*Rosidae*
Order	*Fabales*
Family	*Fabaceae/Leguminosae*
Genus	*Lens*
Species	*Culinaris* Medik.

Medik (ssp. *Culinaris*, ssp. *orientalis* (Boiss.) Ponert, ssp. *odemensis* (Ladiz.) M.E. Ferguson& al and, ssp. *tomentosus* (Ladiz.) M.E. Ferguson& al.); *Lens ervoides* (Brign.) Grande; *Lens lamottei* Czefr; *Lens nigricans* (M. Bieb.) Godron.

Lentil seeds play important role in overcoming the problems of micronutrient deficiencies and malnutrition especially in poor people due to cheapest and excellent source of dietary protein, fibres, energy, lysine amino acid, vitamin, carbohydrates and minerals *viz*., iron (Fe: 73–90 mg/kg), zinc (Zn: 44–54 mg/kg), selenium (Se: 425–673 μg/kg) potassium (K), calcium (Ca) and potassium (P), and also contain oligosaccharides, protease inhibitors and tannins (Kumar et al. 2016; Ray et al. 2014; Thavarajah et al. 2011; USDA-Agricultural Research Service 2015). Their seeds are richest source of high-quality crude protein, ranged from 23.8 to 29.0% with an average of about 26% (Khazaei et al. 2019). However, their seeds are deficient in methionine and cysteine amino acids, but could form a complete protein diet when consumed with cereals (Cash et al. 2001). In addition to these, lentil exhibited low glycemic index which is recommended as a remedy for a number of chronic diseases suffering from diabetes, obesity and cardiovascular problems (Srivastava and Vasishtha 2012).

Lentil is mostly grown as a main crop; although, intercropping can also be done with other crops such as wheat, rice, barley, sugarcane, mustard, linseed and castor bean (Andrews and Mckenzie 2007). It helps in improving soil health by symbiotic nitrogen fixation and net carbon sequestration in cereal-based cropping systems by lowering carbon footprint, i.e. −552 kg CO_2 eq/ha especially in lentil–wheat cropping system as well as provide rotational benefits in controlling weeds, diseases and insect pests effectively (Gan et al. 2014; Kumar et al. 2013). It is a climate resilient cool loving leguminous crop, grown in low annual rainfall (minimum 10 inches) and high temperature (warm temperate, subtropical and high altitude tropical climate) regions of the world (Cash et al. 2001; Muehlbauer et al. 1995). However, these factors could reduce the seed yield if they coincide during flowering and seed setting stages (Muehlbauer et al. 2006). It can be grown in wide ranges of soil with good drainage (Elzebroek and Wind 2008) and broad soil pH condition ranges from 4.4 to 8.2, capable to tolerate moderate alkaline or saline soil conditions (Muehlbauer et al. 2006) but ideally adapted towards neutral range of soil pH from 5.5 to 7 (Elzebroek and Wind 2008). Lentil has an inherent capability to survive along with high yield potential under moisture stress and problematic soil conditions, reveals wide adaptability of this crop (Mishra et al. 2007).

During last few decades, conventional breeding methods along with good agronomic practices in lentil cultivation have significantly improved the crop productivity from 605 to 1038 kg/ha (FAOSTAT 2018). Lentil crop productivity has increased by use of short duration improved cultivars in Bangladesh with the use of BARIMasur series of varieties such as BARIMasur-3, -4, -5, -6, -7 and -8 (Yigezu et al. 2019); in Morocco, yield advancement of 35 kg/ha/year (i.e. around 16–67% over local check from 1989 to 2018) (Idrissi et al. 2019, and in Ethiopia, yield enhancement of 18–28 kg/ha/year from 1980 to 2010) (Bogale et al. 2017). However, currently in developing countries like India, the average crop productivity of lentil is still below the average productivity of 2 ton/ha standard in some countries around the world (FAOSTAT 2019) due to several constraints in lentil yield. The currently identified major constraints in lentil crop are rainfed cultivation under changing climatic and complex edaphic conditions, low annual genetic gain (<0.7%), high drop rate of flowers and shedding rate of pods, compounded with several biotic (Anthracnose, *Ascochyta* blight, *Fusarium* wilt, rust, *Stemphylium* blight, collar and root rots) and abiotic (terminal drought, intermittent heat, cold and frost) stresses during crop cultivation period (Kumar 2016; Sharpe et al. 2013). Moreover, selection for quantitatively inherited traits resulted into poor or no genetic gain in lentil crop due to environmental influences and higher G × E interaction on the phenotypic expression of breeder traits of interest (Kumar and Ali 2006), that make the greatest task for breeder to develop climate-resilient superior performing lentil varieties. The conventional breeding approaches are less efficient and more time consuming in improving of polygenitically inherited several complex yield traits in lentil (Kumar 2016).

The lentil genomic resources are now available to integrate the genomics-assisted breeding approaches in lentil breeding programme which is widely used for genetic relationship/diversity studies (Dissanayake et al. 2020a; Khazaei et al. 2016; Kumar 2016; Lombardi et al. 2014; Wong et al. 2015). Just now, genomics-assisted breeding approaches have become a crucial to break the yield plateau and accelerate the genetic gain in lentil breeding by identification, selection and fixation of superior alleles in breeding populations in more precisely, effectively and rapidly manner than conventional breeding methods (Kumar 2016). In recent times, development of genomic resources with the use of gene-based markers genotyping such as single nucleotide polymorphism (SNP) and/or simple sequence repeats (SSR), and next generation sequencing (NGS) techniques have made possible to access the genomic sequences, transcriptomic sequences and genes/QTLs linked to important yield contributing traits. Thus, there is an urgent need to integrate genomics-assisted breeding approaches for more challenging breeding to develop climate resilient high yielding and resistant cultivars in lentil breeding programme.

In lentil, the earliest genetic map was constructed in early 1980s on the basis of morphological and biochemical (isozyme) markers (Tadmor et al. 1987; Zamir and Ladizinsky 1984). The first comprehensive linkage map was developed using morphological and 177 RAPD, AFLP & RFLP markers with recombinant inbred lines (RILs) population from inter-specific cross (*L. culinaris* × *L. orientalis*) in lentil (Eujayl et al. 1998). Subsequently, the first genomic library developed in a cultivated accession ILL 5588 by using RFLP markers with restriction endonuclease enzyme *Sau3AI* (*Staphylococcus aureus 3A*) and probes *viz.* $(GA)_{10}$, $(GAA)_8$, $(GC)_{10}$, $(GT)_{10}$, $(TA)_{10}$ and $(TAA)_5$ (Hamwieh et al. 2005). For the development of lentil genome maps used diverse PCR-based markers, and among them, SSR markers have contributed more significantly which is now being replaced quickly by DNA chip-based markers, especially with SNPs. The SNPs are plentiful and universally available across the genome of legume crops (Chagné et al. 2007). Transcriptome assemblies are used to identify the expressed sequenced tags (ESTs) derived from SSR and SNP markers, and intron-targeted primers (ITPs). The first EST-based library was constructed from a varietal blend of eight cultivars with changing seed morphology (Vijayan et al. 2009) and the second cDNA library from *Colletotrichum truncatum* inoculated leaflets of a Canadian cultivar "Eston" (Kumar et al. 2014). Thus, the era of genomic tools and technologies has opened new opportunities in lentil breeding programs by deploying the diverse genomic resources of lentil. In this review, the recent developments and future potentials of genomics-assisted lentil breeding have been discussed in detail.

10.2 Origin and Types of Races

Lentil crop was domesticated at Near East in the Fertile Crescent, thereafter; expand to Europe, Northern Africa, Middle East and then the Gangetic plains of India (Ford et al. 2007). The Middle East is considered as primary centre of genetic diversity (Cubero 1981; Zohary 1972). It is originated in Near East from wild progenitor *L. orientalis* (Cubero 1981; Ladizinsky 1979; Zohary 1972). Vavilov centre of origin is the Mediterranean region and southwest Asia (Vavilov 1949; Zohary 1995).

The conventionally cultivated species are grouped into two classes based on seed size by Barulina (Cubero 1981) but this view has been renewed as race by Cubero (1981) and their contrasting traits are defined by Cubero et al. (2009):

1. Micro-sperma race: small podded (6–15 $\times$ 3.5–7 mm) with small seeds (3–6 mm) and 1000 seeds weight $\leq$25 g, small flowers (5–7 mm long) white to violet colour, plant height (15–35 cm).
2. Macro-sperma race: large podded (15–20 $\times$ 7.5–10.5 mm) with large seeds (6–9 mm), 1000 seed weight 25–50 g, large flowers (7–8 mm long) white colour, plant height (25–75 cm).

Europe, North Africa and America are mostly cultivated macro-sperma race (Chilean types), whereas, micro-sperma race (Persian types) is in cultivation of Asia, Egypt and Ethiopia and both of them are grown in the regions of south-eastern Europe and western Asia (Ford et al. 2007).

10.3 Floral Biology (Singh 1991)

Lentil is strictly self-pollinated due to chasmogamy flowering nature (Fig. 10.1) and outcrossing ranged from $<$1 to 6.6% by thrips and other insects (Baum et al. 1997; Wilson and Law 1972).

Inflorescence: Raceme

Flower: Typical papilionaceous

Calyx: 5 sepals

Corolla: 5 petals (one standard, two wings and two keels)

Androecium: Diadelphous (9 + 1) or polyadelphous

Gynoecium: Monocarpellary, sub-sessile ovary, swollen and hairy stigma

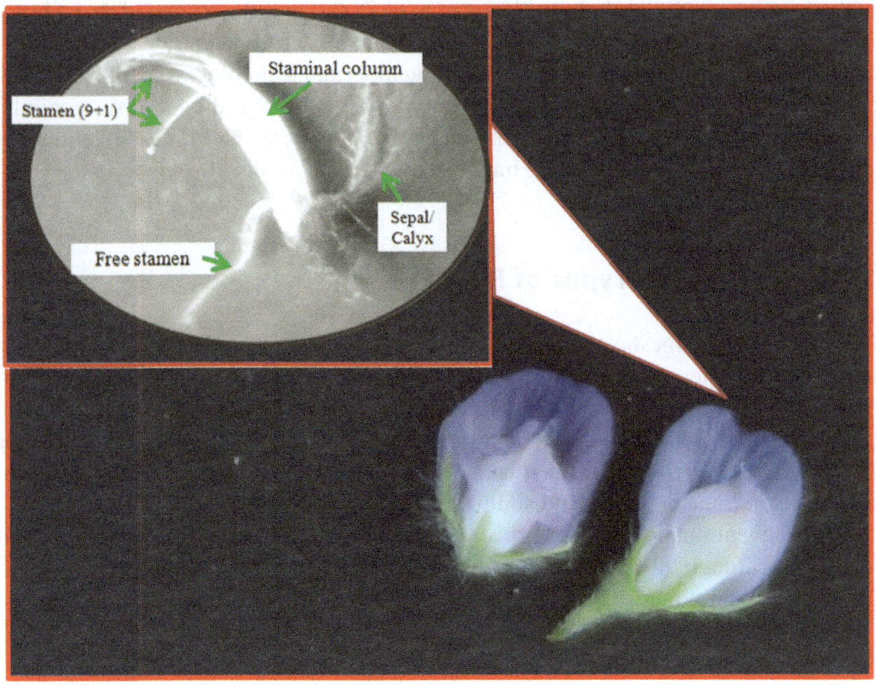

Fig. 10.1 Typical papilionaceous lentil flower and flower organs

10.4 Cytogenetics

Lentil cultivar and its wild relatives have a diploid number of chromosome ($2n = 2x = 14$) with length ranged from 3.0 to 9.2 μ (Mishra et al. 2007). Their chromosome structures have shown to vary from species to species and within species also. Sindhu et al. (1983) represented the chromosomal karyotype of lentil and observed acrocentric (three pairs), metacentric (one pair) and sub-metacentric (three pairs) chromosomes. The lentil species, *L. orientalis*, karyotype was quite similar to *L. culinaris* (Ladizinsky et al. 1984) but varied from *L. nigricans* due to the presence of three inter-changes (Gupta and Bahl 1983). Harlan and de Wet (1971) proposed gene pool concepts on the basis of crossing abilities/hybridization among species. Ladizinsky (1999) allocated *Lens* species into three categories: 1° (primary), 2° (secondary) and 3° (tertiary) gene pools based on the end products of hybridization presented in Table 10.2.

Table 10.2 Ladizinsky (1999) classified lentil gene pool based on hybridization among *Lens* species

Gene pools	Lentil species	Hybridization	F_1 fertility
Gene pool 1	Cultivated races, landraces and wild and weedy relatives: *L. culinaris* ssp. *culinaris*, *orientalis*, *odemensis*	Easy	Fertile
Gene pool 2	Wild and weedy species: *L. ervoides*, *L. Nigricans*	Moderate	Fertile/ sterile
Gene pool 3	*L. lamottei*, *L. culinaris* spp. *Tomentosus*, *L. culinaris* spp. *Orientalis*	Difficult	Sterile

10.5 Breeding Objectives

The milestone goal in breeding of lentil crop is to develop superior varieties with higher yield, better quality, and diseases and insects resistance (Muehlbauer et al. 1995). However, its breeding objectives are varying place to place with the farmer problems and consumer preferences in the specific region of the world. Various stresses of biotic and abiotic causes greatly limit the genetic improvement in seed yield and its component traits (Tomlekova 1998; Yankova and Sovkova-Bobcheva 2009). In addition to theses, agronomic complications such as lodging, pod dehiscence and inadequate crop management are important constraints in a wild progenitor of lentil, *L. culinaris* ssp. *orientalis*, which is cultivated as landrace by farmers. The major breeding objectives in lentil crop are to breed varieties with following:

1. High and stable yielding
2. Early and synchronous maturing
3. Resistance to shattering
4. Resistant to biotic stresses (insect-pest, diseases and weeds)
5. Tolerant to abiotic stresses (terminal drought, heat and frost)
6. Better quality seeds rich in protein content
7. Bold seeded seeds with less cooking time

Biotic Stresses Symbiotic nitrogen fixation enriches the legumes with nitrogen and phosphorus, therefore, these attribute attracts more pests and diseases (Sinclair and Vadez 2012). Lentil suffering from different types of biotic stresses and the important insect pests and diseases are listed in Table 10.3.

Abiotic Stresses The yield of lentil crop is globally hampered by abrupt rise in temperature with exhausting available soil moisture at terminal stages of crop, especially at grain filling stage causes forced maturity and reduction in crop yield. In India, throughout the different production zones of lentil, commonly experiences on an average of 42% yield gap (Ali and Gupta 2012; Reddy and Reddy 2010), due to intermittent or terminal drought, cold and heat stress at seedling and flowering stages of crop (Farooq et al. 2009; Silim et al. 1993; Wery et al. 1994). The drought

Table 10.3 Important insect pests and diseases of lentils

	Insect pests	Diseases
1.	*Aphis cracivora* Koch. (aphids)	*Ascochyta lentis* (*Ascochyta* blight)
2.	*Etiella zinckenella* treit (pod borer)	*Botrytis fabae* and *B. cinerea* (Botrytis grey mould)
3.	*Sitona* spp. (leaf weevil)	*Uromyces viciae-fabae* (Pers.) Schroet (rust)
4.	*Agrotis ipsilon* (cut worm)	*Fusarium oxysporum f. sp. lentis* (wilt)
5.	*Frankiniella* spp. (thrips)	Beet western yellows virus/BWYV bean leaf roll virus/BLRV or subterranean clover red leaf virus/SCRLV
6.	*Cydia nigricana* (pea moth)	(Lentil yellows disease)

and cold are key abiotic constraints of lentil causing major yield loss in crop, although other factors, such as salinity stress, nutrient deficiency and/or toxicity, could also reduce the lentil production and productivity (Muehlbauer et al. 2006; Tivoli et al. 2006).

1. Drought
2. Low temperature (cold and frost)
3. High temperature (heat stress)
4. Salinity
5. Nutrient deficiency and toxicity

10.6 Lentil Genetic Diversity

Lentil breeding started with germplasm collection and evaluation, among them genetically diverse parents for target traits is crossed to breed novel combinations of gene, genetic improvement and widening the genetic base of lentil gene to overcome the limitation of lentil breeding. In the world, different gene banks hold around 58,407 lentil accessions (Khazaei et al. 2016), among them, International Centre for Agricultural Research in the Dry Areas (ICARDA), Lebanon, preserve the largest lentil accessions in the world (About 13,907 cultivated and 603 wild species accessions) and has a prominent intent in the collection and characterization of landraces (Singh and Chung 2016) followed by Australian Temperate Field Crops Collection around 5254 accessions (Malhotra et al. 2019). In India, National Bureau of Plant Genetic Resources, New Delhi (NBPGR) has almost 2655 accessions comprised of around 2083 indigenous collection with 572 exotic accessions (Singh and Chung 2016). The ICARDA landraces were characterized and documented into four major regional groups, namely the Indian subcontinent group, the Ethiopian group, the northern group (Greece, Iran, Turkey, USSR, Chile), the Levantine group (Egypt, Jordan, Lebanon and Syria) around the world

(Erskine et al. 1989). The global germplasm collection further structured into three main groups based on eco-geographical origin by Khazaei et al. (2016): subtropical savannah, Mediterranean and northern temperate. Indian and Ethiopian landraces both come under *pilosae* group which has *pilosae* trait, i.e. the presence of pubescence on the vegetative parts and short/rudimentary tendril at the tip of leaf (Vandenberg and Slinkard 1990). Indian lentils are conventionally micro-sperma type (small seeds), more polymorphous, more pigmented and typically designated with early in flowering and maturity, less in biological yield and height (Erskine et al. 1998), whereas Ethiopian materials with the endemic elongated pod apex. Pakistan germplasm was included into *Pilosae* group for qualitative traits but intermediate between Afghanistan and India for quantitative traits. Afghanistan, Iran and Turkey germplasm were very late types in flowering and maturity among them Afghan germplasm is latest one.

The crop wild relatives (CWR) are ultimate sources of novel traits/alleles used for broadening of genetic base of lentil crop because they are found in adverse agro-climatic conditions and adapted to diversified habitats. They are characterized and documented for diverse agro-morphological, phenological, physiological, biochemical traits as well as different biotic and abiotic stresses (Singh et al. 2020; Malhotra et al. 2019). The core and FIGS (Focused Identification of Germplasm Strategy) sets were developed to display representative genetic diversity for desirable traits for better use and accessibility of available germplasm within genebanks. The FIGS strategy is successfully demonstrated the characterization of several agro-morphological and biotic resistant (insect-pest and diseases) and abiotic tolerance (heat and drought) in lentil at ICARDA using robust geo-graphical datasets. Recently, 162 lentil accessions were screened and identified heat and drought tolerant germplasm using a FIGS set (El haddad et al. 2020), and a core set of 96 accessions developed from 405 accessions of wild *Lens* species which were used to identify the key traits of nutrition, diseases, insect pest or abiotic stresses (Kumar et al. 2018d; Singh et al. 2020). The bi-parental and association mappings were used to detect the QTLs/genes within genetically diverse cultivated and wild germplasm of lentil available for several agro-morphological traits, nutritional quality (Se, Fe and Zn) as well as biotic and abiotic stresses (Kumar et al. 2018d; Khazaei et al. 2018; Polanco et al. 2019). Image-based phenotyping used for rapid screening and precise characterization of useful accessions/traits in lentil germplasm that could accelerate lentil breeding programs. For example, *Aphanomyces* root rots resistant lentil genotypes were identified by screening of lentil germplasm using this technique (Marzougui et al. 2019). The multi-parent breeding populations such as nested association mapping (NAM) and multi-parent advanced generation inter-cross (MAGIC) including both cultivated and wild species increased allelic diversity through novel recombinants in the populations from diverse parents (Scott et al. 2020) that provide a unique platform for selection of desirable recombinants of genes/mapping of QTLs of useful traits in advanced generations (Varshney et al. 2019; von Wettberg et al. 2018). In lentil, efforts for developing a MAGIC population using genetically diverse eight-parents are in progress at ICARDA.

Table 10.4 Genetic diversity in lentil germplasm for yield and its component traits

Sr. No.	Yield and its component traits	Sakthivel et al. (2019)	Kamaluddin et al. (2020)	Satpathy and Debnath (2020)	Iqbal et al. (2021)
1.	Days to 50% flowering	61–80	71–98	53–72	[a]
2.	Days to maturity	105–127	108–124	83–107	[a]
3.	Plant height (cm)	21–28	18–58	20–31	27.38–71.12
4.	Primary branches/plant	2–5	1.32–5.47	1.76–2.43	[a]
5.	Secondary branches/plant	[a]	2.88–29.23	1.42–9.80	[a]
6.	Pods/plant	9–37	18–113	6–44	2–3
7.	Seeds/pod	[a]	1–2	1.22–1.49	1–3
8.	100 seed weight (g)	1.97–3.38	1–3	1.29–1.67	1.05–3.65
9.	Seed yield/plant (g)	0.26–0.77	1.29–4.43	0.82–1.20	[a]
10.	Biological yield/plant (g)	0.66–1.83	2.09–18.54	0.96–1.72	[a]

[a]Data were not calculated for particular trait

The fundamental of any breeding programme is the presence of ample amount of accessible genetic diversity within their crop gene pool. Narrow genetic diversity in lentil crop is bottleneck for the genetic dissection of a particular trait by conventional breeding. Conventional breeding methods in lentil has been successful in improving the crop yield and dealing with the major yield constraints by developing of cultivars resistant to major diseases and insect-pests or tolerant to important abiotic stresses (Materne and McNeil 2007; Muehlbauer et al. 2006; Sarker and Erskine 2006); however, classical lentil breeding programme has certain limitations, i.e. narrow genetic diversity, insufficient genetic information and absence of precise selection techniques which limit the breeders to reach their targets and pursue the achievable supplementary breeding objectives. Therefore, availability of sufficient amount of genetic variability for particular trait is essential for overall genetic improvement. The lentil genetic resources have sufficient variability for numerous key traits, yet not available for some other economically important traits, which limits the plant breeder efforts to address several obstacle. Lentil cultivated in South Asia, exclusively of *pilosae* types, has narrow genetic base, low photoperiod and high temperature sensitivity than the native landraces from West Asia (Erskine et al. 1998). The "daylength bottleneck" restricted the spread of lentil gene pool into the areas of Indo-Gangetic plain compounded with low yield and susceptible to an array of diseases. Currently, the scarcity in genetic variability limits the scope of selection, heterosis, new gene recombination and transgressive segregation. The genetic diversity for important seed yield and its attributing traits were shown in Table 10.4.

List of lentil breeding research institutes and their headquarters:

1. ICARDA: International Centre for Agricultural Research in the Dry Areas, Lebanon, Syria
2. IIPR: Indian Institute of Pulses Research, Kanpur, India
3. NBPGR: National Bureau of Plant Genetic Resources, New Delhi, India
4. Akdeniz, The University, Turkey
5. Bangladesh Agricultural Research Institute, Bangladesh
6. Nepal Agriculture Research Council, Nepal
7. University of Saskatchewan, Canada
8. Washington State University, USA
9. Instituto de Agricultura Sostenible (CSIC), Spain
10. University of Córdoba (UCO), Spain
11. The University of Western Australia, Australia
12. University of Birmingham, UK

10.7 Conventional Lentil Breeding Approaches

The prime objective of any breeding programme is to develop high yielding cultivars of crop plants by exploiting genetic diversity either by selection or hybridization among selected lines. Lentil crop fits remarkably well into genre of a "smart crop", therefore, appropriate breeding strategy is required for genetic improvement in the varieties for higher and stable yields. The major breeding methods in lentil are similar to the breeding methods of autogamous crops includes pure line selection or plant hybridization followed by bulk, pedigree, backcross or single seed descent methods (SSD), or a modification of these procedures, accompanied with mutation breeding for specific purposes (Muehlbauer et al. 2009; Singh and Pedapati 2015; Toker et al. 2007). Pure line selection in landraces is the largely used breeding scheme during initial period of cultivar development in lentil (Muehlbauer 1992). In India, varietal improvement programme in lentil was initiated in 1924 with single plant selection in locally collected mixed seed lots. Subsequently, the lentil breeding programme boosted under All India Coordinated Pulse Improvement Project (AICRP) in 1970s and several important cultivars have been released for various states. The selections led to achieving varieties with erect growth habit, increased seed size and diversity in colour accompanying of decreased dormant seeds and pod dehiscence (Zohary 1996). In the last few centuries, plant selection and hybridization have become the chief assets in crop improvement programme of lentil especially in developed countries; however, heterosis breeding is preferred over selection for rapid development of new varieties with increased global demand.

10.7.1 Plant Introduction

It is the simplest and quickest approach for the genetic enhancement of lentil cultivars by obtaining high yielding widely adaptable lentil cultivars from different geographical regions of within or outside of India. For example, lentil materials

"Precoz (ILL 4605)" introduced from West Asia into the Indo-Gangetic plains which is early in flowering and macro-sperma seeds (4.5 g/100 seeds). The successful plant introduction relies on genetic constitution and adaptability of introduced variety in new agro-climatic condition. Appropriate plant materials in plant introduction are homozygous pure lines due to better adaptability in new environment than heterozygous segregating lines, which necessitate further selection of introduced line with specific desirable traits. Some important lentil cultivars were introduced from ICARDA to different parts of world such as in India (VL-507 & Vipasha), in Nepal (Sikhar, Simal, Khajura Masuro-1 & -2), in Pakistan (Shiraz 96 & Mansehra 89) etc. (Rahman et al. 2009).

10.7.2 Plant Hybridization

It is the breeding procedure of combining desirable traits/alleles from genetically dissimilar parents into a cultivar. The diverse genetic makeup and geographical evolution of micro-sperma and macro-sperma races of lentils are judiciously used in hybridization programmes followed by selection of desirable recombinants in segregating generations by practicing pedigree method for combining desirable traits into developing high yielding, stress (biotic or abiotic stress or both) resistance and wider adaptability varieties for broadening and enhancing the existing genetic base of lentil crop (Chahota et al. 2007; Kumar et al. 2009; Chuni et al. 2000). The selection of parents depend on the objectives, level of genetic diversity and combining ability, however, knowledge of parental genetic relationships during selection of parents is extremely important to achieve genetic gain in lentil crop improvement programs. The efforts have also been made to extend the photoperiod length (18 h) of flowering in lentil at ICARDA to improve the synchronization by extensive crossing of *Pilosae* types with other origin lentils.

10.7.3 Mutation Breeding

The bottleneck in genetic base, variability in traits and low-yielding potential of current cultivars are the major restrictions handicapped the lentil breeding programs for long period of time (Amin et al. 2015; Asnake and Bejiga 2003). Induced mutation would be a quick, cost-effective and complementary breeding strategy for widening the base of lentil. A variety of mutagen (Gamma "γ" rays/n-nitroso-n-methyl-urea "NMU") could be applied to break the major bottlenecks in lentil morphological, phenological or yield-attributing traits and key stresses resistance to broaden the genetic diversity (Erskine et al. 1998; Toker et al. 2007). In recent times, a novel lentil mutant "multipodding (mp)" has been screened and identified which has a huge impact on inflorescence for seed yield and their stability (Amin-Laskar et al. 2018). A number of reports supported the usefulness of combination of physical and chemical mutagenic treatments instead of individual treatment for the progress in lentil breeding (Khan and Tyagi 2010, 2005). Although, the achievement of mutation breeding for genetic enhancement in lentil crop mainly depend upon the efficiency and nature of the mutagenic agents and the precision in screening and identification of mutants (Khan and Tyagi 2005; Manju and Gopimony 2009). Mutations led to development of varieties with greater seed yield potential and quality, consequently advancement in the key agro-morphological traits with greater consumer acceptance (Ahloowalia et al. 2004). In India, "S-256 (Ranjan)" is high yielding and spreading type lentil variety developed in 1981 by X-rays irradiation and "Rajendra Masoor 1", low temperature tolerance and early maturing lentil variety recommended for late sowing developed in 1996 by Gamma rays (100 Gy).

10.7.4 Polyploidy Breeding

The genetic base of lentil cultivars is mainly confined to crossing within the GP-1/ primary gene pools, therefore, exploitation of GP-2/secondary and GP-3/tertiary gene pools are necessary for widening the genetic base of lentil. Ploidy refers to an organism with more than one set of homologous chromosomes in a cell, it is divided into two groups, autopolyploidy: polyploidy developed by doubling of haploid genome from single parent, whereas, allopolyploidy: by doubling of haploid genome from multiple parents (Aversano et al. 2012). Blakeslee and Avery (1937) used colchicine alkaloid for doubling of haploid genome in plants by disrupting the spindle fibre formation during cell cycle (Ye et al. 2010), thereafter, for producing of auto- and allopolyploids (Chen and Ni 2006). Polyploids are generally coupled with several commercial profits such as giant and seedless fruits, more vegetative growth and better resistance to diseases, and in fact the major field and horticultural crops are polyploids in nature including major cereals (Wheat, maize, oats), oilseeds (Groundnut, mustard, rape), forage (alfalfa), fibre (cotton), tobacco, sugarcane, potato, sweet potato, strawberry, banana, pear etc. (Harper et al. 2016; Rehman 2017; Emery et al. 2018). Shahwar et al. (2019) artificially induced phenotypic diversity in lentil with the use of heavy metals (lead and cadmium nitrate) in traits like plant height, plant growth, leaf and flower characteristics and pod size in lentil crop population.

10.7.5 Quality Breeding

In lentil breeding, simultaneous increase in both quantitative traits, i.e. seed yield and protein contents, is a major challenge while maintaining the progress of diseases, insect pest resistance or abiotic stress tolerance because both the traits are negatively correlated (Erskine et al. 1989; Hamdi et al. 1991; Lizarazo et al. 2015). This suggested the independent selection in any one of the traits lead to reduction in one trait by gaining in other during lentil breeding. The seeds of lentil are outstanding source of dietary protein containing approximately 26% high-quality crude protein (Khazaei et al. 2019) with estimated genetic variation of 24.30–30.20% by using a NIR Systems 6500 analyser calibrated in respect of Dumas Combustion method (Wang and Daun 2006); 23.80–29.30% by Dumus Combustion method

(Tahir et al. 2011); 21.80–27.10% by Kjeldahl method (Zaccardelli et al. 2012); 24.60–30.00% by a variety of methods (Heuzé et al. 2014) and 10.50–27.10% using Kjeldahl method (Kumar et al. 2016). These variations in protein content may be variation in locations/agro-climatic conditions, protein estimation methods, genetic cause or G × E interaction for protein content and amino acid composition. Furthermore, seed protein concentration also affected by nitrogen fixing ability of *Rhizobium* species in legumes. The improvement in nutritional quality of lentil could be achieved by upgrading the quality of protein and its composition as well as processing factors as quick expanding human consumption crop globally.

10.8 Drawbacks of Conventional Lentil Breeding Approaches

Since over the years, conventional lentil breeding approaches of selection-recombination-selection have led to fivefold raise in global crop production from 0.85 to 5.73 Mt and productivity from around 60.5 to 119.5 kg/ha (FAOSTAT 2019) during the former five decades through improvement in a number of simple monogenic/oligogenic traits (Nutritional quality, color, shape and taste). Traditional breeding techniques in lentil have been successful in dealing with the important yield constraints by developing superior cultivars, resistant/tolerant to major diseases, insect-pests or abiotic stresses *viz.*, drought, heat, salinity etc. (Materne and McNeil 2007; Muehlbauer et al. 2006; Sarker and Erskine 2006). However, the majority of economically important traits like seed yield and its contributing traits are complex, govern by quantitative/polygenic genes, highly influenced by environment and exhibit wide range of genotypic × environmental interaction (GEI) led to little genetic gain in conventional lentil breeding programme (Kumar and Ali 2006).

Conventional breeding techniques are suffering from narrow genetic base, lack of genomic data and precise selection procedure which confine the lentil seed yield and restricts the lentil breeders to achieve the major breeding objectives (Kumar and Ali 2006). The execution of conventional breeding techniques is time consuming, less precise and difficult in the improvement of agronomically important quantitatively inherited complex traits (Kumar 2016). Thus, there is a necessary to incorporate genomics-assisted lentil breeding approaches such as molecular marker and genomic resources techniques in order to find, select and fix the greater recombination of genes more accurately and efficiently (Kumar 2016). It is quick and more powerful technique in developing high yielding varieties which also helps in genetic diversity assessment, gene-targeted marker-assisted selection (MAS) and gene-specific genetic engineering (GE) in crop breeding program of lentil to enhance the genetic gain and genetic diversity for developing high yielding cultivars.

10.9 Genomics-Assisted Lentil Breeding Approaches

The genomics-assisted lentil breeding is a potent tool, enabled incorporation of genomic tools and high-throughput phenotyping (HTP) with conventional breeding techniques. The presence of existing accessible genetic diversity is the fundamental for mining of beneficial alleles in gene banks which is essential for development of better-quality advanced cultivars with greater performance within defined environments. The accessibility of genomic tools and technologies such as gene-specific markers (SSR or SNP), EST, genome sequences, transcriptome sequences, QTLs maps and NGS have unlocked a novel means to assist the lentil breeding programs by deploying diverse genomic resources (Kumar et al. 2018a, b, c; Singh et al. 2019). The ongoing lentil breeding programs have limited use of genomic resources that cause boundation of the potential implementation of MAS. The slow progress in the development of lentil genomic resources (Kumar et al. 2014), vast genome dimension, low density linkage map, narrow genetic base, shortage and detection problems of candidate genes are the central limiting factors in genomics-assisted lentil improvement. The advent of NGS technologies accelerated the genomic resources development with time and cost effective manner, and changed the legume status from orphan to rich in genomic information (Singh et al. 2017, 2019; Varshney et al. 2005), which led to the development of an initial draft of 23× coverage later refined with further 125× coverage and have been widely used for a variety of purposes mostly for genetic relationship/diversity studies (Dissanayake et al. 2020a); not only in lentil crop but also in diverse legume crops like chickpea, pigeon pea, groundnut common bean, cowpea etc. (Afzal et al. 2020; Bauchet et al. 2019; Varshney et al. 2019).

10.9.1 Drafting the Genome Sequence and Developing of Gene-Specific Markers

The lentil genome sequence has been drafted in V1.2 for the CDC cultivar Redberry using bulk sequencing technique and chromosomal "pseudo-molecules" assembly representing about assembling of seven pseudomolecules and 120,000 scaffolds to a sum total of about 2.6 Gbp. The genes were tagged and annotated for genomics studies in lentil, i.e. open and accessible online on the KnowPulse (http://knowpulse.usask.ca) web portal (Bett et al. 2016). Moreover, the lentil cultivars/accessions have been analysed by re-sequencing to aware of the potential genetic and genomic information and data existing within the lentil germplasm. For example, Australian cultivar "PBA Blitz". genome had drafted, of which included seven pseudo-molecules, a sum of 337.7 Gbp (c. 85× coverage) of top-level genome sequences and its assemblage of about 352,065 scaffolds and 444,011 singletons resulting a sum of about 2.3 Gbp. This drafted genome of lentil cultivar has a resemblance of 99% with respect to genome sequence of CDC cultivar (Kaur et al. 2016). The development of gene-specific markers linked with target traits are being developed in lentil to illustrate the genetic diversity within the available germplasm and their association with diverse agro-morphological traits, such as disease resistance, seed quality and micronutrient concentration (Khazaei et al. 2018). The first locus-specific RFLP markers based on restriction endonuclease enzymes and hybridization with labelled probes was used to construct lentil linkage maps (Eujayl et al. 1998). For example, the first genomic library was created using restriction digestion of genomic DNA of a cultivated accession "ILL 5588" with *Sau3AI* (*Staphylococcus aureus* 3A) and hybridized with radio-labelled probes $(GT)_{10}$, $(GA)_{10}$, $(GC)_{10}$, $(GAA)_8$, $(TA)_{10}$ and $(TAA)_5$ then sequenced with M13 primers on LI-COR DNA sequencer (Hamwieh et al. 2005). However, requirement of high technical skills and difficulties in handling of radioactivity make it impracticable. Later, with the advent of PCR-based markers such as Random Amplified Polymorphic DNA (RAPD), Sequence Characterized Amplified Region (SCAR) and SSR have accelerated the use of corresponding markers in genome characterization of lentil crop (Ates et al. 2018; Singh et al. 2019; Mbasani-Mansi et al. 2019; Polanco et al. 2019). However, these marker techniques are usually labour intensive, time consuming and difficult in large scale application in plant breeding programme. At present, the development of gene-specific SSR markers has become quicker and efficient in cost through transcriptomic analysis or NGS techniques (Kaur et al. 2011; Singh et al. 2019). In current scenario, SNP markers are more popular than PCR-based markers due to even distribution across the genome, high in abundance and high detection ability with high throughput (HTP) techniques (Mammadov et al. 2012).

10.9.2 DNA Sequencing Techniques

The three generations of DNA sequencing techniques are shown in Table 10.5. First-generation/Sanger sequencing technique has been exploited to create ESTs, gene- or

Table 10.5 The generation, basis and techniques of DNA sequencing in lentil

Generations	Basis of sequencing	DNA sequencing techniques
First generation	Sequencing based on cloned DNA fragments into host cells	Sanger sequencing
Second generation	Linker- and/or adapters-based template libraries	Roche 454 pyrosequencing Illumina/Solexa HiSeq and MiSeq sequencing SOLiD (sequencing by oligo-nucleotide ligation and detection) DNA nanoball and ion torrent sequencing
Third generation	Chromosome conformation capture, optical mapping and DNA dilution-based technologies	SMRT (single molecule real time) sequencing Helicos sequencing Nanopore sequencing (MinION and PromethION) NGS by electron microscopy

genome-specific SSR and SNP markers, and re-sequence unigene-derived amplicons (Kumar 2016). For instance, the kompetitive allele-specific PCR (KASP) procedure is utilized to discover the SNPs from existing EST database in lentil (Sharpe et al. 2013; Fedoruk et al. 2013). The various second-generation techniques in lentil has been exploited to draft the sequence of genome and transcriptomes (Kaur et al. 2011; Khorramdelazad et al. 2018; Polanco et al. 2019; Sharpe et al. 2013; Singh et al. 2019). The third-generation DNA sequencing technique can be used to quickly fill the large gaps by generating longer DNA sequence in cost effective manner that is formed in drafting the lentil genome using second-generation techniques and also help in increasing the accuracy of SNPs detection (Stapley et al. 2010). The major limiting factors restricting the progress and purposes of genomics-assisted lentil breeding are narrow genetic base, too big genome size, low density linkage map and lack of candidate genes (Kumar 2016). The NGS and genotyping by sequencing (GBS) tools have unfolded new avenues for quick progress in developing sequence-based markers and sequencing platforms for lentil genome throughout the world (Elshire et al. 2011).

In recent times, NGS techniques have developed quickly which is valuable for identification and validation of unigenes, transcript, functional markers, and enriching genetic diversity, phylo-genetic studies and mapping of QTLs (Li et al. 2010). It has made feasible of SNPs mining in quick and economic manner for gene/QTL mapping and advancement in array-based high-throughput (HTP) genotyping techniques in lentil crop (Kaur et al. 2011; Sharpe et al. 2013). Several studies of NGS technologies were carried out, for example, NGS sequencing data produced on lentil cultivar CDC Redberry (Bett et al. 2014), about 44,879 SNP markers discovered using Illumina Genome Analyzer (Sharpe et al. 2013) and high-density linkage map constructed using 50,960 SNPs in lentil (Temel et al. 2015) which encouraged the progress of Illumina Golden Gate (GG) platforms for genotyping in lentil (Kaur et al. 2014; Sharpe et al. 2013), advancement in linkage maps and detection of genetic diversity and association of markers with traits of economic importance (Ates et al. 2018; Khazaei et al. 2018; Lombardi et al. 2014; Pavan et al. 2019). However, a large number of SNPs from coding regions are discovered in current years through NGS transcripts analysis in the genome of lentil (Kaur et al. 2014; Sharpe et al. 2013; Singh et al. 2019).

10.9.3 Transcriptome Analysis

The analysis of transcriptome assemblies offer an excellent way to identify the ESTs-derived functional markers like SSR, SNP markers and ITPs in lentil (Kumar 2016). In recent past, the classical Sanger sequencing technique has been employed to sequence 150–400 bp cDNA clones at particular crop stage and to generate ESTs from mRNA sequences. The progression in high throughput functional genomics techniques such as NGS and SAGE (Serial analysis of gene expression) has accelerated the designing of ESTs, functional markers based on EST and identification of candidate genes through homologous sequences across

Table 10.6 NGS platforms in lentil used for formation of EST-based functional markers

NGS platforms	SSRs	SNPs	Transcripts	Reads/ESTs	References
454 pyrosequencing	–	44,879	–	1,030,000	Sharpe et al. (2013)
Roche 454 GS-FLX Titanium	2415	–	25,592	1,380,000	Kaur et al. (2011)
Illumina HiSeq2500 platform	9949	8260	–	58,621,121	Singh et al. (2017)
	–	6306	48,453	–	Polanco et al. (2019)
	–	6693	–	46,700,000	Pavan et al. (2019)
			20,009	119,855,798	Verma et al. (2013)
Illumina HiSeq 2000 platform	141,050	194,178	–	26,165,023	Singh et al. (2019)
	–	50,960	–	111,105,153	Temel et al. (2015)
Ion proton sequencer	–	–	317,412	–	Khorramdelazad et al. (2018)

crop species. The first EST library based on seed phenotypes made from a mixture of eight cultivars (Vijayan et al. 2009), subsequently, second cDNA library was set up from the leaflets of "Eston" a Canadian cultivar inoculated with fungal pathogen *Colletotrichum truncatum* (Bhadauria et al. 2017; Kumar et al. 2014). At present, available ESTs for lentil in public domain are approximately 33,371 as accounted by NCBI, June 2020. EST-based databases are valuable genomic assets for the designing of SSR, CAPS, SNP and RFLP markers through different bioinformatics tools such as MISA for SSR detection and Snipper for SNP discovery (Kota et al. 2003; Thiel et al. 2003; Varshney et al. 2005). Kaur et al. (2011) generated 1.38×10^6 unigene/EST assembly, 3470 SNP and EST-SSR markers, contigs (15,354) and single tons (68,715) de novo assembly through transcriptome sequencing by using second-generation Roche 454 GS-FLX Titanium technique from tissue-specific cDNA samples of six lentil genotypes (Indianhead, Northfield, Digger, ILL 2024, ILL 7537 and ILL 6788). The NGS platform used for formation of EST-based functional markers and genomic resources in lentil in quick and cost-effective manner (Table 10.6).

Currently, Illumina GA/GAIIx sequencing technique is a future prospective alternative for development of transcriptome cDNA library. Using this technique Sharpe et al. (2013) developed 1536 SNP Illumina GG array to build a SNP-based linkage map and their population employed for mapping in *L. culinaris*. Similarly, Verma et al. (2013) de novo assembled the transcriptome assemblies of lentil using Illumina GAII and designed SSR markers for diversity analysis. Temel et al. (2015) generated a SNP-based genetic linkage map and their RILs mapping population using Illumina CASAVA pipelines in two lentil cultivars, Precoz and WA 8649041. Singh et al. (2017) assembled 77,346 contigs generated from overall 58,621,121 reads over four treatments of drought tolerant and susceptible lentil genotypes through de novo transcriptome analysis that led to the designing of SSR (9949) and SNP (8260) markers. Subsequently, Singh et al. (2020) generated a sub set of about 50 EST-SSR markers which exploited for the diversity assessment among 234 lentil genotypes and cross species transferability of markers across the legumes. In another study, Singh et al. (2019) generated 96,824 contigs developed from on average 26,165,023 reads across 12 heat tolerant and sensitive lentil genotypes by de novo transcriptome analysis. More recently, Wang and Daun (2006) designed PCR-based KASP markers by converting 78 SNPs out of 276 EST-based SSR markers evaluated from 26,449 EST-SSR and 130,073 SNP markers across 94 accessions of lentil which were developed through RNAseq analysis.

10.9.4 Mapping Techniques

The stepwise flowchart of genomics-assisted lentil breeding has been depicted in Fig. 10.2.

10.9.4.1 Mapping Population (MP)

The ICARDA, Lebnana (Syria) and national lentil programmes in several countries including India were attempted to develop mapping population for genetic dissection of key QTLs/genes associated with target traits of economic significance in order to locate their relative position and genetic distance by using molecular markers (Kumar et al. 2018b, 2011). Intra-specific mapping populations is characterized

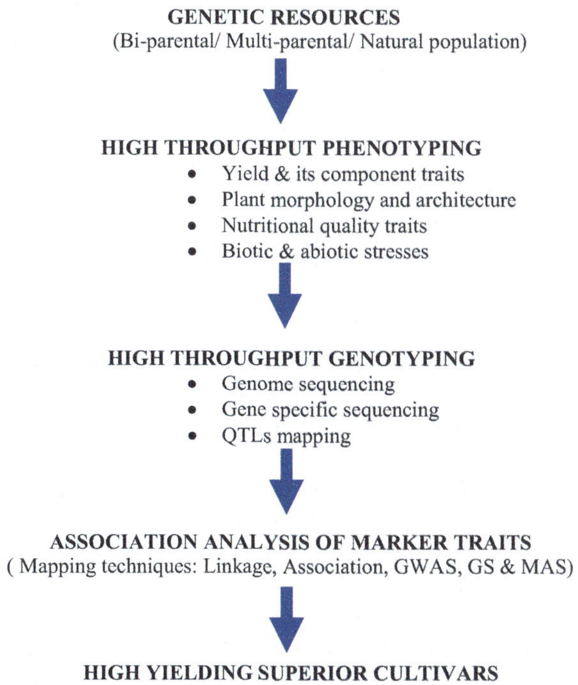

Fig. 10.2 Stepwise flowchart of genome-assisted breeding in lentil

Table 10.7 Mapping populations (MPs) derived from various crosses for different traits in lentil

S. No.	Target traits	Parental cross	MP size	References
1.	Drought	ILL 7946 × ILL 7979 ILL6002 × ILL5888	174 133	Kumar et al. (2016) Idrissi et al. (2015)
2.	Cold	ILL 4605 × ILL 10657	153	Kumar et al. (2016)
3.	Earliness	ILL 7115 × ILL 8009	150	Kumar et al. (2016)
4.	Rust	ILL 5888 × ILL 6002	152	Kumar et al. (2016)
5.	*Fusarium* wilt	ILL 213 × ILL 5883 ILL 5588 × L 692-16-1(s)	150 86	Kumar et al. (2016) Hamwieh et al. (2005)
6.	*Aschochyta* blight	ILL5588 × ILL7537 ILL7537 × ILL6002 *L. culinaris* (Alpo) × *L. odemensis* (ILWL 235)	150 153 78	Rubeena and Ades (2006) Polanco et al. (2019)
7.	Zn content	ILL 5722 × ILL 9888 ILL 9888 × ILL 5480	177 149	Kumar et al. (2016)
8.	Fe content	ILL 9932 × ILL 9951	193	Kumar et al. (2016)
9.	Milling quality	CDC Robin × 964a-46	127	Subedi et al. (2018)

with low genetic diversity lead to less profusely incorporated markers for gene-based maps (Bohra et al. 2012). Therefore, inter-specific RILs population were generated in Lentil by crossing Alpo cultivar (*L. culinaris*) and ILWL 235 accession (*L. odemensis*) for resistance of *Ascochyta* blight (Polanco et al. 2019). The back-cross inbred line (BIL) population was also generated from a cross IPL 220 cultivar (*L. culinaris*) and ILWL 118 accession (*L. orientalis*) to identify the genes for yield traits (Kumar et al. 2019). The RILs (Recombinant Inbred Lines) mapping populations in lentil for key agronomic important and, biotic and abiotic traits *viz*., early growth vigour, earliness, root traits, drought, zinc (Zn) and iron (Fe) were developed at ICARDA, Lebnan (Kumar 2016). In lentil, resistance for *Aphanomyces* root rot (ARR), generated the 189 RILs population by crossing between partial resistant and susceptible lines (Ma et al. 2020). The RILs population for quality traits have been developed from three bi-parental crosses (PI320937 × Eston; ILL8006 × CDC Milestone; and CDC Redberry × ILL7502) to recognize the QTLs for iron (Fe), manganese (Mn), and selenium (Se) uptake (Aldemir et al. 2017; Ates et al. 2016, 2018). The various mapping populations (MPs) derived from various crosses for different traits in lentil have shown in Table 10.7.

10.9.4.2 High Throughput Phenotyping (HTP)

Since plant genomics already revolutionized with the advent of NGS techniques, whereas, plant phenomics are still facing a greatest challenge in practical plant breeding due to technology availability and high initial cost. Conventional phenotyping approaches have done by manual assessments and measurements requires breeder skill and expertise, a lot of energy, labour intensity, time and accompanied with human error. A large-scale phenotyping of complex quantitative

Table 10.8 High-throughput phenotyping platform (HTPP) for various economically important traits

Sr. No.	HTPP technology	Economically important traits	Condition	Reference
1.	Imaging	Growth parameter and chlorophyll fluorescence	Controlled condition	Jansen et al. (2009)
2.	Imaging	Leaf area	Field condition	Montes et al. (2011)
3.	Automated RGB imaging	Cold tolerance, shoot biomass and PS-II efficiency	Controlled condition	Humplík et al. (2015)
4.	RGB and hyper-spectral imaging	*Aphanomyces* root rot (ARR)	Controlled condition	Marzougui et al. (2019)
5.	RGB imaging	Salt tolerance	Controlled condition	Dissanayake et al. (2020b)
6.	Unmanned aerial system and satellite imagery	Vegetation indices	Field condition	Sankaran et al. (2019)
7.	Camera	Growth parameter of leaf	Controlled condition	Massonnet et al. (2010)
8.	Camera	Presence of rice bugs	Field condition	Fukatsu et al. (2012)
9.	Spectro-radiometer	Drought stress	Controlled condition	Lu et al. (2011)
10.	Visual	Root architectural traits	Field condition	Trachsel et al. (2011)
11.	Hydraulic push press	Root distribution pattern and depth	Field condition	Wasson et al. (2012)
12.	Sensor	Height of canopy	Field condition	Andrade-Sanchez et al. (2014)

traits or unbiased scoring of diverse biotic and abiotic stresses is quite difficult in modern plant breeding (Furbank and Tester 2011). Therefore, there is urgent need of HTP technologies for rapid, highly efficient, high quality and accurate measurement and prediction of complex quantitative traits in modern plant breeding (Marko et al. 2018; Zhang et al. 2017). In recent years, rapid advancement in image-based HTP approaches has been furnished excellent opportunities to distinguish the plants for various stresses (Hu et al. 2020; Sun et al. 2018; Wang et al. 2019). For example, advancement in a range of digital imaging sensors have been progressed such as red–green–blue (RGB), fluorescence, hyperspectral and 3D imaging techniques (Light detection and ranging-LiDAR) and unmanned aerial vehicle remote sensing (Pratap et al. 2019).

Quick progress towards high-throughput phenotyping platforms (HTPPs) have been made due to technological evolution in computing and robotics which leads to establishment of plant phenomics. It has potential to non-destructive phenotyping and, high-quality data recording and analysis predominantly for detailed and reliable characterization of plants (Table 10.8). Moreover, an international association established the major plant phenotyping center "International Plant Phenomics Network (IPPN)" for HTP of major plants via robotic and non-invasive imaging technique in short-duration crops/model plants throughout their life cycle (Muraya et al. 2017). The integration of LiDAR, high-resolution camera and hyperspectral imager are able to quick, precise and efficient phenotyping of numerous traits *viz.*, plant height, leaf length, width and angle under controlled condition in the green-house (Guo et al. 2017). Zhang et al. (2017) quantified the dynamic range of growth characters from seedling to tasseling stage in maize RIL population using a HTPP under greenhouse. Undoubtedly, HTP together with genomic selection will increase our understanding of crop physiology for complex traits (e.g., disease resistance, stress tolerance and seed yield) (Reynolds and Langridge 2016). Recently, Marzougui et al. (2019) quantitative evaluated ARR resistance in lentil using image-based phenotyping approaches in controlled condition (351 lines and 191 RILs using digital RGB and hyperspectral imaging) and field (173 RILs using multispectral imaging) conditions. Similarly, Dissanayake et al. (2020b) applied image-based HTP methodology for screening of 276 accessions for salt tolerance in lentils using RGB images on a LemnaTec Scanalyzer 3D phenomics platform.

10.9.4.3 Linkage Mapping

The different marker types and linkage maps have been developed in lentil (Table 10.9). In ancient times, linkage maps have been generated in lentil by using both intra- and inter-specific bi-parental mapping populations (Kumar et al. 2019). For example, a molecular linkage map in lentil was constructed from a inter-specific cross between Precoz × WA 8949041 using 94 RILs population with 166 RAPD and ISSR markers, which comprised of 11 linkage groups covering 1396.30 cM and distance between framework markers ranged from 50.90 to 436.50 cM with an average of 8.40 cM (Tanyolac et al. 2009). Till now, several researches related to genetic linkage map had published, however, the first linkage map generated by exploiting the DNA-based markers during 1989 (Havey and Muehlbauer 1989). Thereafter, development of genetic linkage maps has been progressed with several molecular (RAPD, RFLP, AFLP, ITAP and SSR) and morphological markers (Andeden et al. 2013; Gupta et al. 2012b; de la Vega et al. 2011; Verma et al. 2015) and in fact first such map constructed with the help of these markers and RILs population of a cross between *L. culinaris* and *L. orientalis* (Eujayl et al. 1998). A high-density linkage map constructed by utilizing bi-parental populations with 5385 DArT markers, covered approximately 973.1 cM genetic map distance with 0.18 cM average distance between markers (Ates et al. 2018).

Currently, evolution of NGS techniques has replaced the chip-based method used with SNP markers and made greater availability in number of SNP markers which lead to the generation of high-density linkage maps. Further, low expenses of sequencing guided to the construction of a SNP-based vast number of high-density linkage maps in lentil reflected in Table 10.10 (Bhadauria et al. 2017; Ma et al. 2020; Polanco et al. 2019; Sudheesh et al. 2016). The SSR and SNP markers developed from parts of DNA expressing by NGS-based transcriptome analysis which acts as functional DNA markers and led to construction of high density linkage map from an

Table 10.9 Linkage map in lentil development based on molecular markers and mapping population

Parental crosses	MP size	Molecular markers	Marker loci (number)	Map distance (cM)	References
L. culinaris sp. *culinaris* × *L. c. sp. orientalis*	113 F_2	RAPD, ISSR, AFLP, CAPS, SSR, SRAPS	200	2234	Fratini et al. (2004) de la Puente García et al. (2013)
ILL 5588 × ILL 7537	150 F_2	RAPD, ISSR RGA	114	784	Ford and Taylor (2003)
L 830 × ILWL 77	114 F_2	SSR, ISSR, RAPD	199	3843	Gupta et al. (2012b)
ILL 5588 × L 692-16-1(s)	86 RIL	SSR, AFLP	283	751	Hamwieh et al. (2005)
Eston × PI 320937	94 RIL	AFLP, RAPD, SSR	207	1868	Tullu et al. (2006, 2008)
Precoz × WA 8649041	94 RIL	AFLP, ISSR, RAPD	166	1396	Tanyolac et al. (2009)
ILL 6002 × ILL 5888	206 RIL	SSR, RAPD, SRAP	139	1565	Saha et al. (2013)
Cassab × ILL 2024	126 RIL	SSR, SNP	318	1178	Kaur et al. (2014)
L. culinaris cv. Precoz × *L. culinaris* cv. L830	126 RIL	SSR	216	1183.70	Verma et al. (2015)
Indianhead × Northfield Indianhead × Digger Northfield × Digger	RIL 117 112 114	SNP, SSR	689	2429.61	Sudheesh et al. (2016)
ILL 8006 × CDC Milestone (LR11) CDC Redberry × ILL 7502 (LR8) PI 320937 × Eston (LR39)	RIL 96 120 118	DArT, SNP	50	977.47	Ates et al. (2018)

interspecific (*L. culinaris* × *L. odemensis*) mapping population consisted of 6153 markers grouped into 4682 unique bins, positioned on 10 LGs and covered about 5782 cM length (Polanco et al. 2019). A high-density linkage map formed via transcriptome sequencing with 61 SSR and 264 SNP markers from a lentil EST database communicated to seven linkage groups (LGs) and three satellites covered around 1178 cM with an average of one locus per 3.70 cM (Kaur et al. 2014). A high-density consensus map generated in lentil from three RILs population (Table 10.7) covered 977.47 cM genetic distance and consisted of 9793 SNP markers which were grouped in seven LGs with an average distance between markers of 0.10 cM (Ates et al. 2018). Recently, a linkage map constructed with

Table 10.10 Linkage maps in lentil constructed specifically with SNP molecular markers

Parental crosses	MP size	Total number of SNP loci	Map distance (cM)	Average distance between markers	References
DC Robin × 964a-46	144	542 (included six SSRs)	834.7	1.54	Sharpe et al. (2013)
Precoz × WA 8649041	103	388	432.8	1.11	Temel et al. (2015)
Indianhead × digger Indianhead × northfield Northfield × digger	117 112 114	689	2429.6	3.5	Sudheesh et al. (2016)
L 01-827A × IG-72815 (*L. ervoides*)	94	543	740.9	1.36	Bhadauria et al. (2017)
PI 320937 × eston	96	1784 (Included four SSRs)	1,784	2.3	Ates et al. (2016)
ILL8006 × CDC milestone	118	4177	497.1	0.12	Aldemir et al. (2017)
ILL8006 × CDC milestone CDC Redberry × ILL 7502 PI320937 × Eston	118 120 96	9793	977.47	0.10	Ates et al. (2018)
CDC Redberry × ILL 7502	120	5385	973.1	0.18	Ates et al. (2018)
Alpo (*L. culinaris*) × ILWL 235 (*L. odemensis*)	78	6306 (4682 bins)	5782.19	0.91	Polanco et al. (2019)

265 markers taken in the lentil (433 SSRs) and other legumes genome (250 RAPDs, 145 SSRs and 25 ISSR) distributed on seven LGs with map distance of 809.4 cM and an average of 3.05 cM (Mane et al. 2020).

10.9.4.4 QTLs Mapping

QTLs are genomic regions which is responsible for the variation in the quantitative trait of interest. In a study from Ma et al. (2020) identified and highlighted 19 QTL by QTL mapping and 38 QTL by association mapping linked with ARR resistance in lentil, of which seven QTL clusters and 15 putative genes within these clusters identified on six chromosomes, and also validated these results through expression analysis. Kumar et al. (2018c) identified 24 marker trait associations (MTAs) in nine agronomic important traits of lentil using association mapping at $P < 0.01$ through MTA analysis. The phenotypic variability (%) ranged from 7.30 to 25.80% explaining the marker associated with each specific agronomic trait (days to maturity, secondary branches/plant, pods/plant, reproductive duration, yield/plant and

Table 10.11 QTLs identified for economically important traits using NGS techniques in lentil

Traits	QTLs	Linkage group	1.8 LOD score	Phenotypic variation (%)
Stem pigmentation	SP	LG 1	79.00–145.70	33.96
Time to flowering	TF	LG 6	199.00–222.00	55.73
Flower colour	FC	LG 6	31.30–34.60	84.02
Seed coat spotting	Scp	LG 6	30.07–33.96	85.07
Seed size	Q1, Q2, Q3	LG 1, LG5a, LG5b	81.00–141.00	35.48
Ascochyta blight	AS-Q1, AS-Q2, AS-Q3	LG 6	180.00–222.00	28.46

100 seed weight), among them QTLs for primary branches/per plant explained the leading percentage of total phenotypic variation (23.1–25.8%). Subedi et al. (2018) mapped several QTLs for milling quality traits using 534 SNP markers, seven SSR markers and four morphological markers. In 6 of 7 linkage groups (LGs), the most stable QTLs were grouped on LGs 1, 2, 3 and 7 for dehulling efficiency (DE) and milling recovery (MR), whereas, on LGs 4, 5, 6 and 7 for football recovery (FR). Gupta et al. (2012a) documented three QTLs for *Ascochyta* blight (*Ascochyta lentis*) resistance at seedling and pod/maturity stages. Lately, Khorramdelazad et al. (2018) identified major defence response genes during host-pathogen interactions of *Ascochyta* blight by transcriptome profiling of lentil lines ILL 7537 (resistant) and ILL 6002 (susceptible) via a targeted RNA-Seq technique, while, Sari et al. (2018) concluded the usefulness of designing molecular markers in the gene pyramiding of *Ascochyta* blight resistance through quantitative real time-PCR (RT-qPCR) and RNA-seq analysis. Ates et al. (2018) identified the six QTLs for Mn (Manganese) concentration via composite interval mapping (CIM) which explained total phenotypic variation ranged from 15.30 to 24.10% with LOD scores from 3.00 to 4.42. Polanco et al. (2019) detected the QTLs for economically important traits through RNAseq methodology (NGS technique) in lentil RILs population derived from the inter-specific cross between *L. culinaris* cultivar Alpo and *L. odemensis* accession ILWL 235. There is a list of QTLs identified for yield and other economically important traits of interest in lentil by using NGS techniques (Table 10.11).

10.9.5 Functional Genomics

Functional genomics are comprehensive technique used to identify the functions and their interactions of candidate gene(s)/QTLs responsible for principal trait of interest. The gene cloning approach, resistant gene analogue (RGA) and cross species sequence homology (Synteny and co-linearity) helps in tagging the unknown or targeted genes controlling the trait of agronomic importance or biotic resistance or abiotic tolerance. The lentil genomic resources with cDNA from ESTs and cloned genes could speed up the progress of functional markers to be used in MAS. The functional genomics techniques like SAGE enhanced the cloning of genes in several crops including legumes. In lentil, RGAs have been recognized (Yaish et al. 2004). For example, Sari et al. (2018) identified two RGAs *viz.*, *RGA1* and *RGA71* with putative roles as receptors/transcription factors by NCBI BlastX tool, orthologues of *ABI1*-B and *DDB1-CUL4* involved in the signal transduction pathways and an orthologue of *PRH*, as a transcription factor involved in the transcription activation of *PR* proteins. They identified candidate defence response genes through RNA-seq analysis were "nucleotide-binding site- leucine rich repeat (*NBS-LRR*) and *RLK* gene families" tentatively involved in the *Ascochyta* blight pathogen recognition and resistance, and differentially expressed among the *Ascochyta* blight resistance genotypes "CDC Robin", "964a-46" and susceptible check "Eston". The transcription factor (TF) genes were identified in *Arabidopsis* which is also found in legumes based on cross reference sequence homology (Li et al. 2018; Udvardi et al. 2007). The cross-species distribution of functionally related or differentially expressing TF genes between legume and non-legume or in contrasting genotypes can also be identified or confirmed by functional genomics approaches. Ford et al. (2007) assessed differential gene transcript profiles in lentil resistance (ILL7537) and susceptible (ILL6002) genotypes after inoculation with *Ascochyta lentis*. Functional genomics analysis can be used to identify several defence responsive candidate genes. For example, Cao et al. (2019) identified the function of candidate genes coding for calcium-transporting *ATPase* and glutamate receptor 3.2 for *Stemphylium* blight disease resistance in lentil and validated by BSA (Bulk segregant analysis). Currently, Dissanayake et al. (2020a) identified five candidate genes from Region-1 encoding for metal and ABC transporter, and a mitogen-activated protein kinase 3 (*MAPKK* 3) using cross reference SNP markers.

10.9.6 Comparative Genomics

The synteny among the genomes of lentil and other crops of legume have shown different levels of conservation during course of evolution which has been potentially utilized by comparative genomics approach through designing of PCR-based DNA markers such as SSR and ITAP (Intron-Targeted Amplified Polymorphism) in lentil (Choi et al. 2004; Zhu et al. 2005; Choudhary et al. 2009; Phan et al. 2007). The comparative genomics approaches have improved the cross-species transferability of genetic information among legume crop species (chickpea, pigeon pea, soybean *Trifolium*, common bean and *Medicago*) that has led to the development of SSRs and establishment of phylogenetic relationship among lentil species (Alo et al. 2011; Datta et al. 2011; Gupta et al. 2012a; Phan et al. 2007; Reddy et al. 2010). The synteny and co-linearity among the genome of legume crop species have potential to offer SSR marker transferability to lentil because the latter has limited availability of SSR markers, for example, chickpea-specific STMS (Sequence Tagged Marker

Satellite) primers has 5% transferability to lentil (Pandian et al. 2000); the 62% successful amplification of *Trifolium*; 36% of *Medicago* and 25% of *Pisum* markers in lentil (Reddy et al. 2010); 19 STMS markers transferability from chickpea, pigeon pea, common bean and soybean to lentil (Datta et al. 2011); and also has potential to EST-based SSR marker transferability from model genome *M. truncatula* to lentil crop (Gupta et al. 2012a).

Comparative genomics can also be exploited to detect the cross-species candidate genes responsible for traits of agronomic importance with the help of functional markers such as EST-SSR or EST-SNP or gene-based markers. Weller et al. (2012) identified orthologous gene loci "*ELF3*" responsible for photoperiod response in lentil from pea which is one of the two genes "*HR*" (High response to photoperiod) and "*ELF3*" controlling the difference in photoperiod response between wild and domesticated pea. Kaur et al. (2014) tagged genes for micro-nutrient tolerance of boron in lentil through comparative genomics from the comparative genome of *Arabidopsis thaliana* and *M. trucatuala* by using flanking markers SNP 20002998 and SNP 20000246. In another study by Kaur et al. (2011) developed EST-based SSRs associated with flowering time in lentil led to discovery of genes responsible for flowering time in other legumes through comparative mapping technique (Kumar et al. 2018b). Recently, a study conducted by Singh et al. (2020) screened a subset of 50 out of 9949 EST-based SSR markers developed through transcriptome analysis which is detected different candidate genes associated with different metabolic activities in lentil and 12 legume genera. Thus, these functional markers could be used for identification of candidate genes underlying traits of agronomic interest through comparative mapping in lentil.

10.9.7 Marker-Assisted Selection (MAS)

Conventional breeding methods is largely reliable on selection of desirable plants in F_2 and onwards segregating generation contain appropriate combinations of genes or as a parent or variety based on complex quantitative traits which is highly influenced by the genotype, environment and interaction between them. MAS and DNA-based selection of desirable plant are remarkable ways to introduce QTLs/genes associated with yield enhancement, biotic resistant or abiotic tolerant traits. Currently, the implementation of MAS in lentil-breeding is limited due to insufficient availability, accessibility and slow progress in the enhancement of lentil genomic resources (Kumar et al. 2014). MAS breeding strategy has already been successfully applied and documented in broad range of cereals (rice, wheat, maize, sorghum and barley), pulses (chickpea, pigeon pea and beans), oilseeds (groundnut, mustard and sunflower) and vegetable crops (cassava, cowpea and potato), and other crops as well (Cobb et al. 2019). Till now, the two schemes of MAS, i.e. MABB (marker-assisted backcrossing) and MAP (marker-assisted gene pyramiding) employed extensively for a numerous of complex quantitative traits including abiotic stress (Steele et al. 2006) and grain number (Ashikari et al. 2005). The morphological markers exhibiting monogenic dominant inheritance were classified as qualitative markers such as cotyledon (Yc), seed coat pattern (Scp), anthocyanin pigmentation in stem (Gs), early flowering (Sn), flower color (W), pod indehiscence (Pi), radiation frost tolerance (Rf) and ground color of seed (Gc) in lentil (Duran et al. 2004; Hamwieh et al. 2005; Tullu et al. 2003). Till now, several QTLS have been mapped and documented for various yield traits including disease resistance and abiotic stresses that will facilitate the introgression of theses genomic regions in lentil via MAS. Furthermore, quantitative traits are highly affected by both genetic and environmental factors, thus, RILs or NILs (near-isogenic lines) mapping population are more precise to accurately dissect the specific QTLs.

10.10 Conclusion

The milestone goal in breeding of lentil crop is to develop superior varieties with higher yield, better quality and diseases and insects resistance. In recent past, hybridization and selection have become the chief assets in lentil improvement, however, conventional breeding techniques are suffering from several bottlenecks which restricts the lentil breeders in achieving the major breeding objectives. In recent times, development of genomic resources with the use of gene-based markers genotyping such as SNP and/or SSR, and NGS techniques have made possible to access the genomic sequences, transcriptomic sequences and genes/QTLs linked to important yield contributing traits. Therefore, there is a necessary to incorporate genomics-assisted lentil breeding approaches such as molecular marker and genomic resources techniques in order to identify, select and fix superior recombinants more accurately and efficiently. The genomics-assisted lentil breeding is a powerful tool, enabled integration of genomic tools and HTP with conventional breeding techniques. There is need to establish HTP platform to accelerate the accurate phenotyping the complex quantitative traits. The newly developed genetic and genomic tools and techniques will augment the conventional plant breeding evaluation and selection procedures. The genomics-assisted lentil breeding will broaden the genomic resources, generate a large quantity of next-generation sequencing information, transcriptomes and cross references genomic information among legumes for mining of candidate gene information for traits of agronomic interest, biotic and abiotic stresses. The HTP, high-throughput genotyping and next-generation sequencing platforms pledge to further revolutionize our perceptive of genetic diversity and in designing breeding strategies to utilize precise genomic information in further improvement of lentil crop yield in future breeding programme. MAS and genomics-assisted breeding approaches have great potential to improve lentil breeding at genome level and take crop breeding into new heights of achievements in future. Lentil crop fits remarkably well into genre of a "smart crop", and now considered as target crop for research and development among pulses.

References

Afzal M, Alghamdi SS, Migdadi HH et al (2020) Legume genomics and transcriptomics: from classic breeding to modern technologies. Saudi J Biol Sci 27(1):543–555

Ahloowalia BS, Maluszynski M, Nichterlein K (2004) Global impact of mutation-derived varieties. Euphytica 135(2):187–204

Aldemir S, Ateş D, Temel HY et al (2017) QTLs for iron concentration in seeds of the cultivated lentil (*Lens culinaris* Medic.) via genotyping by sequencing. Turk J Agric For 41(4):243–255

Ali M, Gupta S (2012) Carrying capacity of Indian agriculture: pulse Crops. Curr Sci 102:874–881

Alo F, Furman BJ, Akhunov E et al (2011) Leveraging genomic resources of model species for the assessment of diversity and phylogeny in wild and domesticated lentil. J Hered 102(3):315–329

Amin R, Laskar RA, Khan S (2015) Assessment of genetic response and character association for yield and yield components in lentil (lens culinaris L.) population developed through chemical mutagenesis. Cogent Food Agric 1(1):1000715

Amin-Laskar R, Wani MR, Raina A et al (2018) Morphological characterization of gamma rays induced multipodding mutant (mp) in lentil cultivar Pant L 406. Int J Radiat Biol 94(11): 1049–1053

Andeden EE, Derya M, Baloch FS (2013) Development of SSR markers in lentil. In: Proceedings of plant and animal genome conference Vol 21. Conference XXIP0351, San Diego, CA, p P0351, 12 Jan 2013

Andrade-Sanchez P, Gore MA, Heun JT et al (2014) Development and evaluation of a field-based high-throughput phenotyping platform. Funct Plant Biol 41(1):68–79

Andrews M, Mckenzie BA (2007) Adaptation and ecology. In: Yadav SS, McNeil DL, Stevenson PC (eds) Lentil-an ancient crop for modern times. Springer, Dordrecht, pp 23–32

Ashikari M, Sakakibara H, Lin S et al (2005) Cytokinin oxidase regulates rice grain production. Science 309(5735):741–745

Asnake F, Bejiga G (2003) Breeding lentil for wider adaptation in forage and food legumes of Ethiopia: progress and prospects. In: Kemal A, Gemechu K, Seid A et al (eds) Proceedings of the workshop on food and forage legumes. EIAR and ICARDA, Aleppo/Addis Ababa, pp 80–86

Ates D, Sever T, Aldemir S, Yagmur et al (2016) Identification QTLs controlling genes for Se uptake in lentil seeds. PLoS One 11(3):e0149210

Ates D, Aldemir S, Yagmur B et al (2018) QTL mapping of genome regions controlling manganese uptake in lentil seed. G3 8(5):1409–1416

Aversano R, Ercolano MR, Caruso I (2012) Molecular tools for exploring polyploid genomes in plants. Int J Mol Sci 13(8):10316–10335

Bauchet GJ, Bett KE, Cameron CT (2019) The future of legume genetic data resources: challenges, opportunities, and priorities. Legume Sci 1(1):e16

Baum M, Erskine W, Ramsay G (1997) Biotechnology and genetic resource of grain legumes: lentil and faba beans. In: Watanabe KN, Pehu E (eds) Plant biotechnology and plant genetic resource for sustainability and productivity. RG Landes Company/Academic Press Inc, Austin, pp 117–132

Bett K, Ramsay L, Sharpe A et al (2014) Lentil genome sequencing: establishing a comprehensive platform for molecular breeding. In: Proceedings of international food legumes research conference (IFLRC-VI) and ICCLG-VII, vol 19. Crop Development Center), Saskatoon

Bett K, Ramsay L, Chan C et al (2016) The lentil genome – from the sequencer to the field. In: International conference on pulses, Marrakesh, Morocco, 18–20 April, 2016 Abstract

Bhadauria V, Vijayan P, Wei Y, Banniza S (2017) Transcriptome analysis reveals a complex interplay between resistance and effector genes during the compatible lentil-*Colletotrichum lentis* interaction. Sci Rep 7(1):1–3

Blakeslee AF, Avery AG (1937) Methods of inducing doubling of chromosomes in plants: by treatment with colchicine. J Hered 28(12):393–411

Bogale DA, Mekbib F, Fikre A (2017) Genetic improvement of lentil [*Lens culinaris* Medikus] between 1980 and 2010 in Ethiopia. Malays J Med Biol Res 4(2):129–138

Bohra A, Saxena RK, Gnanesh BN et al (2012) An intra-specific consensus genetic map of pigeonpea [*Cajanus cajan* (L.) Millspaugh] derived from six mapping populations. Theor App Genet 125(6):1325–1338

Cao Z, Li L, Kapoor K et al (2019) Using a transcriptome sequencing approach to explore candidate resistance genes against stemphylium blight in the wild lentil species *Lens ervoides*. BMC Plant Boil 19(1):1–6

Cash D, Lockerman R, Bowman H et al (2001) Growing lentils in Montana. In: Montana state university extension guide MT199615. Montana State University, Bozeman, MT

Chagné D, Carlisle CM, Blond C et al (2007) Mapping a candidate gene (MdMYB10) for red flesh and foliage colour in apple. BMC Genomics 8(1):1–1

Chahota RK, Kishore N, Dhiman KC et al (2007) Predicting transgressive segregants in early generation using single seed descent method-derived micro-macrosperma genepool of lentil (*Lens culinaris* Medikus). Euphytica 156(3):305–310

Chen ZJ, Ni Z (2006) Mechanisms of genomic rearrangements and gene expression changes in plant polyploids. Bioessays 28(3):240–252

Choi HK, Mun JH, Kim DJ et al (2004) Estimating genome conservation between crop and model legume species. Proc Natl Acad Sci U S A 101(43):15289–15294

Choudhary S, Sethy NK, Shokeen B et al (2009) Development of chickpea EST-SSR markers and analysis of allelic variation across related species. Theor Appl Genet 118(3):591–608

Chuni L, Sharma SK, Chahota RK (2000) Combining ability and heterosis for growth parameters in microsperma× macrosperma lentil. Crop Res 19(1):115–121

Cobb JN, Biswas PS, Platten JD (2019) Back to the future: revisiting MAS as a tool for modern plant breeding. Theor Appl Genet 132(3):647–667

Cokkizgin A, Shtaya MJ (2013) Lentil: origin, cultivation techniques, utilization and advances in transformation. Agric Sci 1(1):55–62

Cubero JI (1981) Origin, taxonomy and domestication. In: Webb C, Hawtin GC (eds) Lentils. Commonwealth Agricultural Bureau, Slough, pp 15–38

Cubero JI, Perez De La Vega M, Frantini R et al (2009) Origin, phylogeny, domestication and spread. In: Erskine W, Muehlbauer FJ, Sarker A, Sharma B (eds) Lentil – botany, production and uses. CABI, Wallingford, pp 13–33

Datta S, Tiwari S, Kaashyap M et al (2011) Genetic similarity analysis in lentil using cross-genera legume sequence tagged microsatellite site markers. Crop Sci 51(6):2412–2422

de la Puente García R, García PG, de la Puente CG et al (2013) An improved intersubspecific genetic map in lens including functional markers. Span J Agric Res 1:132–136

de la Vega MP, Torres AM, Cubero JI et al (eds) (2011) Genetics, genomics and breeding of cool season grain legumes. CRC Press, Boca Raton

Dhuppar P, Biyan S, Chintapalli B et al (2012) Lentil crop production in the context of climate change: an appraisal. Indian Res J Ext Educ 2:33–35

Dissanayake R, Braich S, Cogan NO et al (2020a) Characterization of genetic and allelic diversity amongst cultivated and wild lentil accessions for germplasm enhancement. Front Genet 11:546

Dissanayake R, Kahrood HV, Dimech AM et al (2020b) Development and application of image-based high-throughput phenotyping methodology for salt tolerance in lentils. Agronomy 10(12): 1992

Duran Y, Fratini R, Garcia P et al (2004) An intersubspecific genetic map of lens. Theor Appl Genet 108(7):1265–1273

El Haddad N, Rajendran K, Smouni A et al (2020) Screening the FIGS set of lentil (Lens Culinaris Medikus) germplasm for tolerance to terminal heat and combined drought-heat stress. Agronomy 10(7):1036

Elshire RJ, Glaubitz JC, Sun Q et al (2011) A robust, simple genotyping-by-sequencing (GBS) approach for high diversity species. PLoS One 6(5):e19379

Elzebroek T, Wind K (2008) Oil crops. Guide to cultivated plants. CABI, Wallingford, pp 182–225

Emery M, Willis MM, Hao Y et al (2018) Preferential retention of genes from one parental genome after polyploidy illustrates the nature and scope of the genomic conflicts induced by hybridization. PLoS Genet 14(3):e1007267

Erskine W, Adham Y, Holly L (1989) Geographic distribution of variation in quantitative traits in a world lentil collection. Euphytica 43(1):97–103

Erskine W, Chandra S, Chaudhry M et al (1998) A bottleneck in lentil: widening its genetic base in South Asia. Euphytica 101(2):207–211

Erskine W, Sarker A, Ashraf M (2011) Reconstructing an ancient bottleneck of the movement of the lentil (*Lens culinaris ssp. culinaris*) into South Asia. Genet Resour Crop Evol 58(3):373–381

Eujayl I, Baum M, Powell W et al (1998) A genetic linkage map of lentil (*Lens* sp.) based on RAPD and AFLP markers using recombinant inbred lines. Theor Appl Genet 97(1):83–89

FAOSTAT (2018) Food and Agriculture Organization of the United Nations. http://www.fao.org/faostat/en/#data/QC

FAOSTAT (2019) Food and Agriculture Organization of the United Nations. http://www.fao.org/faostat/en/#data/QC

Farooq M, Wahid A, Kobayashi NS et al (2009) Plant drought stress: effects, mechanisms and management. Sustain Agric:153–188

Fedoruk MJ, Vandenberg A, Bett KE (2013) Quantitative trait loci analysis of seed quality characteristics in lentil using single nucleotide polymorphism markers. Plant Genome 6(3):5

Ferguson M (2000) *Lens spp*: conserved resources, priorities for collection and future prospects. In: Linking research and marketing opportunities for pulses in the 21st century. Springer, Dordrecht, pp 613–620

Ferguson M, Erskine W (2001) Lentils (*Lens* L.). In: Plant genetic resources of legumes in the Mediterranean. Springer, Dordrecht, pp 125–133

Ford R, Taylor PW (2003) Construction of an intraspecific linkage map of lentil (*Lens culinaris ssp. culinaris*). Theor Appl Genet 107(5):910–916

Ford R, Redden RJ, Materne M et al (2007) Lentil. In: Pulses, sugar and tuber crops. Springer, Berlin, pp 91–108

Fratini R, Ruiz ML, de la Vega MP (2004) Intra-specific and inter-sub-specific crossing in lentil (*Lens culinaris* Medik.). Can J Plant Sci 84(4):981–986

Fukatsu T, Watanabe T, Hu H et al (2012) Field monitoring support system for the occurrence of *Leptocorisa chinensis* Dallas (Hemiptera: Alydidae) using synthetic attractants, field servers, and image analysis. Comput Electron Agric 80:8–16

Furbank RT, Tester M (2011) Phenomics–technologies to relieve the phenotyping bottleneck. Trends Plant Sci 16(12):635–644

Gan Y, Liang C, Chai Q et al (2014) Improving farming practices reduces the carbon footprint of spring wheat production. Nat Commun 5(1):1–3

Guo D, Juan J, Chang L et al (2017) Discrimination of plant root zone water status in greenhouse production based on phenotyping and machine learning techniques. Sci Rep 7(1):1–2

Gupta PK, Bahl JR (1983) Cytogenetics and origin of some pulse crops. In: Swaminathan MS, Gupta PK, Sinha U (eds) Cytogenetics of crop plants. Macmillan India, New Delhi

Gupta D, Taylor PW, Inder P et al (2012a) Integration of EST-SSR markers of *Medicago truncatula* into intraspecific linkage map of lentil and identification of QTL conferring resistance to *Ascochyta* blight at seedling and pod stages. Mol Breed 30(1):429–439

Gupta M, Verma B, Kumar N et al (2012b) Construction of inter-sub-specific molecular genetic map of lentil based on ISSR, RAPD and SSR markers. J Genet 91(3):279–287

Hamdi A, Erskine W, Gate P (1991) Relationships among economic characters in lentil. Euphytica 57:109–116

Hamwieh A, Udupa SM, Choumane W (2005) A genetic linkage map of *Lens sp.* based on microsatellite and AFLP markers and the localization of *Fusarium* vascular wilt resistance. Theor Appl Genet 110(4):669–677

Harlan JR, de Wet JM (1971) Toward a rational classification of cultivated plants. Taxon 20(4): 509–517

Harper AL, Trick M, He Z et al (2016) Genome distribution of differential homoeologue contributions to leaf gene expression in bread wheat. Plant Biotechnol J 14(5):1207–1214

Havey MJ, Muehlbauer FJ (1989) Linkages between restriction fragment length, isozyme, and morphological markers in lentil. Theor Appl Genet 77(3):395–401

Heuzé V, Tran G, Noziere P et al (2014) Oats. Feedipedia, a programme by INRA, CIRAD, AFZ and FAO

Hu Y, Knapp S, Schmidhalter U (2020) Advancing high-throughput phenotyping of wheat in early selection cycles. Remote Sens 12(3):574

Humplík JF, Lazár D, Fürst T et al (2015) Automated integrative high-throughput phenotyping of plant shoots: a case study of the cold-tolerance of pea (*Pisum sativum* L.). Plant Methods 11(1):1

Idrissi O, Houasli C, Udupa SM et al (2015) Genetic variability for root and shoot traits in a lentil (*Lens culinaris* Medik.) recombinant inbred line population and their association with drought tolerance. Euphytica 204(3):693–709

Idrissi O, Sahri A, Houasli C, Nsarellah N (2019) Breeding progress, adaptation, and stability for grain yield in Moroccan lentil improved varieties. Crop Sci 59(3):925–936

Iqbal MS, Hussain SA, Akbar M (2021) Estimating genetic variability among diverse lentil collections through novel multivariate techniques. https://doi.org/10.21203/rs.3.rs-135543/v1

Jansen M, Gilmer F, Biskup B et al (2009) Simultaneous phenotyping of leaf growth and chlorophyll fluorescence via GROWSCREEN FLUORO allows detection of stress tolerance in Arabidopsis thaliana and other rosette plants. Funct Plant Biol 36:902–914

Kamaluddin RK, Sharma V, Saini HK et al (2020) Variability studies for yield and its component traits in Lentil (*Lens culinaris* Medikus spp. *Culinaris*) under Bundelkhand region. J Pharmacogn Phytochem 9(6):1436–1441

Kaur S, Cogan NO, Pembleton LW et al (2011) Transcriptome sequencing of lentil based on second-generation technology permits large-scale unigene assembly and SSR marker discovery. BMC Genomics 12(1):1

Kaur S, Cogan NO, Stephens A et al (2014) EST-SNP discovery and dense genetic mapping in lentil (*Lens culinaris* Medik.) enable candidate gene selection for boron tolerance. Theor Appl Genet 127(3):703–713

Kaur S, Webster T, Sudheesh S et al (2016) Lentil genome sequencing effort: a comprehensive platform for genomics assisted breeding

Khan MH, Tyagi SD (2005) Induced variation in quantitative traits due to physical (gamma rays), chemical (EMS) and combined mutagen treatments in soybean [*Glycine max* L.) Merrill]

Khan MH, Tyagi SD (2010) Studies on effectiveness and efficiency of gamma rays, EMS and their combination in soybean [*Glycine max* (L.) Merrill.]. J Plant Breed Crop Sci 2(3):55–58

Khazaei H, Caron CT, Fedoruk M et al (2016) Genetic diversity of cultivated lentil (Lens culinaris Medik.) and its relation to the World's agro-ecological zones. Front. Plant Sci 7:1093

Khazaei H, Fedoruk M, Caron CT et al (2018) Single nucleotide polymorphism markers associated with seed quality characteristics of cultivated lentil. Plant Genome 11(1):170051

Khazaei H, Subedi M, Nickerson M et al (2019) Seed protein of lentils: current status, progress, and food applications. Foods 8(9):391

Khorramdelazad M, Bar I, Whatmore P et al (2018) Transcriptome profiling of lentil (*Lens culinaris*) through the first 24 hours of *Ascochyta lentis* infection reveals key defence response genes. BMC Genomics 19(1):1–21

Kota R, Rudd S, Facius A et al (2003) Snipping polymorphisms from large EST collections in barley (*Hordeum vulgare* L.). Mol Genet Genomics 270(1):24–33

Kumar S (2016) Current knowledge in lentil genomics and its application for crop improvement. Crop Breed 25:331–356

Kumar S, Ali M (2006) GE interaction and its breeding implications in pulses. Botany 56:31–36

Kumar N, Chahota RK, Sood BC (2009) Component analysis for seed yield and yield traits in microsperma× macrosperma derivatives of lentil (*Lens culinaris* Medik.). Agric Sci Dig 29(3): 163–168

Kumar J, Choudhary AK, Solanki RK et al (2011) Towards marker-assisted selection in pulses: a review. Plant Breed 130(3):297–313
Kumar S, Barpete S, Kumar J et al (2013) Global lentil production: constraints and strategies. SATSA Mukhapatra 17:1–3
Kumar S, Hamwieh A, Manickavelu A et al (2014) Advances in lentil genomics. In: Legumes in the omic era. Springer, New York, pp 111–130
Kumar J, Singh J, Kanaujia R (2016) Protein content in wild and cultivated taxa of lentil (*Lens culinaris ssp. culinaris* Medikus). Indian J Genet Plant Breed 76:631–634
Kumar J, Basu PS, Gupta S et al (2018a) Physiological and molecular characterisation for high temperature stress in *Lens culinaris*. Funct Plant Biol 45(4):474–487
Kumar J, Gupta S, Biradar RS et al (2018b) Association of functional markers with flowering time in lentil. J Appl Genet 59(1):9–21
Kumar J, Gupta S, Gupta DS (2018c) Identification of QTLs for agronomic traits using association mapping in lentil. Euphytica 214(4):1–5
Kumar S, Choudhary AK, Rana KS et al (2018d) Bio-fortification potential of global wild annual lentil core collection. PLoS One 13(1):e0191122
Kumar J, Choudhary AK, Gupta DS et al (2019) Towards exploitation of adaptive traits for climate-resilient smart pulses. Int J Mol Sci 20(12):2971
Ladizinsky G (1979) The origin of lentil and its wild genepool. Euphytica 28(1):179–187
Ladizinsky G (1999) Identification of the lentil's wild genetic stock. Genet Resour Crop Evol 46(2): 115–118
Ladizinsky G, Braun D, Goshen D et al (1984) The biological species of the genus *Lens* L. Bot Gaz 145(2):253–261
Li F, Liu W, Tang J et al (2010) Rice dense and erect panicle 2 is essential for determining panicle outgrowth and elongation. Cell Res 20(7):838–849
Li Y, He L, Li J et al (2018) Genome-wide identification, characterization, and expression profiling of the legume BZR transcription factor gene family. Front Plant Sci 9:1332
Lizarazo CI, Lampi AM, Liu J et al (2015) Nutritive quality and protein production from grain legumes in a boreal climate. J Sci Food Agric 95(10):2053–2064
Lombardi M, Materne M, Cogan NO et al (2014) Assessment of genetic variation within a global collection of lentil (*Lens culinaris* Medik.) cultivars and landraces using SNP markers. BMC Genet 15(1):1
Lu Y, Hao Z, Xie C et al (2011) Large-scale screening for maize drought resistance using multiple selection criteria evaluated under water-stressed and well-watered environments. Field Crop Res 124(1):37–45
Ma Y, Marzougui A, Coyne CJ et al (2020) Dissecting the genetic architecture of *Aphanomyces* root rot resistance in lentil by QTL mapping and genome-wide association study. Int J Mol Sci 21(6): 2129
Malhotra N, Panatu S, Singh B et al (2019) Genetic resources: collection, conservation, characterization and maintenance. In: Lentils. Academic, Cambridge, pp 21–41
Mammadov J, Aggarwal R, Buyyarapu R et al (2012) SNP markers and their impact on plant breeding. Int J Plant Genomics 2012:728398
Mane R, Katoch M, Singh M et al (2020) Identification of genomic regions associated with early plant vigour in lentil (*Lens culinaris*). J Genet 99(1):1–8
Manju P, Gopimony R (2009) Anjitha-A New Okra Variety through Induced Muta-tion in Inter-specific Hybrids of *Abelmoschus spp*. Induced plant mutations in the genomics era. Food Agric Org United Nations 1:87–90
Marko D, Briglia N, Summerer S et al (2018) High-throughput phenotyping in plant stress response: methods and potential applications to polyamine field. In: Polyamines. Humana Press, New York, NY, pp 373–388
Marzougui A, Ma Y, Zhang C et al (2019) Advanced imaging for quantitative evaluation of *Aphanomyces* root rot resistance in lentil. Front Plant Sci 10:383

Massonnet C, Vile D, Fabre J et al (2010) Probing the reproducibility of leaf growth and molecular phenotypes: a comparison of three Arabidopsis accessions cultivated in ten laboratories. Plant Physiol 152(4):2142–2157

Materne M, McNeil DL (2007) Breeding methods and achievements. In: Lentil. Springer, Dordrecht, pp 241–253

Mbasani-Mansi J, Ennami M, Briache FZ et al (2019) Characterization of genetic diversity and population structure of Moroccan lentil cultivars and landraces using molecular markers. Physiol Mol Biol Plants 25(4):965–974

Mishra SK, Sharma B, Sharma SK (2007) Genetics and cytogenetics of lentil. In: Lentil. Springer, Dordrecht, pp 187–208

Montes JM, Technow F, Dhillon BS et al (2011) High-throughput non-destructive biomass determination during early plant development in maize under field conditions. Field Crop Res 121(2):268–273

Muehlbauer FJ (1992) Use of introduced germplasm in cool-season food legume cultivar development. In: Shands HL, Weisner LE (eds) Use of plant introductions in cultivar development. Part 2. Crop Science Society of America (CSSA) special publication, vol 20. CSSA, Madison, pp 49–73

Muehlbauer FJ, McPhee KE (2005) Lentil (*Lens culinaris* Medik.). In: Singh RJ, Jauhar PP (eds) Genetic resources, chromosome engineering and crop improvement. CRC Press, Boca Raton, pp 219–230

Muehlbauer FJ, Kaiser WJ, Clement SL, Summerfield RJ (1995) Production and breeding of lentil. Adv Agron 54:283–332

Muehlbauer FJ, Cho S, Sarker A et al (2006) Application of biotechnology in breeding lentil for resistance to biotic and abiotic stress. Euphytica 147(1):149–165

Muehlbauer FJ, Mihov MA, Vandenberg A et al (2009) Improvement in developed countries. In: Erskine W, Muehlbauer FJ, Sarker A, Sharma B (eds) Lentil-botany, production and uses. CABI, Wallingford, pp 137–154

Muraya MM, Chu J, Zhao Y et al (2017) Genetic variation of growth dynamics in maize (Zea mays L.) revealed through automated non-invasive phenotyping. Plant J 89:366–380

Ogutcen E, Ramsay L, von Wettberg EB et al (2018) Capturing variation in *Lens* (Fabaceae): Development and utility of an exome capture array for lentil. Appl Plant Sci 6(7):e01165

Pandian A, Ford R, Taylor PW (2000) Transferability of sequence tagged microsatellite site (STMS) primers across four major pulses. Plant Mol Biol Report 18(4):395

Pavan S, Bardaro N, Fanelli V et al (2019) Genotyping by sequencing of cultivated lentil (*Lens culinaris* Medik.) highlights population structure in the Mediterranean gene pool associated with geographic patterns and phenotypic variables. Front Genet 10:872

Phan HT, Ellwood SR, Hane JK et al (2007) Extensive macrosynteny between *Medicago truncatula* and *Lens culinaris* ssp. *culinaris*. Theor Appl Genet 114(3):549–558

Polanco C, Sáenz de Miera LE, González AI et al (2019) Construction of a high-density interspecific (*Lens culinaris* x *L. odemensis*) genetic map based on functional markers for mapping morphological and agronomical traits, and QTLs affecting resistance to *Ascochyta* in lentil. PLoS One 14(3):e0214409

Pratap A, Gupta S, Nair RM et al (2019) Using plant phenomics to exploit the gains of genomics. Agronomy 9(3):126

Rahman MM, Sarker A, Kumar S et al (2009) Breeding for short season environments. The lentil: botany, production and uses. CAB International, Wallingford, pp 121–136

Ray H, Bett K, Tar'an B et al (2014) Mineral micronutrient content of cultivars of field pea, chickpea, common bean, and lentil grown in Saskatchewan, Canada. Crop Sci 54(4):1698–1708

Reddy A, Reddy GP (2010) Supply side constrains in production of pulses in India: case study of lentils. Agric Econ Res Rev 23:129–136

Reddy MR, Rathour R, Kumar N et al (2010) Cross-genera legume SSR markers for analysis of genetic diversity in *Lens* species. Plant Breed 129(5):514–518

Rehman RU (2017) Future prospects in polyploidy research. In: Polyploidy: recent trends and future perspectives. Springer, New Delhi, pp 101–104
Reynolds M, Langridge P (2016) Physiological breeding. Curr Opin Plant Biol 31:162–171
Rubeena TPW, Ades PK (2006) QTL mapping of resistance in lentil (*Lens culinaris ssp. culinaris*) to *Ascochyta* blight (*Ascochyta lentis*). Plant Breed 125(5):506–512
Saha GC, Sarker A, Chen W et al (2013) Inheritance and linkage map positions of genes conferring agromorphological traits in *Lens culinaris* Medik. Int J Agron 2013:618926
Sakthivel G, Jeberson S, Singh NB et al (2019) Genetic variability, correlation and path analysis in lentil germplasm (*Lens culinaris* medik). Pharma Innov J 8(6):417–420
Sankaran S, Quirós JJ, Miklas PN (2019) Unmanned aerial system and satellite-based high resolution imagery for high-throughput phenotyping in dry bean. Comput Electron Agric 165: 104965
Sari E, Bhadauria V, Ramsay L (2018) Defense responses of lentil (*Lens culinaris*) genotypes carrying non-allelic *Ascochyta* blight resistance genes to *Ascochyta lentis* infection. PLoS One 13(9):e0204124
Sarker A, Erskine W (2006) Recent progress in the ancient lentil. J Agric Sci 144(1):19–29
Satpathy S, Debnath S (2020) Genetic analysis of yield and its attributing traits in lentil. J Pharmacogn Phytochem Res 9(2):713–718
Scott MF, Ladejobi O, Amer S et al (2020) Multi-parent populations in crops: a toolbox integrating genomics and genetic mapping with breeding. Heredity 125(6):396–416
Shahwar D, Ansari MY, Choudhary S (2019) Induction of phenotypic diversity in mutagenized population of lentil (*Lens culinaris* Medik) by using heavy metal. Heliyon 5(5):e01722
Sharpe AG, Ramsay L, Sanderson LA (2013) Ancient orphan crop joins modern era: gene-based SNP discovery and mapping in lentil. BMC Genomics 14(1):1–3
Silim SN, Saxena MC, Erskine W (1993) Adaptation of lentil to the Mediterranean environment. I. Factors affecting yield under drought conditions. Exp Agric 29(1):9–19
Sinclair TR, Vadez V (2012) The future of grain legumes in cropping systems. Crop Pasture Sci 63(6):501–512
Sindhu JS, Slinkard AE, Scoles GJ (1983) Studies on variation in *Lens* I. karyotype. LENS 10(1):14
Singh DP (1991) Lentils in India–genetics and breeding of pulse crops. Kalyani Publishers, Ludhiana, pp 217–238
Singh RJ, Chung GH (2016) Landmark research for pulses improvement. Indian J Genet 76(4): 399–409
Singh AK, Pedapati A (2015) Assessment of exchange of crop in view of change climate and International Treaties. J Environ Biol 36(1):309
Singh D, Singh CK, Taunk J (2017) Transcriptome analysis of lentil (Lens culinaris Medikus) in response to seedling drought stress. BMC Genomics 18(1):1–20
Singh A, Dikshit HK, Mishra GP et al (2019) Association mapping for grain diameter and weight in lentil using SSR markers. Plant Gene 20:100204
Singh D, Singh CK, Tribuvan KU et al (2020) Development, characterization, and cross species/ genera transferability of novel EST-SSR markers in lentil, with their molecular applications. Plant Mol Biol Report 8(1):114–129
Srivastava RP, Vasishtha HI (2012) Saponins and lectins of Indian chickpeas (*Cicer arietinum*) and lentils (*Lens culinaris*). Indian J Agric Biochem 25(1):44–47
Stapley J, Reger J, Feulner PG et al (2010) Adaptation genomics: the next generation. Trends Ecol Evol 25(12):705–712
Steele KA, Price AH, Shashidhar HE et al (2006) Marker-assisted selection to introgress rice QTLs controlling root traits into an Indian upland rice variety. Theor Appl Genet 112(2):208–221
Subedi M, Bett KE, Khazaei H et al (2018) Genetic mapping of milling quality traits in lentil (*Lens culinaris* Medik.). Plant Genome 11(2):170092
Sudheesh S, Verma P, Forster JW et al (2016) Generation and characterisation of a reference transcriptome for lentil (*Lens culinaris* Medik.). Int J Mol Sci 17(11):1887

Sun S, Li C, Paterson AH et al (2018) In-field high throughput phenotyping and cotton plant growth analysis using LiDAR. Front Plant Sci 9:16

Tadmor Y, Zamir D, Ladizinsky G (1987) Genetic mapping of an ancient translocation in the genus lens. Theor Appl Genet 73(6):883–892

Tahir M, Lindeboom N, Båga M et al (2011) Composition and correlation between major seed constituents in selected lentil (*Lens culinaris* Medik) genotypes. Can J Plant Sci 91(5):825–835

Tanyolac B, Ozatay S, Kahraman A et al (2009) Linkage mapping of Lentil (*Lens culinaris* L.) genome using recombinant inbred lines revealed by AFLP, ISSR, RAPD and some morphologic markers. J Appl Biol Sci 3(2):179–185

Temel H, Göl D, Akkale H et al (2015) Single nucleotide polymorphism discovery through Illumina-based transcriptome sequencing and mapping in lentil. Turk J Agric For 39(3): 470–488

Thavarajah P, Wejesuriya A, Rutzke M et al (2011) The potential of lentil (*Lens culinaris* L.) as a whole food for increased selenium, iron, and zinc intake: preliminary results from a 3 year study. Euphytica 180(1):123–128

Thiel T, Michalek W, Varshney R et al (2003) Exploiting EST databases for the development and characterization of gene-derived SSR-markers in barley (*Hordeum vulgare* L.). Theor Appl Genet 106(3):411–422

Tivoli B, Baranger A, Avila CM et al (2006) Screening techniques and sources of resistance to foliar diseases caused by major necrotrophic fungi in grain legumes. Euphytica 147(1):223–253

Toker C, Yadav SS, Solanki IS (2007) Mutation breeding. In: Lentil. Springer, Dordrecht, pp 209–224

Tomlekova NB (1998) The mechanism of resistance to (*A. obtectus*) in beans (*P. vulguris* L.). Agric Sci 5:19–21

Trachsel S, Kaeppler SM, Brown KM et al (2011) Shovelomics: high throughput phenotyping of maize (*Zea mays* L.) root architecture in the field. Plant Soil 341(1):75–87

Tullu A, Buchwaldt L, Warkentin T et al (2003) Genetics of resistance to anthracnose and identification of AFLP and RAPD markers linked to the resistance gene in PI 320937 germplasm of lentil (*Lens culinaris* Medikus). Theor Appl Genet 106(3):428–434

Tullu A, Tar'an B, Breitkreutz C et al (2006) A quantitative-trait locus for resistance to *Ascochyta* blight [*Ascochyta lentis*] maps close to a gene for resistance to anthracnose [*Colletotrichum truncatum*] in lentil. Can J Plant Pathol 28(4):588–595

Tullu A, Tar'an B, Warkentin T et al (2008) Construction of an intraspecific linkage map and QTL analysis for earliness and plant height in lentil. Crop Sci 48(6):2254–2264

Udvardi MK, Kakar K, Wandrey M et al (2007) Legume transcription factors: global regulators of plant development and response to the environment. Plant Physiol 144(2):538–549

USDA (2019) Plants database. https://plants.usda.gov

USDA-Agricultural Research Service (2015) National nutrient database standard release. USDA-ARS, Washington, DC. Available at: http://ndb.nal.usda.gov/ndb/foods/show/4782?fgcd=&manu=&lfacet=&format=&count=&max=35&offset=&sort=&qlookup=lentil

Vandenberg A, Slinkard AE (1990) Genetics of seed coat color and pattern in lentil. J Hered 81(6): 484–488

Varshney RK, Graner A, Sorrells ME (2005) Genic microsatellite markers in plants: features and applications. Trends Biotechnol 23(1):48–55

Varshney RK, Pandey MK, Bohra A et al (2019) Toward the sequence-based breeding in legumes in the post-genome sequencing era. Theor Appl Genet 132(3):797–816

Vavilov NI (1949) The origin, variation, immunity and breeding of cultivated plants. Chron Bot 13: 1–54

Verma P, Shah N, Bhatia S (2013) Development of an expressed gene catalogue and molecular markers from the de novo assembly of short sequence reads of the lentil (*Lens culinaris* M edik.) transcriptome. Plant Biotechnol J 11(7):894–905

Verma P, Goyal R, Chahota RK et al (2015) Construction of a genetic linkage map and identification of QTLs for seed weight and seed size traits in lentil (*Lens culinaris* Medik.). PLoS One 10(10):e0139666

Vijayan P, Vandenberg A, Bett KE (2009) A mixed genotype lentil EST library representing the normalized transcriptome of different seed development stages

von Wettberg EJ, Chang PL, Başdemir F et al (2018) Ecology and genomics of an important crop wild relative as a prelude to agricultural innovation. Nat Commun 9(1):1–3

Wang N, Daun JK (2006) Effects of variety and crude protein content on nutrients and anti-nutrients in lentils (*Lens culinaris*). Food Chem 95(3):493–502

Wang J, Badenhorst P, Phelan A et al (2019) Using sensors and unmanned aircraft systems for high-throughput phenotyping of biomass in perennial ryegrass breeding trials. Front Plant Sci 10: 1381

Wasson AP, Richards RA, Chatrath R et al (2012) Traits and selection strategies to improve root systems and water uptake in water-limited wheat crops. J Exp Bot 63(9):3485–3498

Weller JL, Liew LC, Hecht VF et al (2012) A conserved molecular basis for photoperiod adaptation in two temperate legumes. Proc Natl Acad Sci U S A 109(51):21158–21163

Wery J, Silim SN, Knights EJ et al (1994) Screening techniques and sources of tolerance to extremes of moisture and air temperature in cool season food legumes. In: Expanding the production and use of cool season food legumes. Springer, Dordrecht, pp 439–456

Wilson VE, Law AG (1972) Natural crossing in *Lens esculenta* Moench. Am Soc Hort Sci J 97: 142–143

Wong MM, Gujaria-Verma N, Ramsay L et al (2015) Classification and characterization of species within the genus Lens using genotyping-by-sequencing (GBS). PLoS One 10(3):e0122025

Yaish MW, Saenz de Miera LE, Perez De La Vega M (2004) Isolation of a family of resistance gene analogue sequences of the nucleotide binding site (NBS) type from lens species. Genome 47(4): 650–659

Yankova V, Sovkova-Bobcheva S (2009) Studying of bean varieties (*Phaseolus vulgaris* L.) reaction to bean weevil infestation (*Acanthoscelides obtectus* Say). Ann RepBean Improv Coop 52:144–145

Ye YM, Tong J, Shi XP et al (2010) Morphological and cytological studies of diploid and colchicine-induced tetraploid lines of crape myrtle (*Lagerstroemia indica* L.). Sci Hortic 124(1):95–101

Yigezu YA, Alwang J, Rahman MW et al (2019) Is DNA fingerprinting the gold standard for estimation of adoption and impacts of improved lentil varieties? Food Policy 83:48–59

Zaccardelli M, Lupo F, Piergiovanni AR et al (2012) Characterization of Italian lentil (*Lens culinaris* Medik.) germplasm by agronomic traits, biochemical and molecular markers. Genet Resour Crop Evol 59(5):727–738

Zamir D, Ladizinsky G (1984) Genetics of allozyme variants and linkage groups in lentil. Euphytica 33(2):329–336

Zhang X, Huang C, Wu D et al (2017) High-throughput phenotyping and QTL mapping reveals the genetic architecture of maize plant growth. Plant Physiol 173(3):1554–1564

Zhu H, Choi HK, Cook DR et al (2005) Bridging model and crop legumes through comparative genomics. Plant Physiol 137(4):1189–1196

Zohary D (1972) The wild progenitor and the place of origin of the cultivated lentil: *Lens culinaris*. Econ Bot 26(4):326–332

Zohary D (1995) Lentil: *lens culinaris* (Leguminoseae–Papilionoideae). In: Evolution of crop plants. Longman Scientific & Technical, Harlow, pp 271–274

Zohary D (1996) The mode of domestication of the founder crops of the Southwest Asian agriculture and pastoralism in Eurasia. University College London Press, London, pp 142–158

Zohary D, Hopf M (1973) Domestication of pulses in the old world: legumes were companions of wheat and barley when agriculture began in the Near East. Science 182(4115):887–894

Transgenic and Molecular Approaches for Pigeonpea and Chick Pea Improvement 11

Madhuri Arya, S. B. Mishra, and Kamaluddin

Abstract

Pulses are the only major source of protein in vegetarian diet. Among the different pulses pigeonpea and chickpea are the most important pulse crops grown in Asia, sub-Saharan Africa, Australia, Canada, and Middle East. The major objectives of grain legumes improvement include high yield, early maturity, resistance/tolerance to key abiotic and biotic stresses, and nutritional quality of grain. The crop-specific breeding objectives are suitability to machine harvesting and herbicide tolerance in chickpea, development of hybrids in pigeonpea, pod borer resistance in chickpea, and pigeonpea, etc. Until early years of the twenty first century, these pulses remained almost untouched from the genomics interventions. After the advances in genomics of these important pulses that include availability of draft genome sequences, large number of molecular markers, high-density genetic maps, transcriptomic resources, physical maps, and molecular markers linked to genes/quantitative trait loci for the key traits successful examples of introgression of traits through transgenic and molecular techniques are visible. Transgenics events are also evident for pod borer resistance in chickpea and pigeonpea. Advances have also been made in the use of secondary metabolites for the promotion of plant growth, control of insect pests and plant pathogens, and bio-fortification. Seed proteins of leguminous crop species are deficient in the essential sulfur containing amino acid methionine. Therefore, the introduction of DNA sequences encoding methionine rich seed protein via gene transfer

M. Arya (✉) · S. B. Mishra
Departement of Plant Breeding and Genetics, Tirhut College of Agriculture, Dholi, Bihar, India

Kamaluddin
Department of Genetics and Plant Breeding, Banda University of Agriculture and Technology, Banda, Uttar Pradesh, India

Kamaluddin et al. (eds.), *Technologies in Plant Biotechnology and Breeding of Field Crops*, https://doi.org/10.1007/978-981-16-5767-2_11

technologies for the improvement of protein quality of grain legume is an attractive aspect of pulse improvement over the conventional methods.

Keywords

Transgenic · Secondary metabolite · Transcriptomic resources · Physical maps · *Cajanus cajan* · *Cicer arietinum*

11.1 Pigeonpea (*Cajanus cajan* L.)

Pigeonpea (*Cajanus cajan* [L.] Millspaugh) is a multipurpose food legume, grown in the rain-fed dryland areas. It serves as the lifeline to resource-poor farmers in tropical and subtropical regions of Asia, Africa, and Latin America. It ranks sixth in global scenario of pigeonpea production and cultivated in about 4.70 m ha area with an annual production of 3.69 m t. The mean global productivity of this legume grain is 783 kg/ha (Food and Agriculture Organization 2018). India ranks first in annual pigeonpea production with 2.65 m t followed by Myanmar (0.9 m t), Malawi (0.23 m t), Tanzania (0.20 m t), and Kenya (0.09 m t) (Food and Agriculture Organization 2018). India contributes about 80% of total world production.

11.1.1 History

The name pigeonpea was first used in Barbados where pigeon was fed the seeds of *Cajanus cajan* (Plukenet 1692). Vavilov (1951) reported that India is the center of origin for cultivated pigeonpea and is also been widely cultivated in many African countries, Egypt, and a bunch of Asian countries since prehistoric times. Eastern Africa was considered as the center of origin of pigeonpea by some of the workers owing to its occurrence in wild form (Zeven and Zhukovsky 1975). Van der Maesen (1980) inferred India as the primary center of origin and Africa as the secondary center of origin for pigeonpea on the basis of the occurrence and diversity of wild relatives. It is cultivated from 0 to 3000 m altitude 30° North and South Latitude. Pigeonpea is predominantly a crop of the tropical regions and in India; it is mainly cultivated in semi-arid areas of the country. The optimal temperature ranging from 260 to 300 °C in the rainy season and 170 to 220 °C in the post rainy season is congenial. Now it has been acclimatized in several tropical and subtropical countries of the world.

11.1.2 Taxonomy

The genus *Cajanus* is the member of the sub-tribe Cajaninae, tribe Phaseoleae, sub-family Papilionoideae, and family Papilionaceae. Eleven genera, i.e. *Rhynchosia* Lour., *Dunbaria* W., Dunbaria A., *Eriosema* D., *Eriosema* C.,

Reichenb, *Flemingia* Roxb. Ex Aiton., *Paracalyx* Roxb. Ali, *Adenodolichos*, *Baukea*, and *Carissoa* (Mallikarjuna et al. 2011) and 32 species; (18 species are endemic to Asia, 13 to Australia, and one to West Africa De 1974, Van der Maesen 1980) are the closely related genera of the genus *Cajanus*. It is postulated that the cultivated pigeonpea originated from *Cajanus cajanifolius* by selection for size and vigor of the plant, non-shattering pods, and larger seed size. However, the cultivated *Cajanus cajan* differs from *Cajanus cajanifolius* in floral morphology, pod, and seed color, and 100 seed mass (Mallikarjuna et al. 2012). On the basis of the genetic cross-compatibility the species of *C. cajan* is places in primary gene pool (GP1). The germplasms of all available landraces of *C. cajan* along with *C. cajanifolius* are grouped in GP1. It is easily crossable with the cultivated types and produces fertile hybrids. Ten Cajanus species are cross-compatible with *C. cajan* with some difficulty and produces partially fertile offspring from the secondary gene pool (GP2). The rest of the species, which are not cross-compatible with *C. cajan*, is placed in the tertiary gene pool (GP3) (Remanandan 1990). Details of the gene pool types are given in Table 11.1.

11.1.3 Major Breeding Limitations

Insect Pests Ranga Rao and Shanower (1999) reported that more than 200 insect species feed on pigeonpea from germination to harvest. However, a large number of insects attacking the crop do not cause significant loss. Among the different insects feeding on pigeonpea, the pod borer (*Helicoverpa armigera* Hubner) is the most damaging pest worldwide may cause complete crop failure. The total estimate of losses may be caused due to this pest to the tune of more than US $300 million annually. The other most damaging insect is the pod fly (*Melanagromyza obtusa* Malloch), causing an annual loss of US $ 256 million annually which in turn results in 29.8% grain loss (ICRISAT 1992). Management of these insects by genetic means is always welcomed due to its eco-friendly nature. Genetic improvement has always been challenged by limited genetic variability for resistance against these pests in the primary gene pool, whereas there is existence of these genes imparting resistance against the insects in wild species of secondary and tertiary gene pools. However, the transfer of resistance through conventional backcrossing is always a challenge to the breeder.

Disease *Fusarium* wilt, sterility mosaic, and *phytophthora* blight are the major diseases of pigeonpea and can cause loss in yield ranging from the level of infestation from 30% to 100%. Resistance against wilt is governed by single dominant gene (Pawar and Mayee 1986). The first released wilt-resistant variety in 1986 was ICP 8863 (Maruti) followed by ICPL 87119 (Asha) in 1993. Till today, these two varieties are still known sources of wilt resistance in India. International Crop Research Institute for Semi-Arid Tropics (ICRISAT) identified ICPL 20109, ICPL 20096, ICPL 20115, ICPL 20116, ICPL 20102 and ICPL 20094 as wilt-resistant

Table 11.1 Gene pool system of *Cajanus cajan*

Gene pool 1	Gene pool 2	Gene pool 3
1. *C. cajan* (L.) Millsp., *C. cajanifolius*	1. *C. acutifolius* (F. von Muell.) van der Maese 2. *C. albicans* (W. and A.) van der Maesen 3. *C. cajanifolius* (Haines) van der Maesen 4. *C. lanceolatus* (W. V. Fitzg.) van der Maesen 5. *C. latisepalus* (Reynolds and Pedley) van der Maesen 6. *C. lineatus* (W. and A.) van der Maesen 7. *C. sericeus* (Benth. ex Bak.) van der Maesen 8. *C. trinervius* (D.C.) van der Maesen 9. *C. scarabaeoides* (L.) Thouars 10. *C. reticulates* (Dryander) F. von Muell	1. *C. aromaticus* van der Maesen 2. *C. cinereus* (F. von Muell) F. von Muell 3. *C. crassicaulis* van der Maesen 4. *C. crassus* (Prain ex. King) van der Maesen 5. *C. elongatus* (Benth.) van der Maesen 6. *C. grandiflorus* (Benth. ex Bak.) van der Maesen 7. *C. goensis* Dalz 8. *C. heynei* (W. and A.) van der Maesen 9. *C. kerstingii* Harms 10. *C. lanceolatus* (W. V. Fitzg.) van der Maese, 11. *C. lanuginosus* van der Maesen 12. *C. mareebensis* (Reynolds and Pedley) van der Maesen 13. *C. marmoratus* (R. Br. ex Benth.) F. von Muell. 14. *C. mollis* (Benth.) van der Maesen 15. *C. niveus* (Benth.) van der Maesen 16. *C. platycarpus* (Benth.) van der Maesen 17. *C. pubescence* (Ewart and Morrison) van der Maesen 18. *C. rugosus* (W. and A.) van der Maesen 19. *C. villous* (Benth. ex. Bak.) van der Maesen 20. *C. viscidus* van der Maesen 21. *C. volubilis* (Blanco) 22. *C. convertiflorus* F. von Muell.

Source: Kumar et al. (2017). In book: The Pigeonpea Genome

sources (Sharma et al. 2016) after thorough screening of 976 breeding and germplasm lines in wilt-sick plots.

Crop Duration Earlier, pigeonpea cultivation was being avoided by the farmers due to its long maturity period. However, the introduction of short-duration varieties has encouraged more farmers to grow pigeonpea. Promoting short-duration pigeonpea varieties in non-traditional areas has multiple benefits for the farmer in terms of land, labor, and climate. The potential to meet the unfulfilled demand for the pulse can be reduced by medium and short-duration pigeonpea. However, this short-duration pigeonpea is limited to some of the areas only. Due to its longer duration crop is subjected to both abiotic and biotic stresses for longer period of time. Putting all or some of the genes for resistance together in one genotype is very difficult either by conventional or molecular breeding methods. The early varieties can escape from some of the major diseases, pest, and abiotic stress.

The requirement for early maturing germplasm led to the initiation of a research program aimed to develop genotypes that mature in a short time. The first early maturing spontaneous mutant by breeders was detected in 1953 in a farmer's field. This triggered the breeding program for earliness, and by the end of the century, cultivars varying in different maturity periods were bred (Saxena et al. 2019). On the basis of the maturity pigeonpea germplasm is classified into four broad groups, viz.; long duration (>200 to 270 days), mid-early (150–160 days to mature), early (130–140 days to mature), and extra-early (110–120 days to mature). Furthermore, significant gains over earliness were achieved in breeding genotypes which matured much earlier than the extra-early maturing (102 days) over control cultivar ICPL 88039. These genotypes, are grouped into one and nicknamed as "super early" types. There was little variation (48–51 days) in days to 50% flowering was observed within the genotypes of this group but all of them matured in 90 days. It can be used as a source material in breeding early maturing cultivars and also can be introduced for cultivation in new niches or adjusted in a cropping system.

Waterlogging and Salinity Pigeonpea is mainly a rainy season crop affected greatly by waterlogging in high rainfall areas where water stagnation is a normal feature. Pigeonpea cannot afford water stagnation at all. Salinity has also been recognized as one of the major constraints of pigeonpea cultivation. Around 6.727 million ha, i.e. 2.1% of geographical area in India is salt affected, of which 2.956 m ha is saline and rest 3.771 m ha is sodic (Arora et al. 2016; Arora and Sharma 2017).

Drought Deep root system of pigeonpea imparts high degree of drought tolerance; however, moisture stresses at reproductive phase severely damage the crop yield and sometimes reduce to nil. This stress generally reduces *rhizobacterial* growth and subsequently limits symbiotic N fixation and ultimately yield reduction is reported.

Photosensitivity To induce flowering pigeonpea requires shorter days and long hours of darkness. Flower induction is greatly affected by Interaction of day and night temperature with prevailing photoperiod. Adequate day night temperature coupled with required photoperiod for pigeonpea is only available between 30° northern and southern latitudes that restricts it penetration beyond these latitudes showed an Inverse correlation between earliness and photosensitivity was reported by Wallis, both photoperiod and low-temperature sensitivity breaks the expansion of pigeonpea across higher altitudes and latitudes (Turnbull et al. 1981) that narrowed its expansion in new niche (Vales et al. 2012).

Wallis et al. (1981) demonstrated that earliness in pigeonpea was linearly associated with photo-insensitivity; therefore, breeding for earliness can increase wide adaptability. A transgressive segregants for earliness in a cross combination results due to recombination of diverse desirable alleles from the two parents combined together in a single individual. Mostly, these genotypes produced either due to the presence of additive, epistatic or complementary gene action or sometimes chromosomal re-arrangements, mobilization of transposable elements, or DNA-methylation (Rieseberg et al. 1999; Liu and Wendel 2000; Michalak 2009).

Lack of Genetic Diversity The lack of genetic diversity within the primary gene pool of *Cajanus* species is a major challenge for improving the crop to sustain the unprecedented challenges apposed by the nature. This left no option for breeders, rather than utilization of the wild relatives from secondary, tertiary, and quaternary gene pools using sophisticated gene transfer techniques. Despite the high genetic diversity in the wild relatives, its use has been limited as little information on the presence of useful traits is available. Characterization of wild relatives for different characters needed intensive, focused, and robust phenotyping. The combination of poor agronomic traits, incomplete characterization, and limited collections of the wild relatives are major setback to the genetic enhancement of this crop.

Linkage Drag There are many ways that can be adopted for the transfer of desirable gene from donor to the well adapted cultivar. However, the major challenge is linkage or tight association between desirable traits with undesirable characters. For instance, transferring the high protein genes from *C. scarabaeoides* and *Cajanus albicans* to the cultivated type took 12–14 generations for selecting a high protein, productive phenology, and high yield (Saxena and Sawargaonkar 2016).

Lack of Funds Lack of systematic public and minimal or no industrial/private funding support for research and development in pigeonpea is the major reason for the slow development of varieties having low genetic gains (Varshney et al. 2017a, b).

11.1.4 Strategies for Pigeonpea Improvement

There are many ways to fulfill the objective mentioned above. Extensive attempts have been made to meet these objectives by many of the workers. The advancements in pigeonpea can be subdivided into two, i.e. before and after genomic intervention. The former contributed the enrichment of the genetic resource and later into genomic resource of the pigeonpea. Furthermore, these resources can be utilized by the breeders to target the specific issues in the pigeonpea improvement.

11.1.4.1 Before Genomic Intervention

Until a decade ago pigeonpea was considered an orphan legume crop. Knowledge of the genetic basis of yield and other yield attributing traits, quality, and stress tolerance played an important for genetic improvement of pigeonpea. The International Crop Research Institute for Semi-Arid Tropics (ICRISAT) promotes the strategically planned research and development for improving plant genetic resources. Until a decade ago pigeonpea was considered an orphan legume crop. Knowledge of the genetic basis of yield and other yield attributing traits, quality, and stress tolerance played an important for genetic improvement of pigeonpea.

Genetic Resources Development ICRISAT is a global repository for the pigeonpea germplasm has 13,771 accessions in its acquisition through concerted efforts to acquire from different national and international institutes, universities, National Agricultural Research System (NARS), etc. (Gowda et al. 2013; Pazhamala et al. 2015), 11,221 accessions collected at National Bureau of Plant Genetic Resources (NBPGR), India (Pazhamala et al. 2015; Singh et al. 2014), 4116 accessions at U.S. Department of Agriculture (USDA), USA, 1288 accessions at Kenya Agricultural Research Institute's National Genebank of Kenya (KARI-NGBK), Kenya (Singh et al. 2013; Pazhamala et al. 2015) and 433 accessions at National Plant Genetic Resources Laboratory, Philippines (Upadhyaya et al. 2016a, b). Pigeonpea is rich in its genetic resource but very poor in its utilization for crop improvement. The richness of the germplasm treasure on the other hand poses problem in its fruitful utilization too. To solve this ICRISAT attempted and prepared core collection to reduce the drudgery of the breeder by making a representative sub-sets of pigeonpea germplasm comprised of 1290 accessions, mini-core collection of 146 accessions (Gowda et al. 2013; Pazhamala et al. 2015) and genotype-based reference set (Upadhyaya et al. 2006a, b). These collections represent more than 80% of the existing variability present in the entire germplasm collection and are ideal resources for studying genetic diversity, population structure, and association mapping (Reddy et al. 2005; Upadhyaya et al. 2006a, b; Gowda et al. 2013; Pazhamala et al. 2015). Extensive multidisciplinary evaluation of core/mini-core collection at ICRISAT has led to the identification of promising accessions for sterility mosaic disease (24), wilt (6), wilt + SMD (5), pod borer (10), salinity (16), waterlogging (23), high yield (54), high zinc (15) and iron (15) content, whereas NARS identified trait-specific germplasm for early maturity (8), high seed yield (2), wilt (39), SMD + wilt (24) from mini-core collection (Upadhyaya et al. 2016a, b). Some of the wild relatives have been identified that render desirable traits such as tolerance to abiotic stresses, resistance to pests and diseases, high protein content photo-insensitivity, cleistogamy, and cytoplasmic male sterility are presented in Table 11.1. In spite of all those efforts, very few poor marker density genetic maps were developed (Varshney et al. 2010a, b). However, significant contributions have been made in the recent past to develop a good range of genomic resources as given in Tables 11.2 and 11.3.

Table 11.2 Different wild species of pigeonpea useful for rendering valuable traits

Important traits		Wild relatives
Resistance to pests	Pod borer	*C. scarabaeoides*
		C. acutifolius
		C. platycarpus
		C. reticulatus
		C. sericeus
		C. albicans
	Pod fly	*C. reticulatus*
		C. acutifolius
	Bruchids	*C. scarabaeoides*
		C. acutifolius
		C. platycarpus
Resistance to diseases	*Fusarium* wilt	*C. platycarpus*
	Sterility mosaic disease	*C. sericeus*
		C. albicans
		C. volubilis
		C. lineatus
	Phytophthora blight	*C. platycarpus*
		C. sericeus
		C. acutifolius
Grain quality	High protein content	*C. cajanifolius*
		C. sericeus
		C. albicans
		C. lineatus
		C. scarabaeoides
	High seed weight	*C. acutifolius*
Cytoplasmic male sterility		*C. cajanifolius*
		C. sericeus
		C. scarabaeoides
		C. acutifolius
		C. volubilis
		C. platycarpus
Tolerance to abiotic stresses	Salinity	*C. platycarpus*
	Drought	*C. sericeus*
		C. albicans
		C. lineatus
		C. scarabaeoides
	Waterlogging	*C. scarabaeoides*
Plant type	Extra-early flowering maturity	*C. platycarpus*
	Photosensitivity	*C. platycarpus*
	Cleistogamy	*C. lineatus*

Source: Pazhamala et al. (2015)

Table 11.3 Genetic resources available for mapping different traits in pigeonpea

Trait	Type of population	Population
Fusarium wilt resistance	Intra-specific F2	ICPB 2049 × ICPL 99050
Fusarium wilt resistance	F2	GSl × ICPL 87119 GS1 × ICP 8863
Sterility mosaic disease resistance	Intra-specific F2	ICP 8863 × ICPL 20097 TTB 7 × ICP 7035
Fertility restoration	F2	ICPA 2039 × ICPR 2447 ICPA 2043 × ICPR 3467 ICPA 2043 × ICPR 2671
Pod borer	Inter-specific F2	ICPL 8755 × ICPL 227 ICPL 151 × ICPL 87 ICP 28 × ICPW 94
Drought tolerance	Inter-specific F2	ICPL 8755 × ICPL 227 ICPL 151 × ICPL 87 ICP 28 × ICPW 94
Determinacy	F2	ICPL 85010 × ICP 15774
Plant type	F2	TT44-4 × TDT2004-1
Plant height, number of primary and secondary branches, number of pods, days to maturity, and days to flowering	F2	Pusa Dwarf × HDM04-1

Source: Pazhamala et al. (2015)

11.1.4.2 Genomic Intervention

Genomic Resource Development Pigeonpea is a diploid crop having 22 (2n = 2x) chromosomes with a genome size of 833.1 Mbp (Varshney et al. 2012). The first draft of the genomic sequence of a popular variety "Asha" was published in 2011 by Singh et al. The genome was assembled using the sequence reads from 454 GS-FLX

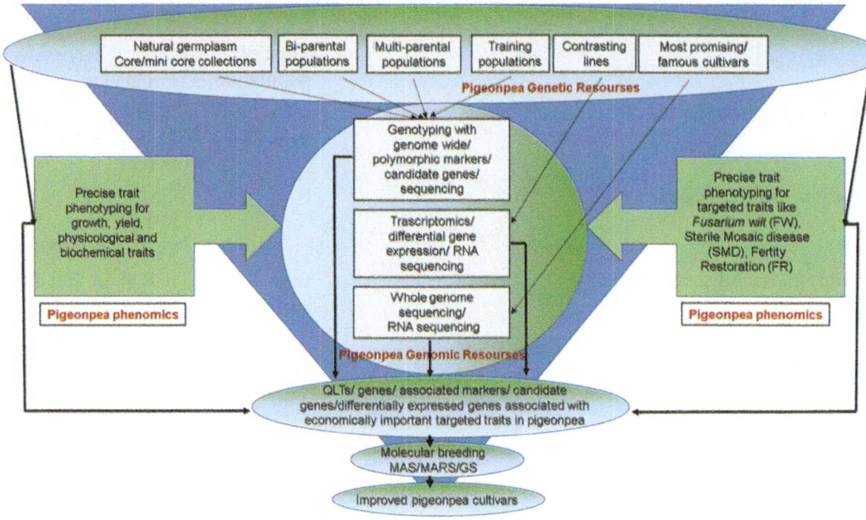

Fig. 11.1 Diagrammatic representation of the modus operandi for the utilization of genomic resource (figure by the courtesy of R. R. Mir et al. 2017)

sequencer with mean read lengths of >550 bp and > tenfold genome coverage, resulting in 510,809,477 bp of high-quality sequence. Total predicted protein coding genes and transposable elements related genes were 47,004 and 12,511, respectively. They identified 1213 disease resistance/defense responsive genes and 152 abiotic stress tolerance genes in this genome. Development of molecular markers tightly linked to the yield and important yield attributing traits is a prerequisite for attempting a molecular breeding program in crops. But most of the agronomic traits on molecular basis in pigeonpea could not be explored due to low level of DNA polymorphism in the primary gene pool and limited number of validated molecular markers (Dutta et al. 2011; Bohra et al. 2011). The last decade witnessed significant achievement in the area of development of genomic resources in pigeonpea to make it "genomic resource rich" crop from the so-called orphan crop (Varshney et al. 2012). Availability of this genome sequence in public domain will accelerate the utilization of pigeonpea germplasm resources in breeding program (Saxena 2008; Varshney et al. 2010b) in the manner described in Fig. 11.1.

Molecular Markers The main goal of genetic mapping is to detect naturally inherited markers in close proximity to the genetic causatives or genes controlling the complex quantitative traits. Genetic mapping can be done mostly in two ways: (1) by the use of experimental populations ("biparental" mapping populations) is called as QTL mapping as well as "genetic mapping" or "gene tagging," or "linkage mapping" and (2) by the use of diverse lines from the natural populations or germplasm collections that is called LD-mapping or "association mapping." Association mapping is one of the powerful tools for genetic mapping of crops, taking

Table 11.4 List of molecular markers developed in pigeonpea

S. no.	Molecular markers	Outcome
First generation molecular markers		
1	Restriction fragment length polymorphism (RFLP)	Developed and employed mainly for diversity studies
2	Random amplified polymorphic DNA (RAPD)	
3	Amplified fragment length polymorphism (AFLP)	
Second generation molecular markers		
4	Genome sequence (gSSRs)	23,410 gSSRs
5	Expressed sequence tags (ESTs-SSRs)	8137 ESTs-SSRs
6	Bacterial artificial chromosome (BAC)-end sequences (BES-SSRs)	6212 BES-SSRs
Next generations sequencing (NGS) technology or third generation molecular markers		
7	DArT arrays	15,360 loci
8	GoldenGate platform	768 SNPs
9	competitive allele-specific polymerase chain reaction (KASPar) assays	1616 SNPs
10	Single feature polymorphisms (SFPs)	5692 SFPs using six parental genotypes of three mapping populations
11	Intron spanning region (ISR)	Developed from the transcriptome assembly of pigeonpea

Source: Pazhamala et al. (2015)

account of nonrandom association of alleles at multiple DNA markers due to their close proximity to one another within a chromosome and co-inheritance. Association mapping has advantages over other mapping techniques in terms of high-resolution, broader allele coverage, and cost-effective gene tagging approach in plant germplasm resources. Molecular marker technologies in pigeonpea have witnessed a transition from the gel-based (RAPD) and hybridization based (RFLP, DArT, SFP) to sequencing-based simple sequence repeat (SSR) and SNP markers (Table 11.4).

11.1.5 Next Generations Sequencing (NGS) Technology

The advancements in genome sequencing techniques reduced the cost of sequencing whole genome. The release of genome sequence draft of pigeonpea in public domain provided large number of SNPs in segregating populations that could lead to the identification of haplotypes. The recombination maps made Genotyping-by-sequencing (GBS) very popular (Elshire et al. 2011). These identified haplotype blocks can then be used as markers for mapping (Deschamps et al. 2012) different traits related to biotic and abiotic stresses. Identification and introgression of genomic segments attributing to disease resistance through genomic assisted breeding

(GAB) program would be an important strategy for the development of disease-resistant pigeonpea varieties (Saxena et al. 2017). New generation mapping populations like multi-parent advanced generation intercross (MAGIC) and nested association mapping (NAM) population best utilize high-throughput genotyping/sequencing platforms and give an edge over, but also offer several advantages in terms of greater resolution and allelic richness aiding in family-based QTL study and Linkage-linkage disequilibrium analysis over conventional (biparental) mapping populations (Bohra et al. 2017). Candidate genomic region CcLG11 was identified as promising QTL for molecular breeding in developing superior lines with enhanced resistance to SMD by genotyping-by-sequencing approach (Saxena et al. 2017). Gnanesh et al. (2011) identified six QTLs explaining phenotypic variation on LG7 and LG9 after robust phenotyping for SMD resistance. Some of the achievements of the NGS techniques in pigeonpea are presented in Table 11.4.

11.1.6 Transgenic Approach

Biotechnology over the years has provided solutions to meet the different challenges posed by biotic and abiotic constraints in crop species. Transgenic technology emerged as a new tool for transferring not only single gene traits but also to engineer multiple genes or plant regulatory machinery that drives the expansion of different stress responsive genes. The advances and standardization in transgenic tools provided scope to attempt for incorporation of several traits such as insect resistance, protein quality, and edible vaccines. *H. armigera* is the most devastating pest of pigeonpea. It is the major constraint in pigeonpea production for which there is no absolute resistance available in the cultivated germplasm. *Bacillus thuringienses* Berliner (Bt) not only produces insecticidal proteins but proteases like insecticidal chitinase. This chitinase plays role in controlling the devastating pod borer by dissolution of the chitin which is an insoluble structural polysaccharide in the exoskeleton and gut lining of insects. Gene pyramiding with two different insecticidal genes and tissue-specific expression to reduce the risk of developing insect resistance is another attractive feature of the genetic engineering approach to obtain durable resistance. Expression of a chimeric *cry1AcF* (encoding *cry1Ac* and *cry1F* domains) gene in transgenic pigeonpea has shown resistance against *H. armigera* (Ramu et al. 2012). *Bt* cry genes have yielded promising results in a collaborative approach of ICRISAT and ICAR-IIPR where a large number of transgenic events are currently being evaluated for their efficacy. These transgenics have shown high mortality of the larva and resisted the damage caused by the larvae.

Another transgenic carrying *dhdps-r1* gene from a mutant *Nicotiana sylvestris*, which encodes a DHDPS enzyme insensitive to feedback inhibition, was used to improve the lysine content in pigeonpea seeds. The *dhdps-r1* coding region governed by a phaseoline or an Arabidopsis 2S2 promoter was successfully over expressed in the seeds of pigeonpea by using *Agrobacterium* transformation and particle bombardment. A twofold to sixfold enhanced DHDPS activity than the wild type was reported in immature seeds at a late stage of maturation in 11 lines. This over expression of *dhdps-r1* led to an enhanced free lysine content in the pigeonpea seeds ranging from 1.6 to 8.5 times compared to wild type. However, this was not reflected in total seed lysine increase. This might be explained by a temporal discrepancy between maximal expression of *dhdps-r1* and the rate of amino acid incorporation into storage proteins. Assays of the lysine degradative enzyme lysine-ketoglutarate reductase in these seeds showed no coordinated regulation of lysine biosynthesis and catabolic activity during seed maturation. All transgenic plants were fertile and produced morphologically normal seeds (Thu et al. 2007).

Work is also going on to develop bio fortified pigeonpea for the enhancement of β-caretone (provitamin A), precursor of vitamin A at ICRISAT. Success in producing transgenic pigeonpea plants with high-levels of β-caretone can add to the nutritional requirement to the malnourished population in the dryland areas of the world. Besides these there have been efforts on developing pigeonpea that can be used as edible vaccines for goat and sheep to rinder pest virus and *peste des petits ruminants virus* by the use of *hemaglutinin* gene of rinder pest virus and *hemaglutinin neuraminidase* gene of peste des *petits ruminants virus* (PPRV-HN) respectively (Prasad et al. 2004).

Similarly, a mutagenized version *P5CSF129A* of wild *P5CS* is introgressed for increased production of proline biosynthesis which is known for conferring enhanced salt and drought stress. Efficient in vitro transformation of embryonic structures of pigeonpea (*Cajanus cajan* (L.) Millsp.) was obtained via *Agrobacterium tumefaciens* strain LBA4404 harboring a modified binary vector *pCAMBIA* 1301 carrying the *hptII* gene for resistance to hygromycin sulfate, GUS reporter gene, encoding β-glucuronidase, and the *Vigna aconitifolia P5CSF129A* genes under a constitutive 35S promoter. When tested for GUS after the first cycle of antibiotic selection after integration of T-DNA into nuclear genome of transformed plants embryonic structures showed blue color and its sexual transmission to the progeny of the transgenic plants are confirmed by PCR amplification of 340 bp *hpt*II, 800 bp *P5CSF129A* fragments and Southern blot hybridization analysis. The resultant primary transgenic plants showed more proline accumulation and a lower lipid peroxidation when subjected to 200 mM NaCl. Levels of proline were also elevated in T1 transgenic plants when grown in the presence of 200 mM NaCl. In addition to their enhanced growth performance, chlorophyll content and relative water content under high salinity, showed lower levels of lipid peroxidation. This suggests that overproduction of proline might play an important role against salt shock and cellular integrity (Surekha et al. 2014).

Thus, application of genetic engineering technology is a viable option to develop transgenics to address complex problem when there is absence of trait/gene-conferring resistance to biotic and abiotic stresses in the primary/cross-compatible gene pool.

11.1.7 Future Prospects

Genetic Diversity via Sequencing/Re-sequencing Limited genetic diversity is the main bottleneck in genetic improvement of pigeonpea. Novel genetic variation can be introduced through the intervening mutation, landraces, and wild relatives in today's systematic breeding programs. However, associated linkage drag with the gene of interest may deviate from the desirable outcome of satisfactory wild gene transfer to cultivated forms. In this context, the available NGS technology and pigeonpea draft genome sequence facilitates a great opportunity for exploring nucleotide-level diversity in cultivated, landraces, and wild species accessions and its relationship with phenotypic diversity (Varshney et al. 2012). Re-sequencing of germplasm accessions will dissect existing genetic diversity, associating gene(s), provide better understanding of phenotypes and opportunity of exploiting natural genetic diversity to develop superior genotypes (Varshney et al. 2017a, b).

Advanced Trait Mapping Approaches The traditional QTL mapping is a time consuming and resource intensive approach for the identification of parental polymorphisms and genotyping of the populations with the polymorphic markers (Abe et al. 2012). Bulk segregant analysis (BSA) on the other hand utilizes the marker screening on the extreme bulks and parents that provides trait-associated markers. Thus, NGS-based BSA approaches could be a future approach for rapid and accurate trait mapping.

Next-Generation Breeding Currently, in pigeonpea, very few genomic inputs like marker-based purity testing in hybrids and parents, DNA fingerprinting, genome assisted breeding for introgression of SMD and FW resistance in elite varieties is employed at ICRISAT (Singh et al. 2017b). The available draft genome, sequencing and re-sequencing data, NGS bio-informatics advances, phenotyping platforms coupled with a significant reduction in marker genotyping cost enabled breeders to select proper and suitable allele combination at early stage of growth that can facilitate successful introgression from wild to elite cultivars without the hindrance of linkage drag (Singh et al. 2017a, b). Genomic selection in the pigeonpea hybrid breeding program would improve yield potential of hybrids and trait-specific parental lines. Conventional breeding with genomic inputs of NGS, high throughput genotyping for early generation screening, marker-assisted backcrossing (MABC), and marker-assisted selection (MAS)can take pigeonpea breeding to the next level in coming days.

11.2 Chickpea (*Cicer arietinum* L.)

Chickpea (*Cicer arietinum* L.) is one of the eco-friendly economically important food legume crops. It is grown in around 17.1 m ha, with a production of 17.1 m t in more than 55 countries of the world (Food and Agriculture Organization 2018). India is the largest producer of the chickpea contributes about 65.5% of the world production. It has been cultivated mainly in the Indian subcontinent, West Asia, and North Africa, but recently large acreages have been introduced in the Americas and Australia). In India Madhya Pradesh is the highest chickpea producing state accounting for 41% of the total chickpea production followed by Maharashtra (18%) and Rajasthan (13%) (Farmer.gov.in). Two main types of chickpea cultivars are grown globally–kabuli and desi, representing two diverse gene pools. The kabuli types are generally grown in the Mediterranean region including Southern Europe, Western Asia, and Northern Africa and the desi types are grown mainly in Ethiopia and Indian subcontinent. *Desi* chickpeas are characterized by angular seed shape, dark seed coat, pink flowers, anthocyanin pigmentation of stem, rough seed surface and either semi-erect or semi-spreading growth habit, whereas kabuli types generally have owl shaped seeds, beige colored seeds, white flowers, smooth seed surface, lack of anthocyanin pigmentation and semi-spreading growth habit (Pundir et al. 1985).

Chickpea is highly rich in essential basic nutrients proteins that vary from 17–22% and 25.3–28.9% as a percentage of total dry seed mass before after dehulling, carbohydrates (62%), fat (2.70–6.48%), iron, folic acid, phosphorus, and dietary fibers (18–22%), minerals (Cu, Fe, Zn, Mn, and Ca), vitamins (A, B1, B2, B3, B5, B6, B12, C, D, E, K, and Biotin), and carotenoids. Chickpea has higher levels of Mn, Zn, and P compared to those present in other legumes. It serves as an important source of protein in human diet and plays an important role in the enrichment of soil fertility.

11.2.1 Taxonomy

The *Cicer* genus belongs to family Leguminoseae, sub-family Papilionaceae and tribe Cicereae. It encompasses 9 annual and 43 perennial wild species. Most of these species are found in West Asia and North Africa covering Turkey in the north to Ethiopia in the south, and Pakistan in the east to Morocco in the west (Table 11.5). It is a highly self-pollinated crop with an outcrossing rate of less than 1%.

Cicer arietinum L. is the only cultivated species among the nine annual species. Van der Maesen (1987) classified the *Cicer* species into four sections based on their morphological characteristics, life cycle and geographical distribution. Eight annual species, namely *C. arietinum*, *C. reticulatum*, *C. echinospermum*, *C. pinnatifidum*, *C. bijugum*, *C. judaicum*, *C. yamashitae*, and *C. cuneatum* were placed in section Monocicer, *C. chorassanicum*, and *C. incisum* (perennial species) in section Chamaecicer, 23 perennial species in section Polycicer and seven woody perennial species in section Acanthocicer.

Ladizinsky and Adler (1976) grouped all the nine annuals of chickpea in different groups on the basis of crossability and fertility of the hybrids (Singh et al. 2008) (see Fig. 11.2).

Table 11.5 *Cicer* species and their distribution

S. No.	Species	Distribution
Annual		
1.	*C. arietinum*	Mediterranean region to Myammar, Ethiopia, Maxico, Chile
2.	*C. chorassanicum*	Afganistan, Iran
3.	*C. bijugum*	Turkey, Syria, Iraq
4.	*C. cuneatum*	Ethiopia, Egypt, Sudan, Saudi Arabia
5.	*C. echinospermum*	Turkey, Anatolia, Iraq
6.	*C. judaicum*	Palestine, Lebanon
7.	*C. pinnatifidum*	Cyprus, Iraq, Syria, Turkey, Former USSR
8.	*C. reticulatum*	Turkey
9.	*C. yamashitae*	Afganistan
Perrennial		
10.	*C. acanthophyllum*	Afganistan, Pakistan, Former USSR
11.	*C. anatolicum*	Turkey, Iran, Iraq, Armenia
12.	*C. atlanticum*	Morocco
13.	*C. balcaricum*	Caucasus
14.	*C. baldshuanicum*	Former USSR
15.	*C. canariense*	Canary islands, Tenerife, and La palma
16.	*C. fedtschenkoi*	Former USSR, Afganistan
17.	*C. flexuosum*	Former USSR
18.	*C. floribundum*	Turkey
19.	*C. graecum*	Greece
20.	*C. grande*	Former USSR
21.	*C. heterophyllum*	Turkey
22.	*C. incanum*	Former USSR
23.	*C. incisum*	Greece, Turkey, Iran, Lebanon, Former USSR
24.	*C. isauricum*	Turkey
25.	*C. kermanense*	Iran
26.	*C. korshinskyi*	Former USSR
27.	*C. laetum*	Former USSR
28.	*C. macrocanthum*	Afganistan, India, Pakistan, Former USSR
29.	*C. microphyllum*	Afganistan, Tibet, India, Pakistan, Former USSR
30.	*C. mogoltavicum*	Former USSR
31.	*C. montbrettii*	Albania, Bulgaria, Turkey
32.	*C. multijugum*	Afganistan
33.	*C. nuristanicum*	Afganistan, India, Pakistan
34.	*C. oxyodon*	Iran, Afganistan, Iraq
35.	*C. paucijugum*	Former USSR
36.	*C. pungens*	Afganistan, Former USSR
37.	*C. rassuloviae*	Former USSR
38.	*C. rechingeri*	Afganistan
39.	*C. songaricum*	Former USSR
40.	*C. spiroceras*	Iran
41.	*C. stapfianum*	Iran

(continued)

Table 11.5 (continued)

S. No.	Species	Distribution
42.	*C. subaphyllum*	Iran
43.	*C. tragacanthoides*	Iran, Former USSR

Source: Chickpea research in India by Massood Ali, Shiv Kumar and N. B. Singh

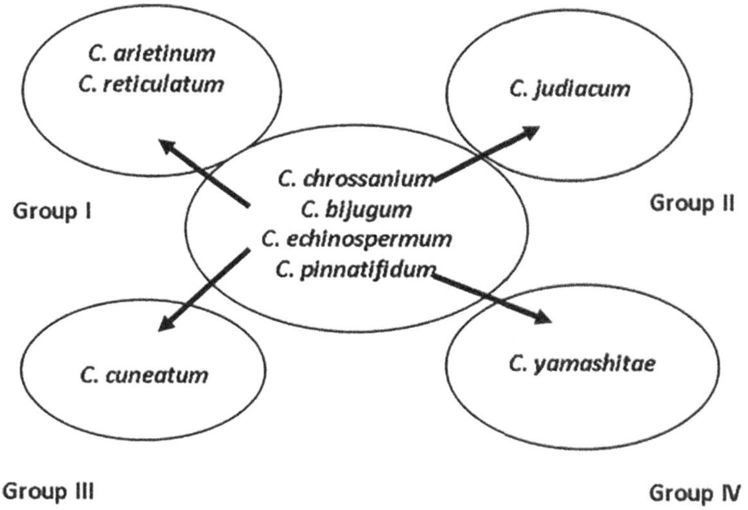

Fig. 11.2 Schematic diagram depicting relationships between nine annual *Cicer* species based on different methods used to establish the relationships. Controversial species are placed in center and arrow indicates the group in which they can be placed (source: Singh et al. 2008; doi: https://doi.org/10.5661/bger-25-267)

11.2.2 Genetic Resources for Trait Discovery and Utilization

The global collection of chickpea is about 100,000 accessions is maintained in 120 national and international genebanks in 64 countries (Upadhyaya et al. 2017). International Crops Research Institute for the Semi-Arid Tropics (ICRISAT) genebank contains maximum accessions 20.8% (20,764) from 59 countries followed by ICAR-National Bureau of Plant Genetic Resources (ICAR-NBPGR) 16% and the International Center for Agricultural Research in the Dry Areas (ICARDA) 15%. These accessions in genebanks have been developed into many subsets like core collection (1956 accessions), mini-core collection (211 accessions), global composite collection (3000 accessions), trait-based FIGS (Focused Identification of Germplasm Strategy) sets and reference set (300 accessions), covering the whole diversity of chickpea. These subset can be used as good source for allelic mining and association mapping to detect QTLs (Upadhyaya and Ortiz 2001; Upadhyaya et al. 2006a, b). Table 11.6 lists wild *Cicer* species with resistance to various biotic and abiotic stresses that can be used a source for gene mining.

Table 11.6 Wild Cicer species with resistance to various biotic and abiotic stresses

Trait	Source of resistance
Ascochyta blight	*C. echinospermum*, *C. pinnatifidum*, *C. bijugum*, *C. judaicum*, and *C. montbretii*
Fusarium wilt	*C. bijugum*, *C. judaicum*, *C. pinnatifidum*, *C. reticulatum*, *C. echinospermum*, and *C. cuneatum*
Leaf miner	*C. chorassanicum*, *C. cuneatum*, and *C. judaicum*
Bruchids	*C. echinospermum*, *C. bijugum*, and *C. judaicum*
Cyst nematode	*C. pinnatifidum*, *C. bijugum*, and *C. reticulatum*
Gray mold	*C. judaicum*, *C. pinnatifidum*
Phytophthora root rot	*C. echinospermum*
Cold	*C. bijugum*, *C. reticulatum*, *C. echinospermum*, and *C. pinnatifidum*

Source: Singh et al. doi: https://doi.org/10.5661/bger-25-267

Until 2005, chickpea was often referred as an orphan crop due to limited availability of the genetic and genomic resources. However, with the further advancement in NGS technology in the last decades, extensive genetic, genomic and transcriptomic resources have been developed to transform chickpea from an orphan crop to a genomic resource-rich crop (Varshney et al. 2009; Nayak et al. 2010; Hiremath et al. 2012; Thudi et al. 2011; Kudapa et al. 2014; Agarwal et al. 2016; Mashaki et al. 2018). Genomics is a successful and effective tool to address the biotic and abiotic challenges towards enhancing yield potential, quality and stability of production through advanced breeding techniques. Varshney et al. in 2013 reported ~738-Mb draft whole genome shotgun sequence of CDC Frontier, a *kabuli* chickpea (*Cicer arietinum*) variety, which contains an estimated 28,269 genes. Resequencing and analysis of 90 cultivated and wild genotypes from ten countries identifies targets of both breeding-associated genetic sweeps and breeding-associated balancing selection. They also identified candidate genes for disease resistance and agronomic traits including traits that distinguish *desi* and *kabuli* and can be used as a resource for molecular breeding of chickpea. For these identified candidate gene different DNA markers has also been identified using the database of whole genome draft. Application of DNA markers to facilitate marker-aided selection (MAS) for crop improvement has proved successful in crossbreeding. Advances in plant genomics further provided means to improve the understandings of crop diversity at species and gene levels, and offer DNA markers to accelerate the pace of genetic improvement (Muthamilarasan et al. 2013, 2014). Robust phenotyping multilocational combined data with genotyping data provides an opportunity for modeling chickpea crop for target environments. To achieve this goal crop physiology can play major role in defining ideotype for enhancing suitability for the target environment. Geographic information systems and passport data are also useful for identification of accessions from stress-prone environments, to improve the target trait. Data available on characterization, DNA fingerprinting and quantitative trait loci (QTL) mapping of trait of interest will assist in selecting promising accessions for modeling the crop specific for particular environment/stress(es). At the same time

Table 11.7 Progress in the development of genetic linkage maps of chickpea

Mapping population	Marker type	No. of markers	Linkage groups	Genome coverage (cM)
F2 population				
C. arietinum × *C. reticulatum*	Morphological and isozyme	29	7	200
C. arietinum × *C. reticulatum* and *C. arietinum* × *C. echinospermum*	Morphological and isozyme	28	8	257
C. arietinum × *C. reticulatum* *C. arietinum* × *C. echinospermum*	Morphological, isozyme, RFLP and RAPD	91	10	550
C. arietinum × *C. echinospermum*	SSR, RAPD, ISSR and RGA	83	8	570
RIL population				
C. arietinum × *C. reticulatum*	STMS	120	11	613
C. arietinum × *C. reticulatum*	SSR, DAF, AFLP, ISSR, RAPD, isozyme, cDNA, SCAR and morphological	354	16	2077.9
C. arietinum × *C. reticulatum*	RAPD, ISSR, isozyme, and morphological	116	9	982
C. arietinum × *C. reticulatum*	RAPD, ISSR and morphological		23	
C. arietinum × *C. reticulatum*	Addition of RGA Potkin 1–2171 to linkage group 5 of Santra et al. (2000)	117	9	–
C. arietinum × *C. reticulatum*	Addition of SSR and RGA marker to map of Santra et al. (2000)	51	9	1175
C. arietinum × *C. reticulatum*	Addition of 47 defense response gene markers to the map of Winter et al. (2000)	296	12	2500
C. arietinum × *C. reticulatum*	SSR and Cyt P450	93	9	344.6
Intraspecific (F2)				
C. reticulatum (ICC1 × Lasseter)	SSR, ISSR and RGA	66	8	534.5
Intraspecific (RIL)				
C. arietinum (ILC1272 × ILC3279)	SSR and Ascochyta blight resistance loci	55	8	–
C. arietinum (ICCV2 × JG62)	SSR, RAPD, ISSR and morphological	111	14	297
C. arietinum (PI 359075 × FLIP 84-92C)	SSR	53	11	318.2

(continued)

Table 11.7 (continued)

Mapping population	Marker type	No. of markers	Linkage groups	Genome coverage (cM)
C. arietinum (Kabuli × Desi)	RAPD, ISSR, SSR and morphological	160	10	427.9

Source: Singh et al. doi: https://doi.org/10.5661/bger-25-267

precise phenotyping for different traits combined with appropriate biometric analysis will help in identifying unique responses of a set of chickpea genotypes to meet the present and future challenges of this crop. These information are further utilized in genomics-aided breeding approaches such as genome-wide selection of promising germplasm for further use in breeding program aiming at both population improvement and cultivar release. A detailed list of genetic maps attempted to develop by different workers have been presented in Table 11.7.

11.2.3 Genomic Resource Development in Chickpea

Until recently, molecular breeding was hampered by the need for very high marker density coverage of the genome. Advancement of next-generation sequencing (NGS) methods has facilitated the development of large numbers of genetic markers, such as single nucleotide polymorphisms (SNP), insertion-deletions (InDels), etc. even in relatively research-neglected crop species. Discovery of novel genes/QTLs/alleles for any given trait was achieved through genotyping-by-sequencing (GBS) approaches. Furthermore, genome-wide association studies (GWAS) could also be utilized to identify the genomic regions pertaining to traits of interest derived from statistical associations between DNA polymorphisms and trait variations in diverse collection of germplasms for which high-density genotyping and robust phenotyping for traits of interest has performed. NGS coupled with GWAS increases the resolution of genetic map for precise location of genes/alleles/QTL (Ma et al. 2012; Liu et al. 2013; Varshney et al. 2014a, b).

Progress made in the development of genomic resources of chickpea during the last decade has brought a paradigm shift in chickpea breeding. Nearly more than 2000 SSR markers (Varshney 2016), thousands of diversity arrays technology (DArT) arrays (Yang et al. 2011; Thudi et al. 2011) for different characters and millions of single nucleotide polymorphism (SNP) and InDel markers have been identified through different approaches in chickpea (Varshney et al. 2012, 2013a, b; Deokar et al. 2014; Das et al. 2015; Kumar et al. 2016). But the cost of these markers was major bottleneck for many researchers. Keeping in view some cost-effective marker genotyping assays have been developed such as Kompetitive Allele Specific PCR (KASP) assays (Saxena et al. 2012; Hiremath et al. 2012), Golden Gate assays, Vera Code assays (Roorkiwal et al. 2013) have also been developed. Very recently, 60K SNP chips have been developed for pigeonpea and chickpea using Affymetrix SNP platform. Many workers have attempted for mapping of the chickpea through

Table 11.8 An overview of sequencing-based trait mapping efforts in chickpea

S. no.	Sequencing technology	Mapping population	Trait mapping approach	Target trait(s)	Significant results
1	GBS	Amit × ICCV 96029	Genetic mapping (3430 SNPs)	Ascochyta blight resistance	QTLs for resistance on CaLG02, CaLG03, CaLG04, CaLG05 and CaLG06 explaining up to 40% of phenotypic variance (PVE)
2	GBS	ICC 4567 × ICC 15614	Genetic mapping (271 SNPs)	Heat tolerance	QTLs on CaLG05 and CaLG06 with a cumulative PVE of 51.89% and 25.84%, respectively
3	GBS	SBD377 × BGD112	Genetic mapping (3228 SNPs)	Seed trait	QTLs explaining up to 29.71% PVE and candidate genes for seed traits
4	GBS	ICC 4958 × ICC 1882	Genetic mapping (743 SNPs)	Drought tolerance-related traits	Refined the "*QTL-hotspot*" region from ca. 29 cM to 14 cM and added 49 SNPs in the region
5	GBS	Cultivated and wild accessions	GWAS, genetic mapping	Photosynthetic efficiency and seed yield per plant	SNPs and candidate genes associated with photosynthetic efficiency and seed yield per plant
6	GBS	Cultivated accessions	Genetic diversity (4349 SNPs)	Seed yield per plant	A pentricopeptide repeat (PPR) gene associated with seed yield per plant
7	GBS	Cultivated accessions	GWAS (16,591 SNPs)	Seed iron and zinc	Genomic loci/ genes (with 29% combined PVE) associated with seed-Fe and Zn concentrations
8	GBS	Cultivated and wild accessions	Genetic diversity (82,489 SNPs)	NIL	Analysis of genetic diversity, population structure and linkage disequilibrium
9	GBS	Cultivated and wild accessions	Genetic diversity (44,844 SNPs)	NIL	Revealed complex admixed domestication pattern, and extended LD decay
10	GBS	Cultivated accessions	Genetic diversity (3187 SNPs)	NIL	Genetic cluster associated with black seeded genotypes
11	RAD-Seq	ICCV 96029 × CDC Frontier	Genetic mapping (604 bins)	NIL	High-density linkage map constructed

(continued)

Table 11.8 (continued)

S. no.	Sequencing technology	Mapping population	Trait mapping approach	Target trait(s)	Significant results
12	Axiom Array	ICC 4958 × ICC 1882 and ICC 283 × ICC 8261	Genetic mapping (13,679 and 7769 SNPs)	Drought tolerance-related traits	Main-effect QTLs for several drought component traits
13	WGRS	ICC 4958 × ICC 1882	Genetic mapping	Drought tolerance-related traits	Delimited "*QTL-hotspot*" region to ~300 kb and identified 26 candidate genes
14	WGRS	ICC 4958 × ICC 1882	Genetic mapping	Plant vigor and canopy conductance	QTLs for plant vigor and canopy conductance traits on CaLG04 and CaLG03, respectively
15	WGRS	ICCV 96029 × CDC Frontier and ICCV 96029 × Amit	QTL-seq	Ascochyta blight resistance	Candidate genes for ascochyta blight resistance
16	WGRS	ICC 4958 × ICC 1882	QTL-seq	100-seed weight (100SDW) and root/total plant dry weight ratio (RTR)	Genomic regions on CaLG01 (1.08 Mb) and CaLG04 (2.7 Mb) linked with 100-seed weight. Two genes (Ca_04364 and Ca_04607) for 100SDW and one gene (Ca_04586) for RTR
17	WGRS	ICC 7184 × ICC 15061	QTL-seq	100-seed weight	Genomic region on CaLG01 harboring six candidate genes for 100-seed weight
18	WGRS	Released varieties and advanced breeding lines	GWAS (144,000 SNPs)	Drought tolerance	MTAs significantly associated with yield and yield-related traits under drought

GBS Genotyping by sequencing, *RAD-Seq* restriction-site-associated sequencing, *WGRS* whole genome re-sequencing (Source: Roorkiwal et al. 2020)

Table 11.9 List of molecular breeding products developed using marker-assisted breeding (MABC) in chickpea

S no	Trait	Donor parent	Recipient parent	Target region	Present status/released variety	Reference
1	Drought tolerance	ICC 4958	JG 11	QTL-hotspot	Released as "Geletu" for commercial cultivation in Ethiopia	EIAR/ ICRISAT unpublished
2			Pusa 372	QTL-hotspot	Released as "Pusa chickpea 10,216" for commercial cultivation in India	ICAR-IARI/ ICRISAT unpublished
3			JG 11	QTL-hotspot	Stable backcross lines under multi-location trials	Varshney et al. (2013b)
4			RSG 888	QTL-hotspot	Under multi-location yield trials of ICAR-AICRP for release in India	Unpublished[a]
5			Pusa 362	QTL-hotspot	Under multi-location yield trials of ICAR-AICRP for release in India	Unpublished[a]
6			JAKI 9218	QTL-hotspot	Under multi-location yield trials of ICAR-AICRP for release in India	Unpublished[a]
7			DCP 92-3	QTL-hotspot	Under multi-location yield trials of ICAR-AICRP for release in India	Unpublished[a]
8			JG 11	QTL-hotspot	Under multi-location yield trials of ICAR-AICRP for release in India	Unpublished[a]
9			ICCV 10	QTL-hotspot	Under multi-location yield trials of ICAR-AICRP for release in India	Unpublished[a]
10	Fusarium wilt resistance	WR 315	Annigeri 1	foc4	Released as "Super Annigeri 1" for commercial cultivation in India	Mannur et al. (2019)
11			JG 74	foc4	Under multi-location yield trials of ICAR-AICRP for release in India	Mannur et al. (2019)
12			Pusa 372	foc4	Under multi-location yield trials of ICAR-AICRP for release in India	Unpublished[a]

(continued)

Table 11.9 (continued)

S no	Trait	Donor parent	Recipient parent	Target region	Present status/released variety	Reference
13			Pusa 391	foc4	Under multi-location yield trials of ICAR-AICRP for release in India	Unpublished[a]
14		Vijay	Pusa 256	foc 2	Under multi-location yield trials of ICAR-AICRP for release in India	Pratap et al. (2017)

[a]*Source*—Annual Report 2018–2019-All India Coordinated Research Project on Chickpea

advanced genomic approaches are listed in Table 11.8. Genetic mapping and QTL analysis are both biparental along with association mapping have accelerated precise understanding of genetic control of agricultural traits and further utilization in Marker Added Selection, introgression of a trait of interest in most popular genotype, gene pyramiding, genetic gain of a character. Here is the list of products developed by the use of marker-assisted breeding in Table 11.9.

11.2.4 Transgenic Development in Chickpea

Recombinant DNA technology has great potential to increase the production and productivity of crops in the future. Judicious application of this technology provides opportunities for alleviating some of the major constraints in chickpea production where conventional breeding is silent. There were two basic requirements for development of transgenic in any crops. First genetic transformation methods and secondly an extensive tissue culture methods. In chickpea particle bombardment and *Agrobacterium tumefaciens*-mediated gene transfer technology have been developed for genetic transformation. Generally, the large-seeded grain legume species have been found recalcitrant to tissue culture. These limitations have been largely overcome in chickpea through extensive tissue culture methods developments for regeneration of transformed plantlets including embryo rescue and micro-propagation. In this context, the standardization of protocols of in vitro plant regeneration and Agrobacterium-mediated gene transfer is fundamental requirement for future genetic manipulation of these crops. The genetic transformation of chickpea began with the initial transformation of callus (Mohapatra and Sharma 1991). Fontana et al. in the year 1993 and Kar et al. in 1996 reported the production of transgenic chickpea plantlets derived from co-cultivation of embryo axes for the *nptll* and *uidA* gene. However, each of these reports could not verify proper inheritance of the introduced gene *nptll* and *uidA*. First stable integration of the *cry1A(c)* gene of chickpea transformation was reported by Kar et al. (1996) for the production of the Bt toxin in ICCV 1 and ICCV 6 genotypes. This was also the first report using biolistic direct DNA transfer to transform immature embryonic axes of chickpea (McPhee et al.

Table 11.10 Reports on genetic transformation of chickpea (*Cicer arietinum* L.)

Sl. No	Name of gene	Chickpea variety/cultivar	Source of the gene	Plant part used to transform	Changes in transformants	Reference
1.	*nptll*, *uidA*	local ecotype (supplied by Consorzio Agrario Provinciale, Rome, Italy)	–	seed-derived embryo axes deprived of the apical meristem	neomycin phosphotransferase II and B-glucuronidase activities	Fontana et al. (1993)
2.	*uidA* and *nptll*	ICCV-1, ICCV-6	–	Embryonic axis without apical meristem	Resistance to kanamycin used as a selectable marker in the transformation	Kar et al. (1996)
3.	*nptll*, *cry1A (c)*	ICCV-1, ICCV-6	*B. thuringiensis*	Embryonic axis without apical meristem of mature seed	*Heliothis armigera*	Kar et al. (1996)
4.	*nptll*, *pat*, *uidA*	PG 1, chafa, turkey, PG 12	–	Embryonic axis	Low reproducibility, *uidA* gene expressed in T_0 transmitted into T_1 (PCR)	Krishnamurthy et al. (2000)
5.	*nptll*, *Bt-cry1A(c)*	Vijay	*B. thuringiensis*	Cotyledons with half embryonic axis	Transgene transmitted to T_1 (PCR for *nptll*)	Das and Samah (2003)
6.	*nptll*, *uidA*, *bar*	P 362, P 1042, P 1043	–	Embryonic axis	Low reproducibility, Gus gene expressed in T_0 transmission to T_1 (PCR)	Tewari-Singh et al. (2004)
7.	*nptll*, *αAl1*	cv. Semsen	*P. vulgaris*	Cotyledons with half embryonic axis	*C. maculates* and *C. chinensis*	Sarmah et al. (2004)
8.	bar, *uidA*, *PGIP*	ICCV-5, H 208, ICCL 87322, K 850	–	Sliced embryonic axis	Good reproducibility, transmission and expression of *uidA* up to T_3	Senthil et al. (2004)
9.	*uidA*, *nptll*	CDC Yuma	–	Sliced embryonic axis	Good reproducibility, transmission and expression of *uidA* up to T_3	Polowick et al. (2004)
10.	*uidA*, *nptll*, *cry1A(c)*	C 235, BG 256, Pusa 362, Pusa 372	*B. thuringiensis*	L2 layer of cotyledonary nodes	Good reproducibility, transmission and expression of *nptll* and *cry1A (c) in T1 was shown, Good cry1A(c)*	Sanyal et al. (2005)

(continued)

Table 11.10 (continued)

Sl. No	Name of gene	Chickpea variety/cultivar	Source of the gene	Plant part used to transform	Changes in transformants	Reference
					activity Resistance to *Heliothis armigera*	
11.	*α-amylase inhibitor (αAI1)*	cultivar K850	*P. vulgaris*	Embryogenic axis	*C. maculates*	Ignacimuthu and Prakash (2006)
12.	*P5CSF129A*	C235	*Vigna aconitifolia*	Axillary meristem	Drought stress	Bhatnagar-Mathur et al. (2009)
13.	*CryIAc*	Cultivar ICCV-2 and A-1	*B. thuringiensis*	Seeds	*Heliothis armigera*	Biradar et al. (2009)
14.	*Allium sativum leaf agglutinin (ASAL)*	cv. ICCV 89314	*A. sativum*	Dissected single cotyledons with half embryos	*A. craccivora*	Chakraborti et al. (2009)
15.	*Cry2Aa*	Cultivars Semsen and ICCV 89314	*B. thuringiensis*	Seeds	*Heliothis armigera*	Acharjee et al. (2010)
16.	*P5CS*	Cultivar Annigeri	*Vigna aconitifolia*	Cotyledonary node	Salt stress	Ghanti et al. (2011)
17.	*Cry1Ab* and *Cry1Ac (Pyrimiding)*	variety P-362	*B. thuringiensis*	Cotyledonary nodes	*Heliothis armigera*	Mehrotra et al. (2011a, b)
18.	*Fused Cry1Ab/Ac*	var. DCP 92–3	*B. thuringiensis*	Embryonic axis	*Heliothis armigera*	Ganguly et al. (2014)
19.	*Cry1Ac*	var. ICCV89314	*B. thuringiensis*	Embryo	*Heliothis armigera*	Chakraborty et al. (2016)
20.	*Cry1Aabc*	var. DCP 92–3	*B. thuringiensis*	Axillary meristem	*Heliothis armigera*	Das et al. (2017)

21.	*Cry1AX*	cv. KAK-2	*B. thuringiensis*	Embryo axes	*Heliothis armigera*	Asharani et al. (2018)
22.	*Lec-RLK*	HC-1	*Oryza sativa*	Seeds	Salt stress	Singh (2018)
23.	*OsRuvB*	HC-1	*Oryza sativa*	Seeds	Salt stress	Preeti (2018)

2007). Considerable progress has been made in developing transgenic addressing different bottleneck in chickpea production particularly biotic and abiotic stresses, bio-fortification, and genetic correction. Here is the list of the genetic transformation attempted by the different workers in Table 11.10 by using different plant part after the standardization of the protocol for tissue culture in chickpea to provide solution for different problems where conventional breeding was taciturn.

Recently, Das et al. in 2021 detailed about the development of four transgenic chickpea lines harboring stress inducible *AtDREB1a* were generated with transformation efficiency of 0.1%. The integration, transmission, and regulated expression were confirmed by Polymerase Chain Reaction (PCR), Southern Blot hybridization, and Reverse Transcriptase polymerase chain reaction (RT-PCR), respectively. Transgenic chickpea lines exhibited higher relative water content, longer chlorophyll retention capacity and higher osmotic adjustment under severe drought stress level 4, as compared to control. The enhanced drought tolerance in transgenic chickpea lines were also manifested by undeterred photosynthesis involving enhanced quantum yield of PSII, electron transport rate at saturated irradiance levels, and maintaining higher relative water content in leaves under relatively severe soil water deficit. Furthermore, lower values of carbon isotope discrimination in some transgenic chickpea lines indicated higher water use efficiency. Transgenic chickpea lines exhibiting better Osmotic Adjustment (OA) resulted in higher seed yield, with progressive increase in water stress, as compared to control. They concluded that *AtDREB1a* transgenic chickpea lines were better adapted to water deficit by modifying important physiological traits on the basis of phenotyping, involving non-invasive chlorophyll fluorescence imaging, carbon isotope discrimination, osmotic adjustment, higher chlorophyll retention, and membrane stability index. These lines can further be utilized in pre-breeding or directly in the development of variety with enhanced drought tolerance.

Productivity of chickpea is thus highly affected by various biotic (*Helicoverpa*, Aphids, *Callosobruchus*, Bruchids *Bromus*, and *Orobanche*) and abiotic (drought and salinity) stresses. Considering quality aspect the legumes are deficient in sulfur containing amino acids, methionine, and cysteine. The possibilities for genetic improvement by marker-assisted breeding and selection approaches are limited in chickpeas due to their sexually incompatible gene pool (Acharjee and Sarmah 2013). Genetically modified chickpeas expressing either of the *cry1Ac/b, cry2Aa* and the *bean α-amylase inhibitor* gene show promise against *Helicoverpa* and bruchids, respectively, but commercialization is still to be achieved. Production and commercialization of transgenic chickpeas containing a single transgene may not give adequate yield advantage as chickpeas are affected by many production constraints both biotic and abiotic in the field as well as in storage. If these constraints anyhow could be managed, then there may be lacking quality trait (protein content, methionine and cysteine, palatability, etc.) hinder its commercialization. Gene pyramiding by incorporating two or more genes may be useful because improving single trait at a time will be time consuming, labor-intensive costly affair. Unfortunately, attempts to generate transgenic chickpeas with increased tolerance to drought and salinity or with increased methionine content have been less successful. The use of modern multi-gene vectors that contain recognition sites for zinc finger nucleases (ZFNs) and homing endonucleases may simplify the incorporation of multiple genes into single genotype (Acharjee and Sarmah 2013) leads to the development of "super chickpeas" that horbor multiple transgenic traits in a single genotype in consortium.

References

Abe A, Kousgi S, Yoshida K et al (2012) Genome sequencing reveals agronomically important loci in rice using MutMap. Nat Biotechnol 30:174–178. https://doi.org/10.1038/nbt.2095

Acharjee S, Sarmah BK (2013) Biotechnlogically generating super chickpea for food and national security. Plant Sci 207:108–116

Acharjee S, Sarmah BK, Kumar PA et al (2010) Transgenic chickpea (*Cicer arietinum* L.) expressing a sequence-modified *cry2Aa* gene. Plant Sci 178:333–339

Agarwal G, Garg V, Kudapa H et al (2016) Genome-wide dissection of AP2/ERF and HSP90 gene families in five legumes and expression profiles in chickpea and pigeonpea. Plant Biotechnol J 14:1563–1577

Arora S, Sharma V (2017) Reclamation and management of salt-affected soils for safeguarding agricultural productivity. J Safe Agric 1:1–10

Arora S, Singh YP, Vanza M et al (2016) Bioremediation of saline and sodic soils through halophilic bacteria to enhance agricultural production. J Soil Water Conserv 15:302–305. https://doi.org/10.5958/2455-7145.2016.00027.8

Asharani BM, Ganeshaiah KN, Raja A et al (2018) Transformation of chickpea lines with Cry1X using in planta transformation and characterization of putative transformants T1 lines for molecular and biochemical characters. J Plant Breed Crop Sci 3(16):413–423

Bhatnagar-Mathur P, Vadez V, Devi MJ et al (2009) Genetic engineering of chickpea (Cicer arietinum L.) with the P5CSF129A gene for osmoregulation with implications on drought tolerance. Mol Breed 23:591–606. https://doi.org/10.1007/s11032-009-9258-y

Biradar SS, Sridevi O, Salimath PM (2009) Genetic enhancement of chickpea for pod borer resistance through expression of *Cry1Ac* protein. Karnataka J Agric Sci 22:467–470

Bohra A, Dubey A, Saxena RK et al (2011) Analysis of BAC-end sequences (BESs) and development of BES-SSR markers for genetic mapping and hybrid purity assessment in pigeonpea (*Cajanus* spp.). BMC Plant Biol 11:56

Bohra A, Jha R, Singh IP et al (2017) Novel new hypervariable SSR markers for diversity analysis, hybrid purity testing and trait mapping in Pigeonpea [Cajanus cajan (L.) Millspaugh]. Front Plant Sci. Article 377. https://doi.org/10.3389/fpls.2017.00377

Chakraborti D, Sarkar A, Mondal HA et al (2009) Tissue specific expression of potent insecticidal, allium sativum leaf agglutinin (ASAL) in important pulse crop, chickpea (*Cicer arietinum* L.) to resist the phloem feeding *Aphis craccivora*. Transgenic Res 18(4):529–544

Chakraborty J, Sen S, Ghosh P et al (2016) Homologous promoter derived constitutive and chloroplast targeted expression of synthetic cry1Ac in transgenic chickpea confers resistance against *Helicoverpa armigera*. Plant Cell Tissue Organ Cult 125:521–535. https://doi.org/10.1007/s11240-016-0968-7

Das A, Samah BK (2003) An efficient in vitro regeneration system for genetic transformation of chickpea (*Cicer arietinum* L.) cultivars. Res J Contemp Conc 1:1–5

Das S, Upadhyaya HD, Srivastava R et al (2015) Genome-wide insertion–deletion (InDel) marker discovery and genotyping for genomics assisted breeding applications in chickpea. DNA Res: dsv020

Das A, Datta S, Thakur S et al (2017) Expression of a chimeric gene encoding insecticidal crystal protein cry1Aabc of *Bacillus thuringiensis* in chickpea (*Cicer arietinum* L.) confers resistance to gram pod borer (*Helicoverpa armigera* Hubner). Front Plant Sci 8:1423. https://doi.org/10.3389/fpls.2017.01423

Deokar AA, Ramsey L, Sharpe AG et al (2014) Genome wide SNP identification in chickpea for use in development of a high density genetic map and improvement of chickpea reference genome assembly. BMC Genomics 201415:708

Deschamps S, Llaca V, May GD (2012) Genotyping-by-sequencing in plants. Biology 1:460–483. https://doi.org/10.3390/biology1030460

Dutta S, Kumawat G, Singh BP et al (2011) Development of genic-SSR markers by deep transcriptome sequencing in pigeonpea [*Cajanus cajan* (L.) Millspaugh]. BMC Plant Biol 11:17

Elshire RJ, Glaubitz JC, Sun Q et al (2011) A robust, simple genotyping-by-sequencing (GBS) approach for high diversity species. PLoS One 6:e19379. https://doi.org/10.1371/journal.pone.0019379

Fontana GS, Santini L, Caretto S et al (1993) Genetic transformation in the grain legume *Cicer arietinum* L. (Chickpea). Plant Cell Rep 12:194–198

Food and Agriculture Organization (2018). http://faostat.fao.org/database

Ganguly M, Molla KA, Karmakar S (2014) Development of pod borer-resistant transgenic chickpea using a pod-specific and a constitutive promoter-driven fused *cry1Ab/Ac* gene. Theor Appl Genet 127(12):2555–2565. https://doi.org/10.1007/s00122-014-2397-5

Ghanti KK, Sujata KG, Kumar V et al (2011) Heterologous expression of *P5CS* gene in chickpea enhances salt tolerance without affecting yield. Biol Plant 55(4):634–640. https://doi.org/10.1007/s10535-011-0161-0

Gnanesh BN, Bohra A, Sharma M et al (2011) Genetic mapping and quantitative trait locus analysis of resistance to sterility mosaic disease in pigeonpea [*Cajanus cajan* (L.) Millsp.]. Field Crop Res 123:56–61. https://doi.org/10.1016/j.fcr.2011.04.011

Gowda CLL, Upadhyaya HD, Sharma S et al (2013) Exploiting genomic resources for efficient conservation and use of chickpea, groundnut, and pigeonpea collections for crop improvement. Plant Genome 6:1–11. https://doi.org/10.3835/plantgenome2013.05.0016

Hiremath PJ, Kumar A, Penmetsa RV et al (2012) Large-scale development of cost-effective SNP marker assays for diversity assessment and genetic mapping in chickpea and comparative mapping in legumes. Plant Biotechnol J 10:716–732

ICRISAT (1992) The medium term plan, vol 1. International Crop Research Institute for the Semi-Arid Tropics, Patancheru

Ignacimuthu S, Prakash S (2006) *Agrobacterium*-mediated transformation of chickpea with alpha-amylase inhibitor gene for insect resistance. J Biosci 31(3):339–345

Kar S, Basu D, Das S et al (1996) Expression of *cryIA(c)* gene of *Bacillus thuringiensis* in transgenic chickpea plants inhibits development of pod-borer (Heliothis armigera) larvae. Transgenic Res 6:177–185. https://doi.org/10.1023/A:1018433922766

Krishnamurthy KV, Suhasini K, Sagare AK et al (2000) *Agrobacterium* mediated transformation of chickpea (*Cicer arietinum* L.) embryo axes. Plant Cell Rep 19:235–240

Kudapa H, Azam S, Sharpe AG et al (2014) Comprehensive transcriptome assembly of chickpea (*Cicer arietinum* L.) using Sanger and next generation sequencing platforms: development and applications. PLoS One 9:e86039

Kumar V, Khan AW, Saxena RK et al (2016) First generation hapmap in Cajanus spp. reveals untapped variations in parental lines of mapping populations. Plant Biotechnol J 14:1673–1681

Kumar CVS, Naik SJS, Mohan N et al (2017) Botanical description of pigeonpea [*Cajanus Cajan* (L.) Millsp.]. In: The pigeonpea genome. https://doi.org/10.1007/978-3-319-63797-6_3

Ladizinsky G, Adler A (1976) Genetic relationships among the annual species of Cicer L. Theor Appl Genet 48:197–193

Liu B, Wendel GF (2000) Retrotransposon activation followed by rapid repression in introgressed rice plants. Genome 43:874–880

Liu S, Wang X, Wang H et al (2013) Genome-wide analysis of ZmDREB genes and their association with natural variation in drought tolerance at seedling stage of Zea mays L. PLoS Genet 9:e1003790. https://doi.org/10.1371/journal.pgen.1003790

Ma Y, Qin F, Tran LS (2012) Contribution of genomics to gene discovery in plant abiotic stress responses. Mol Plant 5:1176–1178. https://doi.org/10.1093/mp/sss085

Mallikarjuna N, Saxena KB, Lakshmi J et al (2011) Differences between *Cajanus cajan* (L.) Millspaugh and *C. cajanifolius* (Haines) van der Maesen, the progenitor species of pigeonpea. Genet Resour Crop Evol. https://doi.org/10.1007/s10722-011-9691-8

Mallikarjuna N, Saxena KB, Laxmi J et al (2012) Differences between *Cajanus cajan* and *C. cajanifolius*, the progenitor species of pigeonpea. Genet Resour Crop Evol 59:411–417

Mannur DM, Babbar A, Thudi M et al (2019) Super Annigeri 1 and improved JG 74: two Fusarium wilt-resistant introgression lines developed using marker-assisted backcrossing approach in chickpea (*Cicer arietinum* L.). Mol Breed 39:2

Mashaki KM, Garg V, Ghomi AAN et al (2018) RNA-Seq analysis revealed genes associated with drought stress response in kabuli chickpea (Cicer arietinum L.). PLoS One 13:e0199774

McPhee KE, Croser J, Sarmah B et al (2007) Development of transgenics in chickpea. In: Yadav SS, Redden RJ, Chen W, Sharma B (eds) Chickpea breeding and management, p 657

Mehrotra M, Sanyal I, Amla DV (2011a) High-efficiency *Agrobacterium*-mediated transformation of chickpea (*Cicer arietinum* L.) and regeneration of insect-resistant transgenic plants. Plant Cell Rep 30(9):1603–1616. https://doi.org/10.1007/s00299-011-1071-5

Mehrotra M, Singh AK, Sanyal I et al (2011b) Pyramiding of modified cry1Ab and cry1Ac genes of Bacillus thuringiensis in transgenic chickpea (*Cicer arietinum* L.) for improved resistance to pod borer insect *Helicoverpa armigera*. Euphytica 182:87. https://doi.org/10.1007/s10681-011-0501-3

Michalak P (2009) Epigenetic transposones and small RNA determinants of hybrid dysfunctions. Heredity 102:45–50

Mir RR, Bhat RI, Ahmad MA et al (2017) Molecular mapping of genes and QTLs in pigeonpea. In: The pigeonpea genome. https://doi.org/10.1007/978-3-319-63797-6_6

Mohapatra T, Sharma RP (1991) Agrobacterium mediated genetic transformation of chickpea, Cicer arietinum L. Indian J Exp Biol 29(8):758–761

Muthamilarasan M, Theriappan P, Prasad M (2013) Recent advances in crop genomics for ensuring food security. Curr Sci 105:155–158

Muthamilarasan M, Bonthala VS, Mishra AK et al (2014) C2H2-type of zinc finger transcription factors in foxtail millet define response to abiotic stresses. Funct Integr Genomics 14:531–543

Nayak SN, Zhu H, Varghese N et al (2010) Integration of novel SSR and gene-based SNP marker loci in the chickpea genetic map and establishment of new anchor points with Medicago truncatula genome. Theor Appl Genet 120:1415–1441

Pawar NB, Mayee CD (1986) Reaction of pigeonpea genotypes and their crosses to *Fusarium* wilt. Indian Phytopathol 30:70–74

Pazhamala L, Saxena RK, Singh VK et al (2015) Genomic-assisted breeding boosting crop improvement in pigeonpea *Cajanus cajan*. Front Plant Sci 6:1–12. https://doi.org/10.3389/fpls.2015.00050

Plukenet L (1692) Phytographia 3, table 213, figure 3.Google Scholar

Polowick PL, Baliski DS, Mahon JD (2004) Agrobacterium tumefaciens-mediated transformation of chickpea (*Cicer arietinum* L.): gene integration, expression and inheritance. Plant Cell Rep 23:485–491. https://doi.org/10.1007/s00299-004-0857-0

Prasad V, Satyavathi VV, Valli SKM et al (2004) Expression of biologically active hemagglutininneuraminidase protein of Peste des petits ruminants virus in transgenic pigeonpea [*Cajanus cajan* (L.) Millsp.]. Plant Sci 166:199–205

Pratap A, Chaturvedi SK, Tomar R et al (2017) Markerassisted introgression of resistance to fusarium wilt race 2 in Pusa 256, an elite cultivar of desi chickpea. Mol Gen Genomics 293: 1237–1245

Preeti (2018) Development and characterization of transgenic chickpea (Cicer arietinum L.) plants with OsRuvB gene for salt stress tolerance. PhD thesis submitted to CCS Haryana Agricultural University, Hisar pp 1–80

Pundir RPS, Rao NK, Van Der Maesen LJG (1985) Distribution of qualitative traits in the world germplasm of chickpea (Cicer arietinum L.). Euphytica 34:697–693

Ramu SV, Rohini S, Keshavareddy G et al (2012) Expression of a synthetic cry1AcF gene in transgenic pigeonpea confers resistance to *Helicoverpa armigera*. J Appl Entomol 136:675–687
Ranga Rao GV, Shanower TG (1999) Identification and management of pigeonpea and chickpea insect pests in Asia. International Crops Research Institute for the SemiArid Tropics, Hyderabad
Reddy LJ, Upadhyaya HD, Gowda CLL et al (2005) Development of core collection in pigeonpea [*Cajanus cajan* (L.) Millsp.] using geo-graphic and qualitative morphological descriptors. Genet Resour Crop Evol 52:1049–1056. https://doi.org/10.1007/s10722-004-6152-7
Remanandan P (1990) Pigeonpea: genetic resources. In: Nene YL, Hall SD, Sheila VK (eds) The Pigeonpea. CAB International, Wallingford, pp 89–115
Rieseberg LH, Archer MA, Wayne RK (1999) Transgressive segregation, adaptation, and speciation. Heredity 83:363–372
Roorkiwal M, Sawargaonkar SL, Chitikineni A et al (2013) Single nucleotide polymorphism genotyping for breeding and genetics applications in chickpea and pigeonpea using the Bead Xpress platform. Plant Genome 6. https://doi.org/10.3835/plantgenome2013.05.0017
Roorkiwal M, Bharadwaj C, Barmukh R et al (2020) (2020) Integrating genomics for chickpea improvement: achievements and opportunities. Theor Appl Genet 133:1703–1720. https://doi.org/10.1007/s00122-020-03584-2
Santra DK, Tekeoglu M, Ratnaparkhe M et al (2000) Identification and mapping of QTLs conferring resistance to ascochyta blight in chickpea. Crop Sci 40:1606–1612
Sanyal I, Singh AK, Kaushik M et al (2005) *Agrobacterium tumefaciens*-mediated transformation of chickpea (*Cicer arietinum* L.) with *Bacillus thuringensis cry1Ac* gene for resistance against pod borer insect *Helicoverpa armigera*. Plant Sci 168:1135–1146
Sarmah BK, Moore A, Tate W et al (2004) Transgenic chickpea seeds expressing high levels of a bean α-amylase inhibitor. Mol Breed 14(1):73–82. https://doi.org/10.1023/B:MOLB.0000037996.01494.12
Saxena KB (2008) Genetic improvement of pigeonpea- a review. Trop Plant Biol 1:159–178. https://doi.org/10.1007/s12042-008-9014-1
Saxena KB, Sawargaonkar SL (2016) Genetic enhancement of seed proteins in pigeonpea – methodologies, accomplishments, and opportunities. Int J Sci Res 4:3–7
Saxena RK, Penmetsa RV, Upadhyaya HD et al (2012) Large-scale development of cost-effective SNP marker assays for genetic mapping in pigeonpea and comparative mapping in legumes. DNA Res 19:449–461
Saxena RK, Kale SM, Kumar V et al (2017) Genotyping- by- sequencing of three mapping populations for identification of candidate genomic regions for resistance to sterility mosaic disease in pigeonpea. PLoS One 7:1–9
Saxena K, Choudhary AK, Srivastava RK et al (2019) Origin of early maturing pigeonpea germplasm and its impact on adaptation and cropping systems. Plant Breed 138(3):243–251
Senthil G, Williamson B, Dinkins RD et al (2004) An efficient transformation system for chickpea (*Cicer arietinum* L.). Plant Cell Rep 23:297–303
Sharma M, Ghosh R, Telangare R et al (2016) Environmental influences on pigeonpea- *Fusarium udum* interactions and stability of genotypes. Front Plant Sci 7:1–9
Singh G (2018) Development of transgenic chickpea (Cicer arietinum L.) plants with lectin receptor-like kinase (Lec-RLK) gene for improving salt tolerance. Ph.D. thesis submitted to CCS Haryana Agricultural University, Hisar, pp 1–76
Singh R, Sharma P, Varshney RK et al (2008) Chickpea improvement: role of wild species and genetic markers. Biotechnol Genet Eng Rev 25(1):267–314. https://doi.org/10.5661/bger-25-267
Singh N, Tyagi RK, Pandey C (2013) Genetic resources of pigeonpea: conservation for use. National Bureau of Plant Genetic Resources (NBPGR), New Delhi, pp 1–49
Singh M, Gautam NK, Rana MK et al (2014) Pigeonpea genetic resources and its utilization in India, current status and future prospects. J Plant Sci Res 1:107

Singh IP, Kumar S, Byregowda CV et al (2017a) Exploitation of heterosis and new plant types in pigeonpea. In: National symposium on pulses for nutritional security and agricultural sustainability from December 2–4th, 2017 at IIPR Kanpur, pp 37–44

Singh VK, Saxena RK, Varshney RK (2017b) Sequencing pigeonpea genome. In: Varshney RK, Saxena RK, Jackson SA (eds) The pigeonpea genome. Springer, Cham, pp 41–54. https://doi.org/10.1007/978-3-319-63797-6-1

Surekha C, Nirmala Kumari K, Aruna V et al (2014) Expression of the *Vigna aconitifolia P5CSF129A* gene in transgenic pigeonpea enhances proline accumulation and salt tolerance. Plant Cell Rep 116:27–36

Tewari-Singh N, Sen J, Kiesecher H, Reddy VS et al (2004) Use of herbicide or lysine plus threonine for non-antibiotic selection of transgenic chickpea. Plant Cell Rep 22:576–583

Thu TT, Dewaele E, Trung LQ et al (2007) Increasing lysine levels in pigeonpea (*Cajanus cajan* (L.) Millsp.) seeds through genetic engineering. Plant Cell Tissue Org Cult 91:135–143

Thudi M, Bohra A, Nayak SN et al (2011) Novel SSR markers from BACend sequences, DArT arrays and a comprehensive genetic map with 1291 marker loci for chickpea (*Cicer arietinum* L.). PLoS One 6:e27275

Turnbull LV, Whiteman PC, Byth DE (1981) The influence of temperature and photoperiod on floral development of early flowering pigeonpea. In: Proceedings of the international workshop on pigeonpeas, vol 1, December 15–19, 1980, ICRISAT Center, India. ICRISAT, Patancheru, AP, 17–223

Upadhyaya HD, Ortiz R (2001) A mini core subset for capturing diversity and promoting utilization of chickpea genetic resources in crop improvement. Theor Appl Genet 102:1292–1298

Upadhyaya HD, Reddy LJ, Gowda CLL et al (2006a) Development of a mini-core subset for enhanced and diversified utilization of pigeonpea germplasm resources. Crop Sci 46:2127–2132. https://doi.org/10.2135/cropsci2006.01.0032

Upadhyaya HD, Furman BJ, Dwivedi SL et al (2006b) Development of a composite collection for mining germplasm possessing allelic variation for beneficial traits in chickpea. Plant Genet Res 4:13–19

Upadhyaya HD, Bajaj D, Das S et al (2016a) Genetic dissection of seed-iron and zinc concentrations in chickpea. Sci Rep 6:24050

Upadhyaya HD, Sastry DVSSR, Vetriventhan M et al (2016b) Mini core collection: means to enhance utilization of germplasm. Open Access repository, ICRISAT

Upadhyaya HD, Dwivedi SL, Sharma S (2017) Managing and discovering agronomically beneficial traits in chickpea germplasm collections. In: Varshney RK, Thudi M, Muehlbauer F (eds) The chickpea genome. Springer, Cham, pp 43–52. https://doi.org/10.1007/978-3-319-66117-9_2

Vales MI, Srivastava RK, Sultana R et al (2012) Breeding for earliness in pigeonpea: development of new determinate and nondeterminate lines. Crop Sci 52(6):2507–2516. https://doi.org/10.2135/cropsci2012.04.0251

Van der Maesen LJG (1980) India is the native home of pigeonpea. In: Arends JC, Boelema G, de Groot CT, Leeuwenberg AJM, Veenman H, Zonen BV (eds) Libergratulatorius in honorem H.C.D. de Wit landbouwhoge school, Miscellaneous paper no 19, Wageningen, pp 257–262

Van Der Maesen LJG (1987) Origin, history and taxonomy of chickpea. In: Saxena MJ, Singh KB (eds) The chickpea. CAB International, Cambridge, pp 11–34

Varshney RK (2016) Exciting journey of 10 years from genomes to fields and markets: some success stories of genomics-assisted breeding in chickpea, pigeonpea and groundnut. Plant Sci 242:98–107

Varshney RK, Hiremath PJ, Lekha P et al (2009) A comprehensive resource of drought- and salinity- responsive ESTs for gene discovery and marker development in chickpea (Cicer arietinum L.). BMC Genomics 10:523

Varshney RK, Penmetsa RV, Dutta S et al (2010a) Pigeonpea genomics initiative (PGI): an international effort to improve crop productivity of pigeonpea (*Cajanus cajan* L.). Mol Breed 26:393–408. https://doi.org/10.1007/s11032-009-9327-2

Varshney RK, Thudi M, May GD et al (2010b) Legume genomics and breeding. Plant Breed Rev 33:257–304

Varshney RK, Chen W, Li Y et al (2012) Draft genome sequence of pigeonpea (*Cajanus cajan*), an orphan legume crop of resource-poor farmers. Nat Biotechol 30:83–89. https://doi.org/10.1038/nbt.2022

Varshney RK, Song C, Saxena RK et al (2013a) Draft genome sequence of chickpea (*Cicer arietinum*) provides a resource for trait improvement. Nat Biotechnol 31:240–246

Varshney RK, Gaur PM, Chamarthi SK et al (2013b) Fast-track introgression of QTL-hotspot for root traits and other drought tolerance traits in JG 11, an elite and leading variety of chickpea. Plant Genome 6:1–9

Varshney RK, Thudi M, Nayak SN et al (2014a) Genetic dissection of drought tolerance in chickpea (*Cicer arietinum* L.). Theor Appl Genet 127:445–462

Varshney RK, Pandey MK, Pasupuleti J et al (2014b) Marker-assisted introgression of a QTL region to improve rust resistance in three elite and popular varieties of peanut (Arachis hypogaea L.). Theor Appl Genet 127:1771–1781

Varshney RK, Saxena RK, Jackson SA (2017a) The pigeonpea genome. Springer, pp 41–54. https://doi.org/10.1007/978-3-319-63797-6-1

Varshney RK, Saxena RK, Upadhaya HD et al (2017b) Whole–genome resequencing of 292 pigeonpea accessions identifies genomic regions associated with domestication and agronomic traits. Nat Genet 49(7):1082–1088. https://doi.org/10.1038/ng.3872

Vavilov NI (1951) The origin, variation, immunity and breeding of cultivated plants. Chron Bot 13(1–6):1–366

Wallis ES, Byth DE, Saxena KB (1981) Flowering responses of thirty-seven early maturing lines of pigeonpea. In: Nene YL, Kumble V (ed) Proceedings of the international workshop on pigeonpeas, vol 2, International Crops Research Institute for the Semi-arid Tropics (ICRISAT), Patancheru, pp 143–159

Winter P, Benko-Iseppon AM, Huttel B et al (2000) A linkage map of the chickpea (*Cicer arietinum* L.) genome based on recombinant inbred lines from a *C. arietinum* x *C. reticulatum* cross; localization of resistance genes for *fusarium* wilt races 4 and 5. Theor Appl Genet 101:1155–1163

Yang SY, Saxena RK, Kulwal PL et al (2011) The first genetic map of pigeonpea based on Diversity Arrays Technology (*DArT*) markers. J Genet 90:103–109

Zeven AC, Zhukovsky PM (1975) Dictionary of cultivated plants. In: Huxby J (ed) The new systematics. Oxford University, London, pp 549–566

Intellectual Property Rights in Plant Biotechnology and Breeding

12

Usha Kiran and Nalini Kant Pandey

Abstract

Plant breeding and innovation in agricultural biotechnology have been very important to increase agricultural productivity and quality which in turn contributed to the economic growth of farmers across the world. High-yielding, diseases resistant, biotic and abiotic stress-resistant crops were the product of advancement in technology and innovations. These plant breeding innovations are the solution to the problem of world food insecurity and ecosystem conservation. The establishment of an effective system of protection, through various policies, rules, and regulations within the national and across the international borders, has been developed for plant varieties and products. Such protection encourages intellectual and financial investments to facilitate the growth of agribusiness, seed industry, and agricultural biotechnology. However, various groups like civil society, researchers, breeders, agribusiness stakeholders, and farm groups approach the law and technological advancements from a different perspective. The development of legal instruments of protection in plant biotechnology and breeding to protect the competing concerns expressed by different groups is a continual process. The chapter attempts to explain the forms and scope of intellectual property rights relevant to plant biotechnology, plant variety, biosecurity, and biopiracy.

U. Kiran (✉)
Vanercia Institute of Technical Education, Gautam Buddh Nagar, Uttar Pradesh, India

N. K. Pandey
Mitakshara IP Services, Ghaziabad, Uttar Pradesh, India

Kamaluddin et al. (eds.), *Technologies in Plant Biotechnology and Breeding of Field Crops*, https://doi.org/10.1007/978-981-16-5767-2_12

Keywords

Plant patents · Utility patents · Plant variety protection act · UPOV · Plant breeders' rights · Sui generis · European patent convention · Indian patent act · Plant variety protection and farmer's right act 2001

12.1 Introduction

With the prediction of the population reaching over nine billion people around the world by 2050, the demand for food is expected to grow substantially. Food security is needed not only in terms of quantity but also nutritional quality. The success story of the green revolution in 1960 was the product of the concerted use of high-yielding seeds, better irrigation methods, and use of fertilizers. This product of industrial society did lead to a quantum leap in food production and transient food security. The intensive cultivation, however, exhausted the soil's natural productivity and jeopardize future productivity. Now the world scientist, breeders, and farmers have a daunting challenge of increasing food production with a depleted agricultural natural resource base, ever-changing climate, and delimited land, water, and energy resources. Innovative plant breeding strategies along with agricultural biotechnology would help scientists develop plant varieties with high productivity, better nutrients, highly resilient to climate and soil conditions. New varieties are now being developed with traits suited to farmers and flower producers' needs, processing industry needs, and to the ever-changing taste of consumers for ornamental flowers.

For centuries, breeders have been able to incorporate suitable traits into plants through selective breeding and cultivation practices. The process of incorporation of the transgene to modify a crop through the conventional method usually take about 10 to 12 years. The biotechnology interventions in agriculture and improvement in research tools led to fast-paced development in the last few decades. Cell and tissue culture techniques, genetic engineering, DNA fingerprinting, and cell transformation techniques showed tremendous advancement. The molecular tools such as zinc-finger nuclease (ZFN) (Shukla et al. 2009; Townsend et al. 2009), oligonucleotide-directed mutagenesis (ODM) (Zhu et al. 2000), and CRISPER-Cas (Srivastava. 2019; Zhou et al. 2018; Yin et al. 2017) led to precise manipulation of genetic material and drastically reduced the time gap for the development of a hybrid and its release for commercial use.

With the advent of affordable high throughput genotyping technologies, the repositories of genome sequences of all major crops were compiled and maintained. These highly useful resources are now being explored and exploited to improve quantitative yields, nutrition, and agronomic performance in breeding programs, world over. Now, genes for the desired traits such as increased vigour and yield, insect resistance, drought, and salt tolerance can be incorporated across the genetic barriers. Thus, these new crops along with the good agricultural practices are not only transforming the way crops are grown but also reducing the use of fertilizers and pesticides, improving water quality and better soil management.

During the early twentieth century, industries gained importance and agriculture became less economically important in western countries, especially the United States and European countries. Governments started diverting funds to industries and progressively reduced financial support for agricultural development and seed distribution to farmers. There was, however, still needed to develop solutions for world food security, agricultural problems, depletion of natural resources, environmental remediation, and ecological imbalance. The private entities or private/public partnerships provide significant intellectual and financial inputs in the research and development for commercial plant breeding and distribution of seeds. With private investments, agricultural biotechnology and plant breeding no longer were the proprietary product solely owned by publicly funded research labs. The plant varieties and plant products such as seed and other propagating material were no longer viewed as a freely exchanged public good but were considered as a product of human intervention that could be owned and protected. This limited the free use of propagating products especially seeds which earlier purchased once can be reused for several generations by farmers (Kameri-Mbote 1999).

Thus, the evolution of agriculture and plant breeding as a new scientific discipline with lots of intellectual and financial inputs required appropriate laws to protect the interest of all stakeholders.

The chapter reviews the relevance of intellectual property (IP) protection in the research and development of plant breeding and the commercialization of agriculture.

12.2 Consortiums for Plant Breeding

To meet the requirement of the ever-increasing world population, new plant varieties are being developed at various sources. Plant breeding is being done at various levels; public institutions, private companies, independent breeders, and farmers. The public-funded institution had always been the major hubs of plant breeding programs in most of the countries. The actual realization of technology for the large-scale commercialization of agriculture needs lots of intellectual and financial inputs. Most of the developing countries, however, have limited scientific infrastructure and lacks financial resources to invest in plant breeding programs. Moreover, the cultivars developed in these public institutions are either released in the public domain, or the finances generated by these programs are not fully invested in these programs again (Shelton and Tracy 2017; Tracy et al. 2016). Thus, it is continually becoming difficult for public institutions to recoup the investment in the plant breeding process. The involvement of private companies in agriculture greatly enhanced the pace of development. BASF Group, Bayer Group, Dupont, Delaware, and Swiss giant Syngenta are some of the companies actively working in plant research. These private companies established their own plant breeding programs to develop and release propriety inbred lines and cultivars. Therefore, now the public along with private companies are active members in plant breeding and cultivar development.

Independent breeders and farmers, who earn living, part or whole, are also the contributors to the commercialization of agriculture (Montenegro DeWit 2017). They select and retain seeds of plants with the best traits for the particular environment for future use or sale. Hence during the repeated selection may discover or invent a new cultivar.

Some international and national non-governmental organizations are also actively engaged in extensive plant breeding programs. These research organizations have active research and breeding programs focusing on specific crops. National and local NGOs research on local crops and release cultivars suited to the farmers and environment of the region. NGOs such as Consultative Group for International Agricultural Research (CGIAR) is the global agricultural innovation network working toward global targets for the transformation of food, land, and water systems across local, regional, and global levels. Initially, CGIAR released its cultivars into the public domain with no associated fees, but now releases cultivars and populations with a standard material transfer agreement (SMTA) to ensure the benefit sharing with the communities contributing to selecting and maintaining seeds over the generations. Apart from this, there are local and national NGOs that engage in farmer-participatory plant breeding programs.

12.3 Intellectual Property (IP) Rights

IP rights enable protection or reward to the creators of new knowledge, by a set of laws. Knowledge, unlike tangible goods, can be used in part or whole, with any number of people. The knowledge when converted to tangible form, can be protected. The tangible products such as inventions, tool kits, publications, plant varieties, etc. require great financial inputs and a considerable amount of time. Once the knowledge becomes tangible, it can be used loaned, gifted, sold, or stolen. Thus, as an instrument of public policy, IPRs provide creator protection against direct copying or use of the process or product without any compensation, over a fixed period of time. The information comes to the public domain to be used freely after the expiry of the protection time, specified under the law. This allows the creator to restrict the use of the IP by other people. No one can use, manufacture, sell the invention without the permission of the creator.

The inventor wants to exercise control over the use of technology developed, research made, or invention done. IPRs promote research and development by providing incentives to investors in the inventive process and enable them to acquire the right to use their inventions, elsewhere. Thus, these laws are instrumental in promoting research collaboration among individuals, public and private investors for driving innovations for benefit of society.

12.4 Importance of IPR in Plant Research and Breeding Program in the Current Context

Before the nineteenth century, publicly funded labs were the sole source of development in agricultural sciences. Commercial plant breeding and cultivation were achieved due to development in biological sciences and fertilizer knowledge during the early nineteenth century. Developing new commercial varieties, however, required lots of innovation in research tools, breeding, and cultivation techniques. The new variety development program is a multi-step research process and carried out by a consortium of people including research scientists, plant breeders, and investors. Public funded institutions generally faced a lack of funds and infrastructure, especially in developing countries. Also, independent breeders usually have very less or no financial resources and are not able to sustain their research work for longer times. The new agri-biotechnology and the breeding programs are being undertaken either by private enterprise or by private/public partnerships that rely on private funding. Also, the capital investments in research and development by these private investors are substantial. These private companies are interested in creating new establishments for the development of new varieties financed through profits from sales of seed. The investors and developers, both want control over the commercialization of new plant innovations as the security for the investment in research and development of plant-related technology (Kloppenburg 2004).

To balance the competing needs of investors, plant breeders, growers, and others who depend upon the health of the farm economy, adequate IPR laws were formulated in the early twentieth century. Thus, the IP protection became more relevant for the commercialization of agriculture and providing an incentive for investment in product research and development. The establishments undertaking the plant breeding program determine the fate of new cultivar released. The cultivar developed in the public-funded institution is generally released in the public domain and usually, no royalty is taken for its use by others. Private companies use IPR-based restrictions on cultivar-use to safeguard their efforts and financial inputs on cultivar development. The revenue generated by charging for the usage of innovations is used up by the companies to fund follow-on plant research and development and support business growth.

12.5 Challenges in Agricultural Biotechnologies and IPR

The research and innovation in agri-biotechnology have 4 major challenges. First, the source of innovations and information is quite geographically dispersed. Major centres for research and development are concentrated in developed countries. Secondly, the development of a variety is a multi-layered process involving a number of biological factors (genes, plants, etc.), tools, and processes. Each part may be developed individually and then collectively used for the generation novel crop variety. Thirdly, the research and innovation in agricultural biotechnology (agri-biotechnology) require lots of investment which is now coming from the private sector, the commercial breeders, and seed companies. Fourth, the perspective on innovation in agri-biotechnology is still evolving and this influences the policies, legislative rules, and regulations governing the industry. The instruments of public policy have been formulated based on developed societies however, the system of agriculture and farming in developing countries is very different from developed countries.

It has been long advocated that a strong IP protection system is vital to stimulate research and development investments. According to this doctrine, a strong IPR system would increase the flow of innovations and products from developed to developing nations and would support local research and innovation. Seed companies and agribusiness stakeholders who own the patents would, however, not like to share seeds without a fee or agreement license to growers or other breeders, to develop their own varieties. This would result in narrowing down of publicly available germplasm pool for breeders to further improve the existing varieties, especially in developing countries. Uncertainty is also expressed that the developed countries may use genetic resources, from industrialized countries, for their product development and may take IPR protection and restrict the use by developing countries. Hence developing countries have to import new biotechnologies which would be a financial burden on their economy. IPRs granted for engineered plant varieties may also lead to loss of biodiversity. Thus, IPR regimes need to be structured according to the current context of research and the requirement of society, to provide opportunities for right holders to generate optimal revenue for the use of their inventions, to promote the dissemination of acquired knowledge, and to commercialize the newly developed technology for benefit of mankind.

12.6 Forms of IPRs

The newly developed intangible property is protected by various IP laws such as brands by geographical indications and trademarks, inventions by patents, designs by design patents or industrial designs rights, and creative work by copyright.

12.6.1 Geographical Indication (GI)

A geographical indication (GI) is a name or tag used on products indicating specific geographical origin and possessing exclusive qualities or characteristics essentially due to its place of origin. According to Data shared for 2019, by 117 national/regional authorities to World Intellectual Property Organization (WIPO), 55,800 protected GIs are in existence.

The protection to GIs is granted by law to avoid misleading the public and prevent unfair competition. These are protected by Article 22 of the TRIPS Agreement which defines that a GI may be assigned or licensed to an individual or entity who is in the area of origin and makes the product according to specified standards. A GI cannot be granted to an individual or entity outside the place of origin or not belong to the group of authorized producers.

GIs defined under Trade-Related Aspects of Intellectual Property Rights (TRIPS) Agreement Article 22.1 as

"Indications which identify a good as originating in the territory of a Member [of the World Trade Organization], or a region or locality in that territory, where a given quality, reputation or other characteristics of the good is essentially attributable to its geographical origin."

Thus, GI rights help the producers to get an economic benefit for their premium product as well as guarantee quality products to consumers. GI also safeguards the regional traditional knowledge and practices for product development. The world-renowned examples of registered GI are Darjeeling tea, Irish whiskey, Café de Colombia, Florida oranges, Vidalia onions, Kampot pepper, Penja pepper, Washington state apples, Napa Valley wines, Idaho potatoes, Toscano olive oil from Tuscany, Roquefort cheese, and Prosciutto di Parma ham from the Parma region of Italy. Usually, names of places are used as GIs however, sometimes non-geographical terms with traditional denominations (geographical connotation) are also, examples include Argane (Morocco) Reblochon (France), and Vinho Verde (Portugal).

The majority of GIs are granted for agricultural products, spirits, and wine. Accumulation of metabolites in plants is typically influenced by geographical location and local factors such as soil status, climate, and specific manufacturing skills and traditions (Belletti and Marescotti 2011). Conferring GI tag reduces market completion at the global level for local breeders and helps them grow economically (Eberlin 2009). This also ensures financial and scientific inputs in the research and development of local varieties for enhanced specific qualities.

Appellation of Origin An appellation of origin is a sub-category of GIs. The indication or tag is generally the geographical name or a traditional geographical connotation which indicates that the designate a product is originated in a particular geographical region and the quality or exclusive characteristics of the product, which has given the product its reputation, are essentially or exclusively because of the prevalent geographical environment (natural and human factors). An appellation of origin differs from other GIs because they represent the strongest link with the geographical area.

Protection of GIs GIs are protected with the country (at the regional as well as national levels) through a variety of legislative approaches. These include a special regime of protection known as Sui generis systems. This system provides protection based on the condition complied within the registration procedure. Sui generis system provides protection against any direct and indirect commercial use or imitation of GIs. The European Union (EU), the African Intellectual Property Organization (OAPI), and other intergovernmental organizations along with many countries use Sui generis systems for GIs protection. GIs are also protected using collective and certification marks under the trademark law. GIs are also protected by consumer protection laws, methods focusing on business practices, unfair competition regulations, and specific laws recognizing individual GIs.

GI Rights Are Territorial in Nature The GI rights are the territorial rights which means that GI rights are obtained in a specific jurisdiction (country or region) and are effective in the territory of that jurisdiction and not outside it. Currently, there is no "international "or "world" right exists. To obtain GI protection in a foreign jurisdiction, GI holders must seek protection under prevailing laws in the jurisdiction, where protection is sought. GI can be protected abroad in four ways; by obtaining protection directly in the country or region concerned, by taking advantage of bilateral agreements (often trade related) decided between them, through WIPO's Lisbon system of international registration of appellations of origin, and through the Marid system of international registration of collective or certification marks.

GIs Can Be Applied The cooperative or association of the producers of the product identified by the GI can request protection. These entities should ensure that the product fulfils the specific requirements which they have agreed upon. In some jurisdictions, a national competent authority (for example, a local government authority) may also request for protection of GI.

GIs Are Granted The GI protection is granted by a national (regional) competent authority upon request. In some countries, a special body responsible for GI protection grants the GI protection while in other countries, the national IP office grants the GI protection. The WIPO has a directory of IP offices listed on its website.

Darjeeling Tea and GI Protection

Darjeeling word and logo were the first Indian GIs to be registered in the name of Tea Board under the Geographical Indications of Goods (Registration and Protection) Act, 1999 in India. Any consumer purchasing Darjeeling tea will expect the tea to be grown and produced in the Darjeeling region of India. They also assume that the tea should have a special flavour and quality due to the traditional and cultural practices used for cultivation and processing prevalent in that geographical location i.e. Darjeeling. GI tag ensured that the word "Darjeeling" could not be used for tea originated in any other region of the world. Darjeeling tea was granted the GI status by India in 2004 and by the European Union in 2011, after authenticating its origin. This led to phasing out of the products, not from Darjeeling. Blenders in the EU however, continued selling their products as Darjeeling tea, by mixing 51% of Darjeeling tea with any tea. Darjeeling tea was the first commodity from India to get protected GI (PGI) status from the German Tea Association and the European Trade Council. The European Trade Union in 2012 decided that packets having the Darjeeling logo and PGI logo and containing 100% Darjeeling tea can only be sold as Darjeeling tea in European Union (Mantrov 2014). Thus, under the protection for GIs, the interest of both the local growers and consumers was protected.

12.6.2 Trademarks

Trademarks like GI, convey the source of a good or service to the customer. Trademark identify a good or service to be originating from a particular company and associate a specific quality or reputation of the particular good or service with the producing company. A trademark is linked to a specific company and not to a particular place therefore, it can be assigned or licensed to anyone, anywhere in the world. Trademark cannot be obtained for plant variety but can be used as a brand mark for the sale of seeds and other propagating material. "Pink Lady" is a Club apple cultivar is registered trademarks are protected under both federal and state law. The apples under the trademarked name, "Pink Lady" could be sold by only those growers who are "in the club." Hence trademark is used in very specific ways in conjunction with plant biotechnology and plant cultivars.

12.6.3 Trade Secret

A trade secret is exclusively held knowledge, a device, pattern, formula, or compilation of information, which gives economic benefit over the competitor, as an incentive to innovate meeting the non-obvious standards of Patent law, and efforts should have been taken to keep it secret. Thus, in absent patents, inventors can rely on trade secrets to protect their discoveries (Friedman et al. 1991). Trade secrets laws have specific legal protection regimes to prevent unauthorized misappropriation (Endres and Goldsmith 2007). In-plant breeding programs, inbred lines used to make an F1 hybrid cultivar are often protected by companies as trade secrets.

12.6.4 Patents

Any invention which is a novel, non-obvious creation of the mind and has beneficial application for mankind, could be protected by patent. These are the strongest rights granted by law to the inventor for exclusive use for a limited period of time (usually 20 years). The invention must not be part or whole, of prior art (should be novel), must be non-obvious to a person with ordinary skills in the particular field of application (should be innovative), and must be a solution of the particular problem (should be useful). Protected by law, the creator or its assignee can use the innovative invention to draw commercial benefits and also can exclude others from imitating or using it. Thus, patent system encourages innovation by giving the creators exclusivity over the invention which can be used to gain economic benefits.

The disclosure of information about innovative inventions to the public on how to create or practice the invention is another core rationale of patent law. The information should be in sufficient detail so that a notional person skilled in that art is able to practice the claimed invention. In most countries, patents are disclosed 18 months after the filing date or earlier. Failing to disclose the invention, a patent on that invention is termed invalid.

The importance of this feature is predicated on the fact that absent patents, inventors can rely on trade secrets to protect their discoveries (Friedman et al. 1991).

12.6.4.1 Benefits of Patents

Economic Benefits The patents are exclusive rights over the novel product or process for a fixed period of time. The knowledge by itself is a pure public good but not the new knowledge derived from mining the existing knowledge. The newly derived knowledge used for the development of new and useful innovative products and processes costs the inventor time and money. The free use of inventors' newly developed knowledge or novel products by the public would not help him to recover his money and time invested in innovation and may discourage the inventor from researching and innovating further. To appreciate the efforts of inventors, patents endow inventors with the property right on their products. These exclusive rights on the invention can be used by the inventor for economic benefits over a fixed period of time.

Promotes New Discoveries The patents grant the inventor with exclusive right to exploit the results of his efforts, economically. The monopoly over the use of patented products or processes ensures the innovator that consumption by other individuals will not reduce his benefits. This is an encouragement to inventors to innovate. Patent protection for exclusive inventor use also leads to disclosure of the new knowledge to the public and likely to increase the innovations.

Dissemination of Knowledge to Avoid Wasteful Innovation A patent filing needs the inventor to agree to disclose new knowledge. Along with the incentive to an inventor, patents also disseminate scientific and technical information. This social payoff property of patents allows scientists and researchers to use the information to avoid duplicating inventions and discoveries. The systematic approach for follow-up inventions or discoveries, avoid waste full innovations and thereby increases the output of research and development (Kitch 1977).

12.7 Protection Under US Legislation

12.7.1 Plant Patent Act (PPA)

Historically, in the United States, private nurseries and breeders specializing in fruits and flowers pioneered the legislative movement towards the creation of exclusive rights over plant varieties. The nurseries were benefiting from the fact that their products were identifiable due to their visible differentiated characteristics, however, were challenged by the competitor engaged in copying and flooding the market with competing products. These growers were not actually breeding but were observing the practice of selecting and saving higher-yielding and better-adapted variety seeds, tubers, and bulbs (vegetative reproductive organs) for the next planting season. Until the nineteenth century, all varieties created by people engaged in growing plants were the "product of nature" and no protection was given to any plant cultivar as all the information was for public consumption. During the nineteenth century, a lot of intellectual efforts and finance went into breeding programs for the development of a new variety. Plant breeders, therefore, stressed the statutory need for safeguarding their work and extend the incentives for agricultural products, as provided under the general patent act. Agriculture gained the same economic status for experimentation and invention that any other industry enjoyed after the legislation extended protection and incentive for a new variety under patent law. The patent system, the strongest form of protection, recognizes the patentee's investment of money and time and protects his invention from easy imitation by others. This protection also encourages the patentee to disclose his invention details to the public.

In 1930, the United States passed the Plant Patent Act (PPA) granting IPRs on plants. Henry F. Bosenberg was the first person to have the patent for his climbing, ever-blooming rose in 1931 under PPA and it was found that nearly half of the 3010 plant patents between 1931 and 1970 were granted for roses (Moser and Rhode 2011). Ornamentals and fruits are the predominant plant types to be protected under plant patents in the United States (Pardey et al. 2013). In FY 2020, 1350 plant patents were granted by the U.S. Patent and Trademark Office.

A patent is granted for the plant if it is a novel variety and the inventor is able to reproduce it by means, other than seeds (asexual reproduction), usually by cutting or grafting.

The most recent amendment to the U.S. plant patent statute (Title 35 United States Code, Section 161) states:

> Whoever invents or discovers and asexually reproduces any distinct and new variety of plant, including cultivated sports, mutants, hybrids, and newly found seedlings, other than a tuber propagated plant or a plant found in an uncultivated state, may obtain a patent, therefore, subject to the conditions and requirements of the title.
>
> The provisions of this title relating to patents for inventions shall apply to patents for plants, except as otherwise provided.

The law gives the right holder or its assignee, exclusive rights for 20 years (25 years for trees or vines) from the date of filing the application to charge a royalty for the use of the new variety or its reproductive material by others. The law states that the inventor or discoverer of a plant is any person who first appreciated the distinctive features of a plant and reproduces asexually the distinct and new variety of plants. The novelty could be due to cultivated sports (a part of a plant that is morphologically different from the rest of the plant), hybrids, mutants, and newly found seedlings. Tuber propagated plants and plants found in the uncultivated states were excluded from the protection under this act. PPA gives the patent owner's right to preclude others from asexually reproducing the patented plant, and or using, offering for sale, or selling the plant so reproduced, or any part of it, in the United States, or importing the same into the United States.

According to the PPA, 1930

1. A living plant is an entity expressing a set of characteristics determined by single genetic makeup or genotype, which can be reproduced through asexual reproduction and cannot otherwise be "made" or "manufactured."
2. Sports, mutants, and hybrids are important and comprehended as; sports or mutants may be spontaneous or induced. Hybrids may occur naturally, from a deliberate breeding program, or somatic in origin. While natural plant mutants may be an act of nature but they must have been discovered in a cultivated area.
3. Algae and macro fungi are regarded as plants, but bacteria are not.

The US legislation under PPA, 1930 only grants protection for the reproduction of the new variety by cloning (asexual reproduction). The new improved varieties bred through sexual reproduction were not included i.e. the legal protection is not provided for seeds. The breeding programs involving new improved varieties by cross-pollinating plants and growing of seed was more complicated and thus, more imperative and defined law was needed. An amendment to PPA in 1954 prevented the patenting of the obvious product of nature by excluding the plants not found in an "uncultivated state" and allowed the patent of "newly found seedlings" other than those found in an uncultivated state (Plant Patent Act 1954 (Amendment)). In many economically important plants, new varieties are a result of spontaneous mutation or sport an orchard tree or mutant varieties in the cultivated state (orchards or farms). The only new cultivated variety which is unique is not found in nature otherwise and could not be reproduced by nature without human intervention are patentable under this act.

In 1980, Dr. Walter F. Beineke ("Beineke"), a well-known plant breeder, filed an application under PPA to obtain a plant patent for improved varieties of white oak trees (Quercus alba L.) with superior genetic traits. He discovered two more than 100-year-old oak trees that appeared to have excellent timber quality and strong central stem tendency, growing in a front yard of a house. Beineke took acorns from each of the trees and grew them for the next few years to confirm the existence of some superior traits in the progeny. After confirming that progeny showed the same traits, he asexually reproduced the trees for a few more generations. After more than 20 years, the asexually reproduced trees were "true to type" i.e. the trees showed the same superior traits, he applied for US plant patent protection for the two "new variety of white oak trees."

The Patent examiner stated that although the claim of being the discoverer and able to asexually reproduce to obtain two new and distinct white oak tree cultivar varieties were met, however, the source trees were not found in cultivated land. The Patent was denied to Dr. Beinke because the oak trees had been discovered in an uncultivated state. He re-appeal for the patent stating that the trees were growing in a cultivated lawn and hence were not in an uncultivated state. PPA legislative defines that the plant discovery resulting from "cultivation" is unique, isolated, and not repeated by nature, nor can it be reproduced by nature unaided by man. The examiner stated that since the lawn was not used for the cultivation of white oak trees and therefore, was an uncultivated land. The claim on cultivars was rejected on the reasoning that the white oak trees were not the result of human intervention and were clearly the obvious products of nature.

12.7.2 Utility Patent

Utility Patents are the most common type of patent recognized by countries the world over. These patents protect the utility or functional aspects of technological products and processes, including machines, compositions, formulations, and anything man-made that has a specific and useful specific function. Utility patents are governed by the Patent Act of 1952, codified in Title 35 of the U.S. Code.42 with a term protection of 20 years. The Patent Act of 1952 grants patent protection to "whoever invents or discovers any new and useful process, the machine, manufacture or composition of matter, or any new and useful improvement thereof." After exploring the text and legislative history of the Patent Act, the US Supreme Court, gave a broad interpretation to the terms "manufacture" and "composition of matter" and included a live, human-made microorganism. The Court reiterated that discoveries in nature are not patentable but genetically modified organisms are a result of human intervention hence patentable. In 1980, the US court in the landmark case of Diamond v. Chakrabarty (447 U.S. 303 (1980)) stated that a genetically engineered microorganism could be patented and after that US started allowing utility patents for plants also. Utility patents have been issued for genetically engineered plants or elements of plants such as plant-based products, fruits, buds, pollens, genes, DNA, etc. The protection under utility patents is now extended to the process for transformation, vectors, markers, and promoters. The prerequisite for the utility patent for the plant is that the plant should be modified by the intervention of the inventor and must be non-obvious, novel, and useful. The claim must define in detail, the modification which resulted in the specific characteristics of the plant. Further, the right of the inventor is considered to be infringed if the plant protected by a utility patent is reproduced either sexually or asexually by anyone, without the consent of the patent holder. A utility patent is a very strong form of the patent therefore, companies working on genetic modification of plants usually apply for protection under it.

Monsanto Canada Inc. v Schmeiser (Federal Supreme Court of Canada 2004), was a leading case on patent rights. Monsanto developed and launched Roundup Ready plants which contained genes for resistance to glyphosate and the Roundup® brand herbicide which is sprayed on the Roundup Ready crop fields. Roundup® brand herbicide kills only the weeds (which do not contain the gene for resistance to glyphosate) and not the Roundup Ready plants which contain the glyphosate-resistant genes. This saves the labor costs for weed control. Monsanto filed the application claim for the glyphosate-tolerant modified plant cell, the genetically modified seeds produced from plants containing such cells, the specific modified genes, and the process of producing the genetically modified plants. United States Patents Nos. 5,352,605 and 5,633,435 were granted to Monsanto and the company launched its first product, Roundup Ready soybeans, in 1996. The agriculture company Monsanto sued Canadian canola farmer, Percy Schmeiser, in Saskatchewan, proclaiming that the farmer was growing a Monsanto genetically modified canola variety without Monsanto's authorization. The seed users were required to enter into a technological agreement with the company stating that they will use the seeds for planting for only 1 year. Percy Schmeiser never purchased Roundup Ready Canola nor did he enter into such an agreement however, his farm was found to have high (95–98%) contamination of genetically modified canola. Monsanto claimed that their patent right has been infringed due to unlicensed use of seed by the farmer however, Mr. Schmeiser pleaded that the contamination may be due to adventitious drift of seeds or pollen from nearby farms which were planted with GM canola. He also pleaded that even though he had GM plants in his field he gained nothing because he did not spray Roundup® brand herbicide to take advantage of the herbicide resistance nature of the modified crops. Also, he claimed instead Monsanto should pay him damages as the seeds or pollen from nearby farms has contaminated his farm, his right over the property has been infringed. Although the contamination source could not be established so the claim was not pursued. The Supreme Court said that the patent right of Monsanto has been infringed which was imperative by the presence of heavy contamination in the field. No damages, however, has to be paid by the farmer as GM seeds were selected unknowingly and the presence of GM plant did not confer any economic benefit to the farmer. The case, however, brought some major concerns for the scientist, breeders, and policymakers especially about the farmer's rights and biosafety concerns. Mere possession of a patented gene or cell without the intention of its commercial use should mean infringement of the inventor's right should be made clear. Also, a patent on gene or cell should be extended to the whole plant is also not clear under the current protection regime. Thus, a strict distinction between the patent of the modified gene and the plant itself should be elaborated under the protection.

12.7.3 Plant Varieties Protection Act, 1970

The research and development of new varieties with cross-breeding resulted in improved cultivars with high yield, disease resistance, and salt and drought tolerance around the mid-nineteenth century. The US legislation under PPA, 1930 only grants protection of the new variety propagated by asexual reproduction. Research and development institutions and plant breeders across the United States demanded to extend the exclusive legislative rights for sexual bred plant varieties also. In 1970, US legislation approved the Plant Variety Protection Act (PVPA) which extent the IPR protection to sexually plants, tubers, and seeds as well. The law facilitated the development of new varieties by ensuring the exclusivity of the variety owner on seed sale and distribution. This act provides plant breeders an opportunity to earn economic benefit for the investment made on research and development of the new variety and re-invest into their breeding programs. This act, however, did not consider the rights of farmers and consumers. PVPA was amended in 1994, to extend the exclusive right of the variety owner. The amended form still prohibits the farmers from selling saved seed without permission from the right holder however, they allow the farmers to save the seeds for their own use (consumption and replanting on his owned land) and sell that same amount to another farmer for their consumption and not replanting.

A "variety" under PVPA is defined as a plant grouping, within a single botanical taxon of the lowest known rank, defined by the reproducible expression of its distinguishing and other genetic characteristics. A certificate is granted under PVPA for the plant variety to a plant breeder that is "new," "distinct," "uniform," and "stable":

New: On the date of filing the application for plant variety protection, the propagating or harvested material of the plant variety has not been sold or otherwise disposed of to any other person, with the consent of or by the owner of the variety for purposes of exploitation of the plant variety, -for more than 1 year prior to the date of application filing in the United States; -more than 4 years prior to the date of application filing in any foreign jurisdiction (outside the United States); -or 6 years prior to the date of application filing in the case of a tree or vine.

Distinct: A variety must be clearly unique or distinguishable from any other publicly known plant grouping, whose existence is a matter of common knowledge, by expression of at least one of the identifiable characteristics such morphological (color, the shape of a leaf, etc.), physiological (disease resistance, drought resistance, etc.) or any other characteristics including characters influencing processing (easy in baking, etc.)or commercial importance, at the time of filing of the application.

Uniform: The variety must be uniform. The variations should be predictable, describable, and commercially acceptable.

Stable: The variety must be stable. The variety when reproduced repeated should remain unchanged with regard to its essential and distinctive characteristics of the variety within a reasonable degree of reliability corresponding with that of varieties of the same group in which the same breeding method has been used.

In the United States, PVPA is a kind of plant breeder's rights and Plant Variety Protection (PVP) certificates are issued by the US Department of Agriculture (USDA). Plant breeder's rights are the protected exclusive commercial rights for a registered variety which gives plant breeder a commercial monopoly over a fixed period of time (20 years). Protection under this act is similar to that provided by the patent with additional qualifiers, the farmer's privilege, and the researcher's exemption. The "farmer's privilege" means that the farmers are can save the seeds of the protected variety for replanting on their own land however, they are not allowed to resell the protected seeds for replanting to another farmer. The connotation "research exemption" means that the use and reproduction of protected varieties for plant breeding and other bona fide research by others e.g. to develop new varieties shall not constitute infringement.

In the Monsanto Co. v. McFarling case (Monsanto Co. v. McFarling, 2002; Monsanto Co. v. McFarling, 2007), a farmer in Mississippi, Mr. McFarling, purchased Roundup® Ready soybean seed in 1997 and 1998 after signing the Technology agreement and paid the license fee for each purchase. The agreement specifically included the clause that the seed would commercially be used for planting for single-season and seeds produced would not be stored for next planting season or supply seeds to any other person for replanting. The farmer saved 1500 bushels of the Roundup® Ready soybean from his harvest during one season and replanted them as seeds for the next season. Farmer did not pay the license fee for this saved soybean seed which retained the genetic modifications of the Roundup Ready® seed. Monsanto filed suit charging infringement of patent and breach of the contract, seeking an injunction. Monsanto appealed that if the farmers were allowed to replant the seed, the company would effectively, and rapidly, lose control of its rights. McFarling challenged Monsanto's claim and raised various counterclaims including patent misuse, antitrust violation, and violation of the Plant Variety Protection Act. He claimed that he did violate the Technology Agreement by replanting saved seeds, but being a farmer, he has the right to save the seeds for replanting of land owned by him under farmers right of PVP act. The Court stated that both patent law and PVP can co-exist and declined to limit the patent law by reference to the PVPA. The Court declared that Mr. McFarling has infringed Monsanto's patent.

12.8 Protection in Plant Breeding and Research Under IPR in Europe

12.8.1 Plant Breeders' Rights Acts in Europe

Various attempts were also made in Europe during the end of the nineteenth century to extend the patent protection to plant varieties. During the beginning of the twentieth century, various breeding programs with on-farm experimentation were carried out for the development of new crop varieties. This resulted in the growth of seed trade in several European countries. The need over the control of selling and distribution of seeds gave impetus to demands for the formulation of IP protection in relation to seeds.

Like in the US, several arguments were put forth to deny the breeder's product protection under patent laws, in Europe (especially in Belgium, Germany, and the Netherlands). According to the commentators, plants were product-of-nature so plant inventions do not qualify for patent protection. This doctrine had many followers in Europe. In Belgium, critics contended that mineral products (such as marble and ivory), plant products (such as fruits, flowers, or vegetables), and animal products occur naturally therefore, they should not be granted patent protection. The objection was based on the faith that these inventions were mainly the result of nature's work with minimal human involvement. In Germany, the "Naturstoff" (the product-of-nature) doctrine had fewer followers but opposed patent protection for culture techniques, breeding methods, and breeder's products. According to them, unlike machines, breeders' products, including those artificially bred, were not the result of a creative process and were "products of nature and hence were not inventions or, as the Germans put it Nicht-Erfindungen." Thus, inventive steps along with industrial applicability were the main category of non-compliance of the legal requirement of patentability. After some failed attempts by France and Germany, the Netherlands was the first country in Europe to introduce PVP legislation in 1942, followed by Germany in 1953. Congrès Pomologique de France in 1911 suggested the need for special international protection for the plant variety. This suggestion became the basis of the establishment of the International association of plant breeders for the protection of plant varieties (ASSINSEL) in 1938.

12.8.2 Union pour la Protection des Obtentions Végétales (UPOV, 1991)

Unlike the US where nurseries influenced the legal system of plant patents, in Europe commercial breeder's pressure led to the adoption of a plant variety protection regime. Since the seed companies were the main influencers, the design of the legislation was focused on seed regulations (distinctness, homogeneity, and stability) and ensured the identification and quality of seeds. Thus, the national PVP laws inspired an initiative aiming at the formulation international sui generis system on the matter. A sui generis system, "one that is of its own kind" is the formulation of new international law granting specific privileges to reward the generation of plant-based innovations but with essential environmental safeguards and securing farmers' livelihoods. ASSINSEL, in June 1956 stressed the need to adopt an international system for the protection of new plant varieties. In response to the ASSINSEL call, the French government invited European countries to a diplomatic conference in Paris to formulate an international instrument on plant variety protection. In the Paris conference, it was agreed upon, that a plant variety should be granted protection if the variety is distinct from pre-existing varieties and is sufficiently uniform and stable in its essential characteristics. This was an important step towards the establishment of an international system of plant variety protection (PVP).

The Lisbon Conference, 1958 was held to discuss the Revision of the Paris Convention for the Protection of Industrial Property concluded with a suggestion that a "special law" was needed to protect new plant varieties. The conference held in Paris for the negotiation of an international instrument on plant variety protection in 1961 resulted in the establishment of the International Convention for the Protection of New Varieties of Plants. Several European countries signed the convention and created a Union for the Protection of New Varieties of Plants defined by the French abbreviation UPOV (Union pour la Protection des Obtentions Végétales). UPOV Convention 1961 called for the adaptation of new existing breeders' rights regulations in contracting member states where plant protection law was already into place. The UPOV Convention also introduced a structured framework for plant breeders' rights protection in contracting member states which did not yet have special legal provisions for plants. The convention was adopted in 1961 at the Paris conference and came into force by 1968. Multiple revisions were adopted thereafter (1972, 1978, and 1991) to remodel legislation to adjust new developments in plant breeding.

12.8.3 European Patent Convention

Second-Generation Patent Acts were established in Europe with the European Patent Convention (EPC) a regional patent treaty adopted in 1973. EPC provided the legal framework for granting European patents before the European Patent office. A single application, in one language, may be filled at any EPO and solitary examination procedure ensures the protection of the invention in all contracting member states which signed the EPC.

The most important article of EPC with respect to a patentable matter is Article 52(1) entitled "Patentable inventions" which states that the "European patents shall be granted for any new inventions which are susceptible to the industrial application that involves an inventive step." The requirements in this article recognized the fundamental provision of first-generation national patent which governs the patentability of inventions. The exclusionary provision of plant varieties from patent protection under Article 53(b) of the EPC states that "European patents shall not be granted in respect of plant or animal varieties or essentially biological processes for the production of plants or animals."

This provision was aimed to exclude the patent protection for EPC subject matter from patentability that was also protectable under independent national plant variety acts. Thus, the excluded subject matter, namely the protection to plant varieties, was identical with the subject matter that was protected under the UPOV and corresponding national plant breeders' rights laws.

Provision of EPC, Article 53(b) prohibited the patenting of plants or their propagating material in the genetically fixed form of the plant variety, but did not prohibit the patenting the plant per se i.e. the exclusion did not apply to plants which either do not belong to the classification unit taxonomically higher than that of the variety or do not meet the profile of a variety. This means according to the provision of EPC, Article 53(b) protection was possible for the potato (*Solanum tuberosum*) but was not possible for the potato variety Charlotte.

According to Article 4(1) of the Biotechnology Directive, "plant and animal varieties and essentially biological processes for the production of plants or animals shall not be patentable." The "essentially biological processes" means a non-microbiological process for the production of plants that consist entirely of natural phenomena of sexual crossing and selection and is excluded from patentability. For the processes or product to be patentable at least one additional technical step must be added or performed within the steps of sexual crossing the whole genome of the plant. The subsequent selection should result in the introduction of a trait into the genome or modify a trait in the genome of the plant produced such that the introduction or modification of that trait is not an obvious result of the mixing of the genes of the plants chosen for sexual crossing. Deliberate genetic manipulation or the production of the transgenic plant through chromosomal incorporation with human intervention is not included in "essentially biological processes."

Further, according to Article 4(2) of the Biotechnology Directive, "inventions, which concern plants or animals, shall be patentable if the technical feasibility of the invention is not confined to a particular plant or animal variety." In 1990, EPO issued patent protection to Plant Genetics Systems (PGS) including claims to plant cells, plants, and seeds that are herbicide-resistant due to herbicide-resistance gene integrated into the plant genome. In this unique interpretation of Article 53 of the EPC, a patent was granted for a transgenic plant having a foreign sequence (herbicide resistance) in its genome. It was stated that although the plant varieties were not patentable, however, if the technical feasibility of an invention with respect to the plant is not limited to a particular variety and can be extrapolated to more than one variety, then such an invention shall be patentable. According to this, a patent claim, wherein the specific plant variety is not claimed individually, is not excluded from patentability under the provision of EPC, Article 53(b) even though the plant falls in per view of variety. In the case of Novartis, the Enlarged Board of Appeal held that the genetically created herbicide resistance would not create a plant variety as this resistance could also be transferred to other varieties. Thus, the fact that the technical feasibility of the invention was not limited to only one plant variety. The Board, therefore, concluded that the variety is not directly patentable but the method to produce such variety is not excluded from the scope of patentability. In 1995, EPO policy, however, stated that the plant cells could be patented whereas claims on plants per se for protection could not be granted.

Currently, in Europe, plant inventions and plant varieties are protected under the patent law and plant variety rights law, respectively. Article 52 (b) of the European Patent Convention excludes the plant varieties from patent protection. The exception that plant varieties as such cannot be protected by a patent has been confirmed in a directive of the European Community (the "Biotech Directive"). At the Community level, however, the Biotechnology Directives allow plant varieties to be patented indirectly because the biotechnological process that developed them, is patentable.

12.8.4 Plant Varieties Act 1997

The Plant Varieties Act 1997 is defined as "An Act to make provision about rights in relation to plant varieties; to make provision about the Plant Varieties and Seeds Tribunal; to extend the time limit for the institution of proceedings for a contravention of seeds regulations, and for connected purposes." The right delegated by this act is known as plant breeders' rights. The protection under Plant breeders' rights is given to all varieties of all plant genera and species.

The Act defines the "variety" as "a plant grouping within a single botanical taxon of the lowest known rank." The grouping can be

a. defined by the expression of the characteristics resulting from a given genotype or combination of genotypes,
b. distinguished from any other plant grouping by the expression of at least one of those characteristics, and.
c. considered as a unit with regard to its suitability for being propagated unchanged.

12.8.5 The Community Plant Variety Protection System

The scope of Community plant variety rights is to a high degree similar to that of a utility patent even though there are some differences due to the specific nature of the subject matter of plant variety rights, plant varieties. The scope of rights is mentioned in Article 13 (2) of the Basic Regulation and includes the exclusive right for the holder to produce and reproduce conditions for the purpose of propagation, sell, market, import, and export to the Community and stocking. In addition to acts effected in relation to the protected variety itself, acts effected in relation to essentially derived varieties, other indistinct varieties, and hybrid varieties dependent on the protected variety for their production may also constitute infringements (Article 13(2)4 read in conjunction with Article 13(5)).

The Community Plant Variety Protection (CPVP) system, 1995 was formed by the Council Regulation (EC No. 2100/94) (the "Basic Regulation"). Both the 1997 Act and CPVP are based on the 1991 UPOV Convention (Jordens 2005). Community Plant Variety Protection right (CPVR) is the IP regime granting protection to all "new" botanical genera and species, including their hybrids, provided that the varieties meet exactly the same requirements as outlined under the UPOV Convention. The community system (CPVR) in European Community exists in parallel with the national plant variety protection schemes aligned to the UPOV Convention in individual European countries. The protection for the same plant variety cannot be granted under both the Community and a national PVP system at the same time. In the case where a CPVR is granted in relation to a variety for which a national right has already been granted, the CPVR protection overrides the national right for the duration of the CPVR. The CPVR is valid (or cancelled) across all EU contracting countries and not in individual countries therefore it provides an added value to breeders to get a return on investment (ROI) throughout the entire European Union.

Application for Plant Variety Protection; Where and Who Can Apply

The Community protection system is managed by Community Plant Variety Office (CPVO). An application for plant variety protection can be made in any of the 20 official languages of the European Community direct to the CPVO or to one of the national agencies in a Member State, which in turn will take the necessary steps to send it on to the CPVO.

Any individual or company whose domicile or headquarters is located in the European Union. Individuals or companies from a state which is a member of the International Union for the protection of new varieties of plants (UPOV) but not a member of the European Union can also apply, provided that an agent domiciled in the Community has been nominated.

12.9 Impact of UPOV 1991 in Developing Countries

In the early 20th the agricultural and seed production in Europe was dominated by commercial breeders and seed companies. The demand for a legal framework for the protection of IPRs over the new plant varieties was proposed and supported by these commercial breeders and business organizations involved in plant breeding. The International Convention for the Protection of New Varieties of Plants (UPOV) for the agricultural systems and the modalities of seed production was developed to accommodate protection rights for newly developed plant variety and seed usage control in developed countries. The UPOV Convention was adopted in 1961 with an aim to extend the new model of plant variety protection internationally.

The enactment of special legal regimes of plant variety protection (PVP) under UPOV was at that time a privileged legal package for influential stakeholders, notably commercial breeders and seed companies. The participation of other relevant stakeholders, notably farmers and consumers were ignored. Also, the UPOV legislature was designed based on the farming systems in developed countries especially, Europe and the US, which was significantly different from those predominant in developed countries. The resource-poor farmers in developing countries save seeds and propagating material for replanting as compared to developed country farmers who were dependent on commercial breeders and seed companies for seed supplies and this difference still exist. Saving seeds from farms for replanting, exchange, and sale among the farmers is a general practice in developing countries that ensures farmers access to affordable seeds and protects them from uncertainties of the commercial seed supply. Further, the livelihood of farmers depends on income generated by selling seeds and propagating material. Saving seeds for replanting is also crucial for preserving a diversified supply of seeds, adapted to local conditions and a changing environment. The implementation of UPOV imposes the obligation of commercial seed use which would narrow down the genetic diversity in the fields.

The first two revisions in 1972 and 1978 did not result in a change in the system of protection but the third revision in 1991 was significant in expanding and strengthening the plant breeder's rights along with giving the limited right to farmers (agricultural exemption/farmers' privilege) to save and use farm-saved seeds.

UPOV 1991 (Article 14) states that:

(1)(a) Subject to Articles 15 and 16, the following acts in respect of the propagating material of the protected variety shall require the authorization of the breeder:

i. production or reproduction (multiplication),
ii. conditioning for the purpose of propagation,
iii. offering for sale,
iv. selling or other marketing,
v. exporting,
vi. importing,
vii. stocking for any of the purposes mentioned in (i) to (vi), above.

(b) The breeder may make his authorization subject to conditions and limitations.

World Trade Organization (WTO) members bound by the agreement on TRIPS are required to provide some compatible legal system for IP protection for plant varieties (Janis 2014). In response to the obligation created by the TRIPS agreement for free trade, some developing countries joined UPOV 1991 however, others established sui generis PVP regimes depending on the physiognomies of their agriculture and seed supply systems. Many of the developing countries however are still struggling to design plant variety protection regimes adequate to their seed and agricultural systems due to constraints of limited expertise, institutional capacity, and experience. In the Sixth Meeting of the Governing Body of the International Treaty on Plant Genetic Resources for Food and Agriculture in Rome, 2015 an initiative was undertaken to the extent the context of UPOV, 1991 taking into account the needs of the farmers in developing countries and alternative sui generis systems for plant variety protection that exist in some developing countries.

UPOV 1991 however, still does not prevent the misappropriation of farmers' varieties by commercial companies. Thus, the absence of the provision for appropriate farm management practice of freely saving, exchanging, and selling seeds or propagating material contributing to growth and concentration of commercial seed supply enterprises, ultimately endangering biodiversity, livelihoods, and food security were the main concerns for developing countries against joining PVP system under UPOV. Thus, modifications in PVP regimes are needed to recognize the changes in biotechnology and plant breeding methods over the years and fair competition rules are required to be aligned to the changes in technology (Janis and Smith 2007).

12.10 Protection in Plant Breeding and Research Under IPR in India

12.10.1 Indian Patent Act, 1970

The patent in India is provided for an invention that is novel, useful, has industrial applicability, and should not fall under the purview of Section 3. Indian Patent Act, 1970, Section 3 provides a list of inventions that cannot be patented in India. According to Section 3(h) of the Act, a conventional method performed on actual open fields should be considered as a method of agriculture/horticulture and is not patentable. A technique of spraying insecticide for controlling pests is a method of agriculture and not considered a patentable subject matter.

No protection to plant variety is provided under Indian patent law. According to Section 3(j) "plants and animals in whole or any part thereof other than microorganisms including seeds, varieties animals" is not patentable. Thus, the mere act of isolating a gene from a living thing, is not patentable, however, isolating, purifying, and then chemically modifying to get a new modified nucleic acid sequence with the industrial application is patentable. Simply isolating sequence is seen as "discovery" of naturally occurring DNA sequence and hence not patentable.

However, when a gene, DNA, or amino acid sequence is genetically modified using an inventive step into a novel product for a specific industrial application, it is patentable. Gene, DNA, or amino acid sequence with unidentified functions and without any industrial application are not patentable. Recombinant DNA-based vaccines and artificially developed microorganisms are, therefore, considered to be patentable.

Section 3(j) also states that apart from naturally occurring embodies, living entities of artificial origin such as transgenic plants and animals and/or parts thereof are not patentable. Plasmids, recombinant DNA, and the process of forming the same, however, are patentable if an inventive step is employed.

12.10.2 Plant Variety Protection and Farmer's Right Act, 2001

India, being a member of the WTO is bound by the agreement on TRIPs which requires WTO Members to provide protection to plant varieties either by patent or by sui generis system or by both. Indian patent system does not provide plant variety protection therefore, the Indian Government opted for a sui generis system by introducing the Plant variety bill in 1999 for plant variety registration and protection of commercial plant breeders and farmer's rights (Brahmi et al. 2004).

The Plant Varieties and Farmers Rights Act, (PPVFR Act), 2001 facilitates the protection of new plant varieties, encourages the development of new plant varieties, accelerates agricultural development, and stimulates financial investments in plant breeding programs. These programs facilitate the higher growth of the seed industry and facilitate the availability of high-quality seeds and propagating material to farmers.

Any variety to be registered as a new plant variety under the Protection of Plant Variety and Farmers Right Act, 2001 is required to fulfill the criteria of being:

Novel: If at the date of filing of the application for registration of plant variety, the propagating or harvested material of such a variety has not been sold or otherwise disposed of by or with the consent of its breeder or his successor for the purposes of exploitation of such variety:

a. in India, earlier than 1 year.
b. or outside India, in the case of trees or vines earlier than 6 years, or, in any other case, earlier than 4 years, before the date of filing such applications, then plant variety is considered novel.

Distinct: A variety is said to be distinct if it is clearly distinguishable by at least one essential characteristic from any other variety whose existence is a matter of common knowledge in any country at the time of filing an application.

Uniform: A variety is said to be uniform if it is sufficiently uniform in its essential characteristics.

Stable: A variety is termed as stable if the essential characteristics of the variety remain unchanged after repeated propagation or, in the case of a particular cycle of propagation, at the end of each such cycle.

Plant Breeder's Right The breeders are granted the rights of commercialization for the variety developed by him under this act. The legal framework safeguards his right including the right to reproduce, sell, market, distribute, export, or import the variety. The infringement of breeders' rights invites substantial punishment as fines and jail terms under this act. The breeder could execute his commercial right either in person or through anyone designated by him.

Researchers' Right (Exemption) PPV&FR act, 2001, allows scientists and breeders to freely use registered varieties for research purposes. The registered variety can also be used as an initial source for the purpose of creating other, new varieties.

Farmer's Right (Farmers' Exemption) PPV&FR act, 2001, allows farmers to use, save, sow, and resow, share, exchange or sell seeds registered under this act. A farmer is not a cultivator but a breeder who selective grows and breeds varieties suitable for the prevalent environmental conditions. The act recognizes the farmer's efforts and has provisions for the registration of such farmer's varieties. To safeguard themselves by being scavenged by formal sector breeders, farmers with the help of NGOs can register the plant variety. Such an arrangement allows the farmer to sell seed in the way he has always done, with the restriction that this seed cannot be branded with the breeder's registered name hence protecting both farmers' and breeders' rights.

12.11 IPR as Custodian of Traditional Knowledge: The Controversial Patent Cases

The IPR regimes were drafted and implemented during the industrialization era in western countries according to the perceived needs of their technologically advanced societies. In most developing countries, traditional knowledge is important as a dependable source of indigenous people for health, livelihood, and well-being. Traditional knowledge is the know-how, skills, practices, innovation, teachings, and learning in the community that is developed, sustained, and passed on from one generation to another within the community. Since this knowledge has ancient roots, often informal and oral, and is part of community culture, and hence not easy to protect it by the current IP right system. However, when community members innovate within traditional knowledge and traditional cultural expression (folklore) framework, they may need it to be protected. This need prompted some countries like India, China, etc. to implement sui generis protection systems for innovation-based traditional knowledge. WIPO members established an Intergovernmental Committee on Intellectual property and Genetic Resources, Traditional Knowledge and Folklore (IGC) in 2000, to develop an international legal instrument that would effectively protect traditional knowledge, genetic resources, and traditional cultural expression. Bio-piracy or the misappropriation of traditional knowledge with an intention to gain IPR protection especially the patent, over the knowledge, is the biggest challenge faced by the IPR system in its current form (Gavin 2004).

12.11.1 The Enola Bean Plant Biopiracy Case

Mexican diet includes beans and yellow beans like azufrado and mayocoba which have been cultivated in Mexico, for centuries. An American, John Proctor, the president of seed company PODNERS, LLC, brought back some beans from Mexico to Colorado. After planting, he allowed them to self-pollinate and selected yellow beans for several generations. This resulted in the yellow bean population in which the colour of beans was stable, uniform, and remain unchanged by season. He was granted a patent by the United States Patent Office (USPO) (U.S. Patent No. 5,894,079 1999). With this Proctor got an exclusive monopoly on yellow beans for 20 years and excluded the importation or sale of any yellow bean displaying the yellow shade of the Enola beans in the state. The patent was challenged by the International Center for Tropical Agriculture (CIAT) arguing that the patent claims are invalid because it does not meet the criteria of novelty, non-obviousness. As six beans identical to Enola were already in existence so the patent claims disregarded the prior art existing in the country (Patent Act, 35 U.S.C. §§ 102–103 2002). Proctor's claims linked to Enola bean were rejected by USPO in 2008 and were a landmark decision to raise concerns about biopiracy.

12.11.2 Battle for Basmati

The researchers at Texas-based Company RiceTec claimed that they have invented novel varieties of basmati with high yield and quality and filed a patent with USPO. RiceTec was granted the patent (US5663484) in 1997. The patent gave RiceTec a monopoly with regards to plant and seeds, over various rice lines including some having characteristics similar to Basmati lines and this sparked the "Battle for Basmati." Rice is the staple food in most Asian communities. Basmati is an aromatic rice variety grown mainly in India and Pakistan agricultural belt. Farmers in this region have produced, developed, and conserved varieties based on taste, flavour, aroma, and needs, for centuries. In its claim, RiceTec also acknowledged this fact. The Indian government made an appeal in 2000 against the patent stating that novelty in the patent application made by RiceTec is questionable. Some claims were withdrawn by the company and some claims were cancelled by USPTO in 2000. A US patent was, however, granted for the three strains developed by RiceTech which were obtained by crossing the strain with a western strain of grain. Although the withdrawal of claims was accomplished it also exposed the problems of the lack of concerns for the appropriation of traditional knowledge and associated resources, especially originating in developing countries.

12.11.3 Turmeric

Turmeric (*Curcuma longa*) a perennial herb of the ginger family is native to the Indian subcontinent. It has been used as a colouring and flavouring agent in various Indian gastronomic dishes and is an integral ingredient in medicines for cough, cold, gastric problems, cut, wound, inflammation, infertility, skin infections prepared under the Indian system of medicine and many ethnic and folk medicinal practices (Ibáñez and Blázquez 2020). The US patent no. 5,401,504 on turmeric was granted to two Indian nationals from the University of Mississippi Medical Centre in 1995 for wound healing property. The subject matter of claim was the use of "turmeric powder and its administration," both topical and oral, for wound healing. The exclusive right was granted for selling and distribution. The Indian Council of Scientific and Industrial Research (CSIR) appealed to USPTO to re-examine the claim as turmeric was used from time immemorial by the Indian community for wound healing and skin infections. CSIR provided the documentary pieces of evidence of the prior art, published in Sanskrit, Hindi, and Urdu languages, and a paper published in 1953 in the Journal of the Indian Medical Association to USPTO. USPTO stated that the claims by patentees were not novel but obvious and anticipated and approving that the use of turmeric as part of traditional knowledge for healing wounds. USPTO revoked the patent and it became the landmark case where traditional knowledge belonging to a developing country was safeguarded.

12.11.4 Hoodia Cactus

Ancient inhabitants of Southern Africa, the San people, from thousands of years have been using Hoodia cactus (*Hoodia gordonia*) to quench thirst and hunger on long hunting trips across the desert. A Dutch anthropologist in 1937 listed the use of Hoodia while studying the Sans. The South African Council for Scientific and Industrial Research (CSIR) studies the plant for its potential as an appetite suppressant. CSIR in 1995 was granted a patent for the appetite-suppressing element (P57), the active component of the Hoodia. Later, in 1997 CSIR gave an exclusive license to the British company, Phytopharm, to commercialize Hoodia's active components. Later on, American pharmaceutical company Pfizer acquired the license from Phytopharm for a milestone payment to develop and market P57 as a potential slimming drug and cure for obesity. Throughout the whole process, however, the institution did not recognize the San's knowledge. In 2001 The South African San Council along with some NGOs claimed the CSIR has not taken any prior-informed-consent (PIC) from Sans, and that their traditional knowledge has been stolen mounting to biopiracy. They also threatened to take legal action against CSIR. In 2002 mutually acceptable benefit-sharing agreement was signed between

South African CSIR and San people. This agreement recognized Sans as the custodians of traditional knowledge associated with the Hoodia plant therefore, they would receive 6% of CSIR royalties and 8% of CSIR milestone payments derived from the sale of the Hoodia products. Although the percentage share of eventual sale for Sans was very less, however, the potential market size for such drugs could mean a substantial amount. Although the things were settled in favour of Sans, however, the Hoodia cactus case emphasizes that the position of stakeholders of traditional knowledge is vulnerable and needs much-defined laws for the protections to safeguard their interests. The local communities should be made partners sooner in the process so that they can get the benefit of the commercialization of their traditional knowledge and practices.

12.12 Conclusions

The possible benefits of the development of new varieties using modern biotechnology in research and development of quality, quantity, and variety of agricultural products are tremendous. Currently, the world needs to develop new biotechnologies and varieties that perform well in the existing environments. This should be a collaborative effort of public and private scientists, researchers, breeders, farmers, and investors. In any country, the areas of common interest and opportunity between the private sector and public sector institutions should be identified and worked upon to realize the maximum benefit of advancement in technology and generated knowledge base. IP rights are very important to secure the economic returns for the intellectual and financial investments in research and developments. Inventions and discoveries resulting from the applications of biotechnology are of different nature and kind and therefore, overly broad IP rights should not be granted. IPRs should be narrowly tailored to be able to recognize the actual scope of new inventions and discoveries, and hence support continual research, innovation, and development. An appropriate IPR system should be designed to safeguard the interest of all stakeholders' public-funded labs, private institutions, breeders, seed companies and private investors, farmers, and consumers.

The IPR protection system has emerged as the essential institution for asserting IP in the field of biotechnologies and plant breeding and variety development. The protection to get incentives on innovations has given agriculture a similar status as enjoyed by any other industry. This has given impetus to the ongoing innovation and development of products based on plants for the benefit of society.

References

Belletti G, Marescotti A (2011) Origin products, geographical indications and rural development. In: Barham E, Sylvander B (eds) Labels of origin for food. CABI, Wallingford, pp 75–91

Brahmi P, Saxena S, Dhillon BS (2004) Protection of plant varieties and farmer's rights act of India. Curr Sci 86:392–398

Eberlin R (2009) GIs role in rural development. FAO regional expert meeting: livestock based geographical indication chains as an entry point to maintain agro-biodiversity

Endres AB, Goldsmith PD (2007) Alternative business strategies in weak intellectual property environments: a law and economics analysis of the agro-biotechnology firm's strategic dilemma. J Intel Prop Law 14(2):237–268

Federal Supreme Court of Canada (2004) 1 S.C.R. 902, 2004 SCC 34

Friedman DD, Landes WM, Posner RA (1991) Some economics of trade secret law. J Econ Perspect 5:61–72

Gavin S (2004) Biopiracy within the pharmaceutical industry: a stark illustration of how abusive, manipulative and perverse the patenting process can be towards countries of the South EIPR 17

Ibáñez MD, Blázquez MA (2020) *Curcuma longa* L. rhizome essential oil from extraction to its Agri-food applications. A review. Plants (Basel) 28(1):44

Janis MD (2014) Patenting plants: a comparative synthesis. In: Okediji RL, Bagley MA (eds) Patent law in global perspective. Oxford University Press, Oxford, pp 213–242

Janis MD, Smith S (2007) Technological change and the design of plant variety protection regimes. Chic Kent Law Rev 82(3):1557–1615

Jordens R (2005) Progress of plant variety protection based on the international convention for the protection of new varieties of plants (UPOV convention). World Patent Inform 27(3):232–243

Kameri-Mbote AP (1999) Commission on genetic resources for food and agriculture, possible formulas for the sharing of benefits based on different benefit-indicators, 8th edn, Rome, pp 19–23

Kitch EW (1977) The nature and function of the patent system. J Law Econ 20:265–290

Kloppenburg J (2004) First the seed: the political economy of plant biotechnology. In: Still the seed. University of Wisconsin Press, Madison, WI, pp 1492–2000

Mantrov V (2014) EU law on indications of geographical origin. In: Theory and practice. Springer, New York

Monsanto Co. v. McFarling, 302 F.3d 1291 (Fed. Cir. 2002)

Monsanto Co. v. McFarling, 488 F.3d 973 (Fed. Cir. 2007)

Montenegro DeWit M (2017) Beating the bounds: how does 'open source' become a seed commons? J Peasant Stud 46(1):44–79. https://doi.org/10.1080/03066150.2017.1383395

Moser P, Rhode PW (2011) Did plant patents create the American rose? In: Lerner J, Stern S (eds) The rate and direction of inventive activity revisited. The University of Chicago Press, Chicago, pp 413–438

Pardey PG, Koo B, Drew J et al (2013) The evolving landscape of IP rights for plant varietal rights in the United States, 1930–2008. Nat Biotechnol 31(1):25–29. https://doi.org/10.1038/nbt.2467

Patent Act, 35 U.S.C. §§ 102–103 (2002) (providing the statutory requirements for novelty and non-obviousness)

Plant Patent Act (1954) Amendment, 3 Sept 1954

Shelton AC, Tracy WF (2017) Cultivar development in the U.S. Publ Sect Crop Sci 57:1823–1835

Shukla VK et al (2009) Precise genome modification in the crop species *Zea mays* using zinc-finger nucleases. Nature 459:437–441

Srivastava. (2019) CRISPR applications in plant genetic engineering and biotechnology. In: Khurana S, Gaur R (eds) Plant biotechnology: progress in genomic era. Springer, Singapore. https://doi.org/10.1007/978-981-13-8499-8_19

Townsend JA et al (2009) High-frequency modification of plant genes using engineered zinc-finger nucleases. Nature 459:442–445

Tracy WF, Dawson JC, Moore VM (2016) Intellectual property rights and public plant breeding recommendations. In: Fisch J (ed) Proceedings of a conference on best practices for intellectual property protection of publicly developed plant germplasm, Raleigh, NC, pp 12–13

U.S. Patent No. 5894079 (issued April 13, 1999)
Yin K, Gao C, Qiu JL (2017) Progress and prospects in plant genetic editing. Nat Plants 3(8):17107
Zhou J, Wang G, Liu Z (2018) Efficient genetic editing of wild strawberry genes, vector development and validation. Plant Biotech J 16:1868–1877
Zhu T, Mettenburg K, Peterson DJ et al (2000) Engineering herbicide-resistant maize using chimeric RNA/DNA oligonucleotides. Nat Biotechnol 18:555–558

turer's authorised representative in the EU is Springer
ner Service Centre GmbH, Europaplatz 3, 69115 Heidelberg,
u have any concerns regarding our products, please
Safety@springernature.com

d by CPI Group (UK) Ltd, Croydon, CR0 4YY
30/09/2026
02225301-0006